Birkhäuser

Nikita N. Moiseev

How Far It Is to Tomorrow...

Reflections of an Eminent Russian Applied Mathematician 1917-2000

Translated by Robert G. Burns and Iouldouz S. Raguimov

Birkhäuser

Nikita N. Moiseev
Moscow, Russia

Translated by
Robert G. Burns
Toronto, ON, Canada

Iouldouz S. Raguimov
Toronto, ON, Canada

ISBN 978-3-030-96650-8 ISBN 978-3-030-96651-5 (eBook)
https://doi.org/10.1007/978-3-030-96651-5

This book is published under the imprint Birkhäuser, www.birkhauser-science.com by the registered company Springer Nature Switzerland AG
The registered company address is: Gewerbestrasse 11, 6330 Cham, Switzerland

For Antonina—wife and friend

Preface

On rereading the book *How Far It Is to Tomorrow*, I sensed once more the charming personality of my Teacher, and again took pleasure in his life story.

I became Nikita Nikolaevich's graduate student in 1960 following a dramatic—for me—exam in hydrodynamics with him as examiner: I wasn't able to give his interpretation of Zhukovsky's Theorem on the wing vortex, and he, at his own initiative and with his expansive nature and sense of humour in evidence as always, announced: "I'll take you on as my graduate student and make a man of you!"

It is for me a great honour to write down my recollections of NN for a wide audience of academically oriented readers. When I was asked to do this, in my mind I glanced back over the 40 years I spent working alongside him, and felt anew a great emotional lift at contemplating the grandeur of his life's work. I participated in much of NN's work in the Computing Centre of the Academy of Sciences—for example, work on the computation of changes in a satellite's orbit by means of low-thrust engines, on numerical methods of solution of optimal control problems, on hierarchical game theory, water resources and agrarian projects, and the information theory of hierarchical systems. Moiseev's demeanour was rarely severe or care-ridden, his mien and gait always reflected a continual movement ahead, as if taking flight. Academician A. A. Petrov, a former student of his, characterized him with the phrase: *Sturm und Drang*. This showed through in his scientific *oeuvre*, and in the way he was forever organizing new scientific groups and research collectives. His presence tended to fill whatever room he was in. A courteous host, he always helped us students with our coats when we visited him at home. "Only lackeys don't help each other with their coats," he used to say.

I would say that of all the students of Nikita Nikolaevich, I am perhaps the most conscientious, or at least that's how I feel: insofar as time and my own nature permitted, I aped his approach to life and science both.

* * *

August 23, 2017, marked a 100 years since the birth of academician Nikita Nikolaevich Moiseev, distinguished scientist, pedagogue and citizen. His scientific works in mechanics, control theory and computational mathematics, his educational and popular essays, and his invention of the philosophical concept "universal evolutionism" remain relevant to this day, and his ideas continue to be developed further by scientists of various generations, both those engaged in theoretical research and those working in applied mathematics.

NN was endowed with what might be called a universal nature; he was many-sided.

Thus, in his youth he was an outstanding sportsman, attaining a professional level in alpinism and cross-country skiing, where he earned the rank of "master." At one time he was an instructor at a sports camp. He was a very tenacious competitor and untiring in ski relay races. Doubtless the strenuous physical activity of his youth toughened him up for what was to come. After a year's study at a pedagogical institute, he even considered a professional career as sportsman as a more agreeable alternative, but his unquenchable thirst for knowledge decided him otherwise.

And in his youth, NN attended a course at a literary institute and began writing verse, which opened up the possibility of pursuing a writing career. One might say that he realized this option, in a way, through the many scientific and educational books and philosophical essays he wrote.

His path through life might be characterized by the sentence (composed once again by A. A. Petrov): "Nikita Nikolaevich Moiseev represents the fate of our nation in the fate of a scientist."

If a titan may be allowed faults, here are two of Nikita Nikolaevich's. While, as a true Russian patriot, he passionately propagandized the cultural values of Russian society, I would say he underestimated the depth of human beings' acquisitiveness and greed, and overestimated the degree of inertia of society's cultural consciousness—although, on the other hand, he believed in the adaptability of world society as a whole. A second caveat: although as it has turned out one of the political scenarios he predicted has been realized, he nevertheless saw the difficult transitional period as the harbinger of catastrophe for Russian civilization, and lost heart.

* * *

The present book describes, in particular, those of NN's scientific achievements having greatest impact on the scientific community and the government of the time. For the series of papers on the dynamics of a rocket with tanks of liquid fuel and computations of cosmic trajectories, he was made a corresponding member of the Academy of Sciences of the USSR and awarded a notable governmental prize. He also published a long series of articles on cybernetics and mathematical methods of optimization, and worked tirelessly at popularizing applications of control theory and cybernetics to society. In view of the official Party line that cybernetics was, along with genetics, a pseudoscience, this represented a brave attempt to enhance education and societal culture in the USSR. This work led him into problems of mathematical modelling of socio-economic and ecological systems, and then to a general approach to imitative modelling of natural and technological systems. NN's contribution to the automation of aircraft design was recognized with his appointment to full membership of the Academy and the award of another prestigious government prize.

The "nuclear winter" scenario of the aftermath of nuclear war, consciousness of which among world leaders led to a halt in the apparently runaway arms race, is also closely associated with the name of Moiseev. The computations of NN's research group in this regard were crucial to rendering iron-clad the conclusions of the Washington conference held on October 31, 1983, with Carl Sagan presiding, as to the catastrophic transformation of nature that must follow a nuclear world war.

Together with his students, he contributed to the setting up of a research programme aimed at investigating the mutual interactions of the biosphere and anthropogenic processes. Starting from the ancient mechanisms of coevolution of nature and human societies, he sought a new developmental path for our civilization into a future where we might live in harmony with nature. With this as his aim, his wide-ranging intellect led him to a philosophy combining Teilhard de Chardin's concept of the Universum and Vernadsky's concept of the noosphere with the anthropic principle and ideas of the fundamental role that uncertainty and chance play in the universe.

Finally, there is his substantial contribution to the development of information theory in hierarchical systems. Reflecting on the central problem of the management of an economy—that of centralization *vs.* decentralization—he, in collaboration with Yu. B. Germeier, advanced the thesis that a hierarchy arises in an economic system precisely when the central organ is unable to assimilate the information circulating throughout the system.

* * *

When I look for words to use in a general summing-up of NN's personality, the ones that come to mind are: OCEAN, BRIDGE and PROPHET.

Influenced by certain of Stanislaw Lem's images, I express my comprehension of NN as an OCEAN by means of the following haiku-like form of words reflecting NN's clearest character traits:

In the Style of a Haiku

I was approached and asked to write
My recollections of my Teacher.
I remember thinking:
Any wretch may come to the Ocean and see himself in its
Waves' exuberance,
Dolphins' playfulness,
Flickering gulls,
Tsunamis' anger,
Wind's smell,
Foam's sparkle,
Shifting light,
Chattering pebbles,
Water's bitterness.
I can't take leave of my reflections.
I preserve my Teacher in myself.[1]

[1] Composed in Bobrov, Voronezhsky Province, August 23, 2001.

BRIDGE. In his literary and scientific *oeuvre*, N. N. Moiseev accumulated observations on a variety of socio-economic and political processes, and gave them precise, though complex, scientific formulations, thereby joining practicalities to science. All of his work fits naturally into a general process aimed at laying, for Russia as for the world at large, the intellectual foundations of a future coevolution of humankind and nature.

PROPHET. The final chapters of the present book are devoted to hypotheses as to what the near future of Russia will be like, and these predictions have turned out to be quite close to what we today observe around us. The reason for their accuracy is to be found in his systematic and rigorous approach to such questions. Although the various possible future scenarios he envisaged are only described in words, had he continued in the direction his research was taking him, he would have realized, as he did in automating aircraft design and in producing his scenario of a post-war nuclear winter, those imitative systems his students had begun working on aimed at producing well-founded scenarios of peaceful worldwide development or possible armed conflict between the leading world nation-players. Even the tentative initial results of these computations have made a strong impression on some political decision-makers. These might, therefore, furnish a basis for the transformation of Russia's socio-economic and political systems.

* * *

What might be of particular interest to the Western reader in NN's book?

In the first place, his wide, one might say universal, intellectual grasp of the main problems facing humankind, his formulations of these and the insights emerging from his theoretical worldview—which, given that he was a realist, are likely to be as close to the truth as humanly possible.

Secondly, the story of his life. NN absorbed the spirit and contradictions of his epoch. He was educated as a member of the Russian intelligentsia. His grandfather was the director of the Far-Eastern Railway and Deputy Minister of Communications, and his father was a *privatdozent* at Moscow University. But NN would experience in full measure the harshness of the epoch. The twists and turns of his fate might furnish the plot of a detective story set in repressive times: the dismissal, arrest and death of his father, the banishment to the Gulag of his stepmother, his own dismissal and ostracism, and, finally, war, the whole of which he spent at the front, taking part, in particular, in the terrible battles in the Sinyavsky Marshes.

But his great spirit enabled him not only to survive such vicissitudes but go on to remarkable scientific achievements—to the benefit of Russia, as well as enhancement of belief in life's value and in himself.

Ascribing great importance to the education of the up-and-coming generation, NN advanced his particular concept of *Teacher* and adhered to it. Among his former students were 4 full and 2 corresponding members of the Academy and over 50 Doctors and candidates of science. He was invited to become dean of the Aero-Mechanical Faculty at the Moscow Institute of Physics and Technology ("Phystech," the Soviet equivalent of MIT) in the 1950s. Here, among other achievements, he created the Faculty of Applied Mathematics and Control Theory in the 1970s. This work took its toll both spiritually and physically—in the latter case not least because each day he commuted a total of 100 km on the roads around Moscow by means of a Fiat which had seen better days. His physical fitness, preserved since his youth, stood him in good stead in this regard.

NN was a great Russian patriot. He used to say: "Russia is like a painful wound to me." He was in contact over many years with M. S. Gorbachov, and doubtless influenced the actions of that historic leader to some extent. As a member of B. N. Yel′tsin's Presidential Council, although failing to influence the decisions taken, he shamed those members of the council concerned only with their narrow ambitions and the attainment of power, rather than serving Russia.

Sharing Toynbee's opinion that Russia is an idiosyncratic civilization combining the cultures of East and West, he considered the collapse of the USSR to be a catastrophe on a planetary scale. He warned against the mistake of regarding Russia as politically second rank and continually stressed the necessity of maintaining equilibrium among the world's political systems.

NN considered himself a stakeholder in international intellectual culture. He formed personal and professional attachments with L. Zadeh, A. B. Rapoport, R. Bellman, J. Forrester, D. Meadows and several French intellectuals, among others. He maintained contact with The Club of Rome, the International Institute of Systems Analysis in Vienna and other scientific organizations in the USA and France.

Nikita Nikolaevich Moiseev was a personage of historic dimensions.

Moscow, Russia Felix Ivanovich Ereshko

Contents

Introduction

Nikita Nikolaevich Moiseev (1917–2000) here presents the reader with reflections on his life, or what amounts to a very thoughtful autobiography. Since his life spanned the whole of the period of existence of the Soviet Union, we get an insider's view of the history of that regime from its founding in October 1917 to its collapse in 1991, as well as a little of its aftermath—the view of an eminent applied mathematician educated and then working in some of the Soviet Union's top institutions of higher learning.

We see vividly the precariousness of life just after the October revolution, his happy family life during the years 1921–1928 of the New Economic Policy (NEP) (following Lenin's re-introduction of a free market and limited capitalism as a temporary measure to forestall economic collapse); the destruction of his family by Stalin's regime following the ending of the NEP; his trials as a social outcast, in particular, difficulties in being admitted to Moscow State University; his experiences as a Soviet Air Force engineer in World War II, including sorties as gunner and a brush with an NKVD agent; post-war euphoria, marriage and further ostracism; and then the vicissitudes of a highly varied academic career. Along the way we meet many famous Soviet engineers and scientists: Korolev and Pobedonostsev, pioneers of Soviet rocket science, the physicist Tamm and the engineer Venttsel′, and the mathematicians Gel′fand, Keldysh, Kurosh, Lavrentiev, Petrovsky, Sobolev, Tikhonov and Vinogradov, among many others. We also have descriptions of encounters with Western thinkers: Bellman, Zadeh, Forrester, Meadows, Bellamy and others. The last several chapters are devoted more to wide-ranging reflections on God, philosophy, science, communism, modelling the biosphere, the threat of nuclear winter, and, finally, to the author's thoughts concerning the impending and then final collapse of the USSR in 1991, and hopes for Russia's future. Of particular interest are the author's comments on Gorbachov, whom he knew personally, and the *perestroika* programme. (Note that there is essentially no overlap with V. V. Putin's rule, since the latter became president in 2000.)

* * *

It may not be entirely out of place to offer to Western readers a word of warning concerning a few passages in the book. N. N. Moiseev was most emphatically a Russian, even Russian imperial, patriot, and he makes this clear at intervals throughout the book. There are places where he tends to trumpet a touch jarringly, perhaps, the virtues and achievements of the Russian people, especially the Russian intelligentsia, which figure significantly throughout. (The appendix, in particular, is an essay on that set of people and its history. The author defines a member of the Russian intelligentsia to be any kind, empathetic individual whose interests go beyond merely personal ones, embracing, in particular, those of their country. Thus, one such need not be an intellectual or even have any formal education.) At the few junctures in the book where this patriotism becomes a little chauvinistic, one might say, we have deemed it appropriate to insert footnotes as correctives. It is our hope that the reader won't find these overly obtrusive.

In conclusion we thank Ms. Ekaterina Kurapova and Professor F. I. Ereshko for permission to include the photographs.

Robert G. Burns
Iouldouz S. Raguimov

To the Reader

I have written this book for myself, my wife, my children, and perhaps for others close to me. Originally I hadn't thought of further possible readers, but gradually came to realize that what I was writing might very well be of interest to a wider set of people.

Although this book is essentially about the sort of work I did and my country, I have been unable to keep more personal considerations separate from these. Thus, the book has turned out to be above all about me. I have tried to write honestly and sincerely throughout—and to adhere strictly to the facts.

Of everyone mentioned here of those whose path crossed mine I have retained their actual names. Alone certain dates may be in error. Note that this is not, however, a memoir in the standard sense, but a record of reflections based on my recollections.

I was born in 1917, a few months before the October revolution, and lived through the whole tragic period of the formation, triumph, and fall of the socialist government.[2] My personal story is also nontrivial; in some sense it is even exceptional: of my vast extended family, formerly spread through all of Russia's cities, I alone of my own and earlier generations survive to this day. And in my case there occurred a great many metamorphoses—to use a biological term—no less indicative of the singular nature of that era than historians' treatises.

Perhaps my "Free Reflections" on the precariousness of the paths and bridges over the abyss of those decades will help to preserve the memory of the past and so throw a bridge over to succeeding generations.

[2] Of the USSR.

The Beginning (Instead of a Preface) Mist ... Taman′...

Mist ... Taman′...
The solitary landscape is hearkening unto God.
—How far it is to tomorrow!
And Lermontov comes out onto the road alone,
His silver spurs ajingle.
(Georgy Ivanov, 1918)[3]

The idea that I would perhaps some day write this book occurred to me over 50 years ago, in June 1942.

We had just extricated ourselves from a ridiculous trap, from which, as we realized soon enough, we may easily not have escaped. For several days we had trudged knee-deep in muck through a former peat works somewhere to the south of the Voibokal station,[4] careful not to risk moving to dry land: above us a German scout plane, which we called "the frame,"[5] was continually reconnoitring, but we were invisible from above against the peat. At that time we hadn't heard of Myasnoy

[3] Taman′ is a town on the eastern side of Kerch Strait, which joins the Sea of Azov to the Black Sea. Here the second line and part of the fourth are taken from the famous poem "I came out onto the road alone" by the great Russian poet Mikhail Lermontov, who, in his novel *A Hero of Our Times*, described the town Taman′ disparagingly. Note also that the Russian word for "mist" is "tuman."

[4] Voibokal is a town some 90 km east of St. Petersburg on the railway line between that city and Volkhov.

[5] The German twin-engine, twin-boom tactical reconnaissance aircraft *Focke-Wulf Fw 189* was called a "frame" by Soviet soldiers because its twin fuselage made it look from below like a rectangular frame.

Bor[6] or that General Vlasov[7] had surrendered to the Germans. . . . All we wanted was to find the front line, but there was no front line in the Ladoga marshes.[8] Eventually we arrived at Voibokal station without ever finding the front line and, to our great relief, without meeting a single German.

There followed a bath, clean underwear, and more or less edible food.

And there I am seated on the shore of Lake Ladoga. Before me is a fringe of reeds with paths cut through it for boats and then the smooth whitish surface of the lake stretching over the horizon. And all about me extraordinary silence reigns. I don't know what time it is: on those flatlands in June, evening gives way imperceptibly to morning. And in any case does it matter, when the war has moved away, at least temporarily? It likewise had gone beyond the horizon. Does the lake even have a further shore? Does this peaceful watery plain drowsing in the early morning come to an end somewhere?

Somehow it doesn't occur to me to think about the future. In a month the remains of the regiment will be sent to Alatyr′[9] where my comrades will begin familiarizing themselves with the new planes being produced at the Kazan′ factory, with which in June 1943 we will surprise the Germans near Mtsensk.[10] By midwinter I also will turn up in Alatyr′. But first I'll have to get through the autumn on the outskirts of Shlissel′burg,[11] whence I'm to be posted with my armourers.

There various adventures await me, with the chances of survival perhaps not so good. But my "luck" will hold. One day, sometime after the raising of the blockade of Leningrad, a clod of frozen earth kicked up during a shelling will strike me on the spine and I'll be sent to recover in a hospital near Volkhov, and returned thence to my former division stationed in the glorious town of Alatyr′. As for my armourers, they will all remain in the 14th Air Army, where they will function as both armourers and rear gunners in IL-2's.[12] I don't know if any of them survived to the end of the war. I never again met any of them. It makes one think about what it is that

[6] Following earlier successes, in April 1942 the Soviet Second Shock Army was surrounded by German forces in the region of Myasnoy Bor (Meaty Pine Forest), a village in the Novgorod municipality. An operation to liberate the army mounted in May and June resulted in horrendous Soviet losses.

[7] Andrey Andreevich Vlasov was a skilful Soviet general who, in the course of a failed attempt to lift the siege of Leningrad, was captured by the Germans in July 1942, and in September 1944 raised a Russian Liberation Army to fight against the USSR. He was captured by the Soviets in early 1945 and hanged in Moscow on August 1, 1946.

[8] Lake Ladoga is a large lake just to the northeast of St. Petersburg.

[9] A town in the Nizhny Novgorod district some 400 km east of Moscow.

[10] This may refer to the "battle of the Kursk salient" of July and August 1943, when the Germans were forced to make a significant tactical retreat. Mtsensk, Oryol, and Kursk lie, in that order, on a line running roughly south from Moscow. Mtsensk is about 300 km south of Moscow.

[11] A town at the head of the river Neva where it emerges from Lake Ladoga. Schlüsselburg in German.

[12] The Ilyushin IL-2 was a Soviet ground-attack aircraft produced in large numbers during World War II.

determines one's fate! Perhaps it was that clod of frozen clay, which caused me to suffer thenceforth from radiculitis.

But all that was yet to come; in the meantime, I was enjoying the peacefulness, gazing out onto the glassy smoothness of the lake and listening to the whispering of the reeds. I recited a few lines of poetry to myself, either made up or from memory. I still remember them, 50 years later. Here they are:

How bright they are,
These moonless white nights.
The reeds at the edge of the grove
Are coated with a grey shimmer
From the wave's motion.

I lay down near the roots of an old birch tree, looked out over the watery expanse and thought idly of the future. It didn't occur to me that I might perish. No. The war will end and decades will pass. There will be much for me to do. I sensed so much energy and strength within myself—a youthful sentiment like the belief that nothing bad could happen to me! Of course I understood that there would inevitably be very hard times, but I was absolutely sure that I would be able to cope with those, and some day perhaps would write an Afterword to all I do and all that befalls me. Or maybe the other way around, a Foreword?

I already then understood that life is nothing but a fragile bridge between two non-existences. All the same I wanted the book that I felt almost certain I would write to be a Foreword. But if an Afterword, then an Afterword to a collection of poems as symbols of something marvellous.

But that was not ordained to come to pass. It was an impossibility, in fact. At that time, I understood neither this nor the fact that a person's strength and time are severely limited, and as a rule his intentions not realized. And in any case life is nothing like poetry. As for the idea of a Foreword, it is now difficult for me to recall what I had in mind back then.

For all that, here before you is the promised book. No "vista of a freely created novel,"[13] but a vista of free reflection and a gift of memory. An idiosyncratic document about its author, his occupation and his country. About the infinitely complicated, contradictory and difficult twentieth century, which in one of my publications I called a century of warning. I would like to think that this book will be of interest to someone and even fulfil a need he feels on his path across the teetering bridge. For such a person, perhaps it will function more like a Foreword.

The hours I spent on the shore of Lake Ladoga, their concentrated solitariness, have remained with me throughout my life. I have grasped to myself the delight of those few hours. I had come to the lake for a dip, but in the end didn't enter the water. As I walked away I nonetheless felt myself baptized into a new faith. Those hours have fixed themselves in my innermost being: in times of grief and joy I often retreat into myself, and then I again see with my mind's eye the pale, limitless, and soothing

[13] In Russian, *dal′ svobodnogo romana*, an oft-quoted phrase from A.S. Pushkin's novel in verse *Evgeny Onegin*.

expanse of the lake, and in the rustling of the reeds sense an unseen power lending me support in times of adversity. Perhaps during that interval by Lake Ladoga, by its tranquil grey vastness, I learnt in my solitude a defence against loneliness.

It would be more appropriate to title this book "Certain episodes from my ruminations by Lake Ladoga"—about the past, the present, and perhaps the future.

Chapter 1
On a Knife Edge

Converging Contingencies

I imagine that many of those who have achieved success in some enterprise or other or have inadvertently avoided foundering on reefs directly in their way, have spontaneously wondered what portion of their good fortune—what they had obtained or what life had given them—was really theirs to claim in the sense of being earned or deserved. Where in the accomplishment did they themselves figure? Perhaps the outcome had been arranged by fate or dumb luck was all there was to it. In my own life there have been many occasions prompting such reflections.

Good fortune has indeed played a leading role in my life. The mere fact of my being here to speak of this is essentially due to a run of good luck.

Having studied self-organizing systems extensively, I know that the whole world, the Universum, is pervaded by indeterminacy and randomness, from the microphysical processes of the elementary particles to the spiritual life of a human being. Nevertheless, the chain of chance events leading up to my present involvement in the writing of this book sometimes seems to me fantastically improbable. It is well known that when Napoleon asked Laplace about the place of God in his cosmogony, Laplace answered laconically "I had no need of that hypothesis." If I were to explicitly marshal the unique succession of events constituting my life, I would doubtless find it difficult to take the Marquis de Laplace's view. But on the other hand, *every* event, considered as a unique occurrence, is highly improbable! To put it briefly, I have voyaged through life along a knife edge, and fate has preserved me "without nannies or French tutors."

Here is just one link of that chain of events.

I began my narrative by recalling how, by damaging my spine, a clod of frozen clay was in all probability responsible for saving my life—since no one tempting fate as a rear gunner in an Ilyushin IL-2 lasted very long. Thus what looked like misfortune turned out to provide the means to a long life.

N. N. Moiseev, *How Far It Is to Tomorrow…*,
https://doi.org/10.1007/978-3-030-96651-5_1

A few months prior I had had another narrow escape. Just before graduating from the Zhukovsky Academy[1] it was proposed that I fly to America as a member of a group of specialists tasked with organizing the delivery of certain aviation technology under the Lend-Lease program.[2] Some knowledge of foreign languages, good references from one's instructors, and, oddly enough, skill at sport: these were the qualifications carrying most weight with those responsible for choosing a team of graduates of the Academy. And although I was not a *komsomolets*,[3] who cared anything about that in April 1942? The proposal was tempting, and everyone congratulated me and envied me. But I refused outright. For me, the front and only the front! So I was posted to the 14th Air Army on the Volkhov front, as senior technician in a squadron responsible for arming airplanes.

As it turned out, with this decision I had preserved my life, although I discovered this only much later.

The "American team" was brought up to its full complement, and under the leadership of a certain colonel arrived safely on the West Coast of the US. There for four years they worked not out of fear but in good conscience. Their subsequent history was, however, tragic. On the return flight via Alaska and Siberia, during one of several stopovers—in Magadan[4] perhaps, or Khabarovsk[5]—practically the whole team was dispatched to a place whence in those years hardly anyone returned. It seems that the colonel alone returned safely to Moscow. In any case I have never heard anything further of the fate of any of the "lucky ones." And there was a rumour that towards the end of the 1940s, the colonel in question was found shot dead in his Moscow apartment.

The war spared me several times. Above my right eye, 3 or 4 mm above the socket, there is still visible a scar left by some "forest brother." I received this mark in early May 1945 far in the rear of the action, standing on an airfield near the town Augustovo,[6] on the border with East Prussia. Although the machine-gun bullet must have been almost spent, all the same had it struck me just a few millimetres lower this book would never have been written.

During the war other life-threatening situations arose which I more or less successfully managed to extricate myself from. These were, however, more of an adventuresome sort than fateful, having more to do with the usefulness to the young of a serious commitment to sport than with the role played by chance in my life.

---

[1] A military institute of higher education in Moscow, specializing in training engineers for the Soviet (now Russian) Air Force.

[2] A program whereby the US supplied, largely *gratis*, allied nations with food, oil and *matériel* between 1941 and 1945.

[3] Member of the Communist Youth League, stepping stone to full-fledged membership in the Communist Party of the USSR.

[4] City in far north-eastern Siberia.

[5] City in eastern Siberia, close to the border with China.

[6] Now Augustów, in north-eastern Poland.

On one occasion, however, chance, in the form of sloppiness or lack of initiative on the part of some functionary, once again saved my life. The details of this episode were revealed to me only much later—and completely by accident—in the safe years of the 1950s. . . .[7]

Non-exploding Bombs and the Kiss of Judas

In 1955 I was appointed dean of the Faculty of Aeromechanics and Flight Engineering of the Moscow Institute of Physics and Technology (MIPT), responsible for the education of specialists for the aerospace industry. Our graduates were taken on by the most prestigious—and top secret—design departments, where the work demanded a very thorough training. It has to be said that the instruction we gave our students was well up to the required level of rigour! Even taking into account the tremendous competitive screening which in those days was standard at the MIPT, the reputation of our graduates—one of ability combined with high professionalism—was generally recognized. In the following years I participated in a multitude of seminars and conferences abroad, and often gave lectures at prestigious foreign universities, so that I can objectively compare the abilities of young Western specialists with our own. I attribute our technological successes of the 1950s and 1960s most of all to the excellence of our engineering and technical personnel. The extremely good preparation of our young specialists compensated in many ways for bad organization, departmental monopolism, and the slackness of our administrative apparatus (no worse, incidentally, than the incompetence I had to deal with in America and France). And recalling all that, I have more than once wondered: if then, back in the 1950s, all that intellectual power and energy had only been in good hands. . . .

In those years the position of dean of "Aeromech"[8] entailed having access to top secret information about aerospace and nuclear technology. However, we now know that the biggest government secret of that era had nothing to do with technological secrets but with the quantity of resources wasted on propping up the communist regime. Be that as it may, in order for me to be able to fulfill my duties, I needed to obtain appropriate official permission to have access to said information, which involved the institute's administration's applying to the relevant security organs.

The necessary documents were indeed prepared by the MIPT and submitted to the appropriate office, but. . . there was no response! I began doing the work required of me despite my superiors' alarm over this as constituting a direct breach of iron protocol: working without appropriate formal authorization.

The rector of the MIPT was at that time Lieutenant-General Ivan Fedorovich Petrov, formerly one of the sailors who had stormed the Winter Palace, former

[7] After Stalin's death in March 1953.

[8] That is, of the Aeromechanical and Flight Engineering Faculty of the MIPT.

famous airman, former head of CAHI,[9] former air commander of the Northern Fleet, formerly in command of the Northern Sea Route, etc., etc. He had been removed from every position he at one time held. He once rather trustingly confided in me (this was, it's true, after the 20th Congress[10]): "Everyone is wondering why I was never arrested. I myself don't understand it. But I've thought of a reason: I was removed from each posting in good time." For all his sailor's rough manners and the fact of his being a true product of his era, I. F. Petrov was absolutely a person worthy of respect.

He did much good for those institutions he was put in charge of—perhaps that explains why they kept dismissing him. In particular, he arranged for CAHI to move out of its cramped premises on Radio Street, and created in the Zhukovsky a modern centre of aviation science. (They still speak of the pre-Petrov and post-Petrov CAHIs.) His chief virtue, however, was the sincerity with which he held his convictions, so that people trusted him despite the sizeable dollop of peasant cunning present in him. He was good at recruiting suitable people and could be counted on to defend them once hired, thanks to which he had many genuine friends, and is to this day recalled favourably by many. For instance, as head of CAHI he had hired M. V. Keldysh, later to become president of the Academy of Sciences of the USSR and chief theoretician of Soviet space technology, even though Keldysh, being the son and grandson of generals, was at the end of the 1930s not permitted to work on classified projects.[11] And here was I, now in the 1950s, in a similar situation. However, Ivan Fedorovich decided, at great risk to himself, to allow me to begin working without the necessary clearance, even though this may well have caused him any number of problems.

So one day, when further delay seemed to the top brass at the MIPT to augur possible serious unpleasantness, Ivan Fedorovich went in person to the Lubyanka.[12] He knew whom to talk to there and the line to take, and managed to get permission to read my dossier, in which, so he said, he saw a standard denunciation recorded—a denunciation by a "special type," a senior lieutenant and head of the SMERSH[13] detachment of the air regiment in which I had served throughout the war. I don't now recall his name, but I remember clearly how ingratiatingly he had tried to become a friend to me. He used to drop in on me often. I gave him alcohol, of which I had a plentiful supply. I also had morsels of food to take the edge off, my sergeant-major at the time being a very resourceful fellow.

---

[9] The Central Aero- and Hydrodynamic Institute, founded in Moscow in 1918 by the pioneer of Russian aviation pioneer Nikolai Egorevich Zhukovsky.

[10] In February 1956, at the 20th Congress of the Communist party of the Soviet Union, Nikita Krushchov denounced the dictatorship and personality cult of Joseph Stalin.

[11] The years 1937, 1938 were those of Stalin's great purge, whereby, in particular, a great many high-ranking members of the Soviet armed forces, together with their families, were arrested and sent to the Gulag or executed.

[12] Infamous prison and headquarters of the NKVD and then KGB in Moscow.

[13] Short for *Smert' shpionam* (Death to spies), Stalin's special umbrella organization for the various Soviet intelligence agencies.

Although I was the regimental armaments engineer, all the same I preferred not to bed down with the regimental staff. I usually set myself up closer to the airplanes with my sergeant-major Eliseev, formerly a driver on a kolkhoz,[14] a kind-hearted man and highly skilled handyman. He was my driver, clerk, and at once mother and father. He was exactly twice my age, and his attitude towards me had much of the paternal in it. Occasionally he would forget himself and call me "sonny." To get to the point: Eliseev, as I addressed him, could not stand our "special type." "Ugh, that louse. Loves to get something for nothing." (The expression "free ride"[15] wasn't in use back then.) He was free with alcohol, since it was on the government. But getting an onion or chunk of bread out of him was like getting blood out of stone.

The "special type" never got drunk, and he and I had long and, generally speaking, friendly chats. He was incredibly ignorant and with apparent pleasure and interest used to quiz me about all sorts of things. We discussed Russian history and literature. Throughout the war I kept in my rucksack, together with a sweater from home and woollen socks, a book titled "An Anthology of Russian Poets," which I had purchased in the town of Troitsk, in the Chelyabinsk region, just before boarding my flight to the front. The "special type" and I sometimes read the poems out loud to one another. We both especially liked *The Bacchic Song*.[16] Occasionally I myself tried to compose poetry—and I think he did too. In any case, it's true that he was genuinely sensitive to the music of Russian poetry. There was one memorable time when I had just returned from a week-long assignment as front-line communications officer for our air division. Eliseev, glad to see me back, found a can of pork stew somewhere, and we were about to celebrate my safe return, when the "special type" burst into my dugout.[17] However on that occasion his arrival on the scene could not spoil even Eliseev's mood. As I remember it, we drank gloriously.

During my tour of duty at the checkpoint, from where in case of need I had to maintain contact with the air command, I had managed to compose the following lines:

Since morning fluffy winter
Has dressed in holiday wear
Forest and field. Through the window
Hilly expanses are visible.
A carpet spread with silver,
Brilliant with the joyful glitter
Of myriad grains of light.
The breastworks are covered with snow,
And the white hulk of a tank
Is snowed in on the minefield.

---

[14] Collective farm.

[15] The two Russian expressions are *darmovshchina* and *na khalyavu*. The latter is the one said not to have been in use at the time in question, i.e., during the war.

[16] A poem by Pushkin.

[17] In Russian *zemlyanka*, a crude dwelling half dug into the ground.

A final shell bursts,
And once again the world is quiet.
Radiant, splendid, and clear
Is the smile of the winter dawn.

The "special type" was the first and perhaps the last person to whom I read these lines. He listened attentively and then said some words to me that seemed sincere and well-meant. I trusted him—especially following the evening when we three, captain, senior lieutenant and sergeant-major, had washed down American pork stew with a good dose of government spirits. I began to draw closer to the "special type," and we practically became friends. I even remember saying to my bosom friend Senior Lieutenant Volodya Kravchenko, a superior pilot and a decent and courageous person: "Even among the 'special types' there are real people." For Volodya, however, getting close to such a one was not on the cards. More than once he said to me: "It's impossible for a normal young person to get inside the skin of a 'special type'. He is expected to expose a prescribed quota of spies, and he'll denounce you just to remain in good stead with his superiors." Influenced by such words, I resolved after all to tread carefully around my unlooked-for "special" friend, avoiding opening my heart to him. And, as it turned out, not without reason!

In 1944 our regiment began receiving trophy dive bomber bombs, which, unlike ours, had the detonator located on the side. The German bombs employed electrically activated detonators so were without the arming vanes[18] of ours, which armed the detonators mechanically. In our bombs the axis of the arming vane was perpendicular to the side of the bomb. This type of fuse—the so-called Oranovsky fuse—was used in the Russian army in World War I and, a rather large quantity of these detonators having been accumulated in army warehouses since the time of the airplane "Ilya Muromets,"[19] it was our dubious good fortune that they had begun to turn up for use in the air regiments. However, in practice the use of Oranovsky fuses didn't work out so well. Very often the bomb failed to explode on impact, even though the detonators were certainly serviceable.

Those in command became worried and began to issue angry orders, ascribing the failure to explode, naturally, to the armourers. These orders all involved the same accusations: slackness in the preparation of aircraft equipment and failure to follow the appropriate technical instructions in such preparation. We poor armourers were threatened with harsh punishment. My immediate superior Colonel Tronza, chief arming engineer of the 15th Air Army, was especially enraged. A pedantic, rigorous Latvian, he had been an active participant in the 1917 Russian revolution. One day a U-2[20] arrived out of the blue with Tronza and his own mechanic on

[18] These had the form of miniature propellers. Air resistance along the side of the falling bomb was supposed to cause the arming vane to turn and thereby become unscrewed, arming the detonator.

[19] The Sikorsky Ilya Muromets was a class of four-engine heavy bomber used during World War I by the Russian Empire. Named after a folk hero of Kievan Rus′.

[20] The Polikarpov U-2, nicknamed "corncob," was a general purpose Soviet biplane used for low-cost ground attack and aerial reconnaissance during World War II.

board. He demonstratively had several bombs armed, took off with them in the U-2 in which he'd arrived and dropped them over a neighbouring marsh. Every one of them exploded!

Tronza then publicly accused me of betraying the worker-peasant government (not the Motherland but the government!), removed me from my duties and ordered that I go to trial, adding that he had long before been thinking of replacing me with a new regimental engineer. Everyone knew what all this meant for me: a death sentence, in effect.

I was at a loss to understand what was going on. When *we* had prepared the bombs we had done exactly the same as the mechanic of the colonel who had so come down on us, yet *our* bombs had not exploded! In my agonizing search for an explanation, on which my very life depended, Eliseev unexpectedly came to my aid. He had been sitting at the far end of our hut staring gloomily at the ground. Suddenly he turned to me with an expression of enlightenment: "Comrade captain, perhaps the reason lies in the fact that they dive-bombed using the U-2." Light dawned at once.

Our airplanes flew five times faster than the famous "corncobs." It followed that the air resistance acting on the blades of the arming vane would be twenty-five times greater. Thus the force on the arming vane as a whole would also be twenty-five times greater. Clearly such a force would simply bend the shaft of the arming vane, flattening it against the surface of the bomb. It would therefore fail to turn and the detonator would not get armed. That was all! All that was needed was to take a pair of pincers and snip off all the blades of the arming vane except for a pair of diametrically opposite ones. One hardly needed to be an engineer to understand this.

Later on I would be thanked for finding this solution—no, not by Colonel Tronza, with whom fate would never again bring me into contact, but by the army commander Lieutenant-General Naumenko. The procedure of "snipping off excessive blades" became widely used in other regiments, and dropped bombs thenceforth exploded as they should. But that was some time later. Back then I ran straight to the regimental commander, who got my drift immediately, and emitted violent curses aimed, in particular, at me and Tronza and our respective mothers. We at once drove over to the airfield, where I myself prepared several bombs, cutting off the excess blades, hands trembling with anxiety. The commander's airplane went aloft and all of the bombs exploded on hitting that same marsh!

As the commander exited the aircraft, Tronza unexpectedly appeared on the scene. He had been about to board his plane to leave the regiment, when he heard the explosions. A threatening growl was heard: "Lieutenant-colonel, who gave you permission? I had Captain Moiseev removed from duty. You'll answer for this!"—and so on and so forth in this vein. But all this had no further relevance!

Well, my "special type" had described this episode in his denunciation, but without the ending, that is, without mentioning the army commander's gratitude. Copying Colonel Tronza's words, he called me a traitor of the Motherland and recommended immediate arrest. However, someone had scrawled something illegible across his report, and following these indecipherable words was the word "delay" or "postpone" with an equally unclear signature. And that is how that denunciation got into my dossier. But at the Lubyanka they had in any case decided not to permit me to engage in classified work.

When in the spring of 1946 I left active army service as engineer in the air division, the "special type," who had also risen in rank, came to see me off. He embraced me—I did not then know that this was the kiss of Judas!—and showered me with good wishes.

The episode I have just related might easily have cost me my life. If Sergeant-Major Eliseev hadn't twigged to a possible explanation, if not for the laziness or negligence of someone among my "special type's" bosses.... Or perhaps, as Captain Kravchenko surmised, the requisite plan for dealing with exposed traitors of the Motherland had not been delivered to the divisional agency of SMERSH or the old plan had already been fulfilled and they had put aside the denunciation to keep in reserve!

And when Ivan Fedorovich Petrov had grasped the essence of the matter, he wasn't slow in putting everyone in their place: Stalin was already dead, Beria shot, and the 20th Congress of the Party was imminent. The atmosphere had changed in an essential way. I soon got level-one access, that is, at the highest level, to classified work, and even more: I thenceforth never encountered difficulties with arranging flights to the testing ground or field trips abroad.

Flight, Turning into Triumph

The last of the episodes which might have turned my life upside down took place at the beginning of the 1950s.

My stepmother, who had for a quarter century worked as a junior high school teacher in Skhodnya[21] was unexpectedly arrested under Article 58 as an active participant, if one could believe it, in a group preparing an armed uprising. They sentenced her to ten years' imprisonment and dispatched her to a camp[22] near the town Taishet. Generally speaking, this was not at all an unusual occurrence in those times. For me in particular, it had extremely painful repercussions and may have ended in real tragedy, if not for ... if not for yet another lucky turn of events. But I shall come to that in due course.

Towards the end of 1948, following demobilization I began working right away in two places. My main workplace was at Scientific Research Institute No. 2 (SRI-2) of the Ministry of the Aviation Industry, where I had been appointed as a "theoretician" in a group under Dillon, chief designer of rocket-propelled aviation torpedoes. Although he was suffering from consumption and was very weak, Dillon got through a surprising amount of work, and was always full of various ideas and

[21] Formerly a town in the Moscow district, absorbed in 2004 into the city of Khimki.

[22] Of the infamous Gulag, so-called.

initiatives. His inventiveness was legendary: had he lived longer he'd have produced many more technological innovations.

As it turned out, I was assigned to the same group as my friend Yury Borisovich Germeier.[23] We had met and become friends way back in Grade 10, in the mathematics circle[24] conducted at the Steklov Institute[25] by I. M. Gel′fand,[26] then still a dozent[27] at Moscow State University (MSU). In our student years I shared a room with Yura in the student residence on Stromynka St., and we graduated from the same department of MechMat[28]—that of the theory of functions and functional analysis—and had the same supervisor, Professor D. E. Men′shov. We have lived our whole lives, essentially, in parallel to one another. Later, when I began working at the Academy of Sciences, I brought Germeier over to the Computing Centre, where he set about organizing a department of operations research. He went on to found a similar department in the Faculty of Computational Mathematics and Cybernetics, one of the most interesting of that faculty.

But to return to 1948: Germeier had not gone to the front. As someone with a German name, he was not admitted to the army, and, even though his mother was Russian, he was, like everyone with a German surname, subject to resettlement. At first he was sent to Stalingrad, where he was assigned to factory work. Then at the time of the German advance on Stalingrad, in the midst of the frenzy and tumult leading up to the heroic Stalingrad saga, for some unknown reason someone had him sent back to Moscow. There was, however, nowhere for him to come back to. Finally it was proposed that he work in one of the most secret engineering facilities in Moscow, namely that where the first katyushas[29] had been built. Thus it was that once again Yura and I found ourselves sharing a room: no longer in a student residence, but in Scientific Research Institute No. 2. He was working on problems of efficiency, and I on the dynamics and ballistics of aviation torpedoes.

Our section worked with enthusiasm: such was the general mood of the immediate postwar years. The work went quickly and very successfully. The head of the institute was then Major-General P. Ya. Zalessky, a fine engineer, a poor mathematician, and, like all Odessans, very witty. When he had to participate in meetings where he might need to discuss complex computations of some kind or other, he took me along with him. He would announce me publicly as "the governor's scientific Jew," although he himself was the Jew as well as the "governor." To put it briefly, the work at the institute was not only interesting, but took place in a creative atmosphere, as they like to say nowadays. The work moved

---

[23] Hermeyer in German.

[24] A special mathematics class for gifted students.

[25] A prestigious mathematics research institute attached to the Academy of Sciences of the USSR.

[26] Later to become a pre-eminent Soviet mathematician, contributing to several branches of mathematics.

[27] Academic rank roughly equivalent to lecturer or associate professor.

[28] Short for Mechanical-Mathematical Division.

[29] Soviet rocket launchers. *Katyusha* is a diminutive form of Ekaterina.

ahead by leaps and bounds; our section and in fact the whole institute was on the rise.

I combined research with teaching, which was not a whit less fascinating. I was appointed acting dozent in the Department of Rocket Technology of one of the best institutes of technology in the whole of the USSR: The Moscow Higher Institute of Technology (MHIT), from which in the distant nineteenth century my grandfather had graduated and for which I had from childhood been accustomed to holding in the highest esteem. When I took up my teaching duties there, the Department of Rocketry was led by Professor Yury Alexandrovich Pobedonostsev, a legendary persona.

First of all, he was one of those very few fathers of Soviet rocket science who had escaped imprisonment at the time just before the war when Stalin engineered the destruction of the whole of Soviet rocket technology, nurtured for so long by Tukhachevsky, shot dead for his pains. Yury Alexandrovich told me that he had spent two years in daily expectation of arrest. Although, like I. F. Petrov, he could not think of any reasonable basis for his arrest, even of the most minor sort, he nonetheless kept a valise in readiness next to his bed containing all an arrestee might need.

He considered his greatest scientific achievement to be his invention of slow burning explosive powders whose rate of combustion was constant over a very wide range of natural conditions: temperature, humidity, etc. It was this, in fact, that accounted for the success of our "katyusha," that fearsome weapon of the Great Patriotic War.[30] Yury Alexandrovich correctly claimed that Kostikov[31] had essentially robbed him of the credit by appropriating to himself all laurels as the "inventor of the katyusha."

In 1949 Professor Pobedonostsev was at the zenith of his career. He was chief engineer, that is, essentially scientific director of the famous Scientific Research Institute No. 88 (SRI-88), in one of whose design offices the still not fully rehabilitated S. P. Korolev[32] had begun working. On the eve of the 1950s Yury Alexandrovich was not only head of SRI-88, but in effect in charge of the team of engineers and scientists taking shape at that time that would lay the foundations of modern space science and technology.

---

[30] World War II.

[31] From 1941 Andrey Kostikov was for a few years director of the Soviet Rocket Research Institute. It is thought that earlier he may have had a hand in arrests made in that institute.

[32] Sergey Pavlovich Korolev, the so-called "father of Soviet space technology," was the leading Soviet rocket engineer and spacecraft designer during the space race between the Soviet Union and the United States in the 1950s and 1960s. He and others working at the Rocket Research Institute were arrested in mid-1938 as "wreckers," possibly denounced by their colleague Andrey Kostikov. Korolev was imprisoned in the Lubyanka, where he was tortured, before being sent to a Gulag camp in Siberia in 1939. In late 1939 he was returned to Moscow to work in a *sharashka*, a special penitentiary where imprisoned scientists were forced to work on prescribed projects, and in 1942 he was moved to another *sharashka* in Kazan′. He was allowed to see his family again in mid-1944, but the charges against him were dropped only in 1957.

During those years he created at MHIT a Department of Reactive Power Technology, later headed for many decades by Professor Fedoseev. Pobedonostsev assembled in his department a very interesting, albeit apparently totally incompatible, collective of people. There was the aforementioned Sergey Pavlovich Korolev, future Chief Designer of Rocket and Space Technology, who, although without a formal academic degree, was given the rank of dozent, and then a youthful Professor Chelomey, who gave superb lectures, and also Barmin, future member of the Academy and a rather gloomy and ungracious type, among many others to whom the country owed the creation of its rocket technology. Later on they would disperse to their separate institutions, but at the end of the 1940s they were all together in one place.

Pobedonostsev himself was, alas, in personal decline during those years. His interest in science had diminished, and his thoughts were more directed towards his family and his garden, which he loved passionately. His lecturing was careless: he didn't bother preparing his lectures thoroughly, and he often had young instructors give them in his place. Thus he sometimes deputized me to lecture on the combustion of powders, a topic I understood poorly. He had also lost interest in the day-by-day business of the department. Once as a joke I hung up a poster bearing the slogan "Brothers, we *will* lift a finger!"[33] in the room where departmental meetings were held. Yury Alexandrovich was a decent person and had a good sense of humour, and he laughed spontaneously when he saw the slogan, and asked that it be left in place. I should mention that our collective was so constituted that research and teaching continued by inertia along well-worn tracks, despite the fact that sometimes Yury Alexandrovich didn't attend departmental meetings in person, preferring to run them by telephone.

I was, however, unexpectedly rebuked ... by the secretary of the MHIT Party Committee, not for anything to do with my work, and not even for the satirical text of the slogan, but for hanging up a poster without clearing the text with the Party Committee.

One day in the instructors' cafeteria over lunch I began enthusiastically describing to Yury Alexandrovich something to do with the special guidance peculiarities of a certain rocket system I was working on. He listened politely for a while and then, interrupting me, suddenly said:

"Nikita, you also live out of town I believe."

"Yes, in Skhodnya."

He thereupon turned his face towards me—he seemed to have grown suddenly younger—and continued animatedly:

"You know, I have a small apple tree," and he extended his hands above the floor to show how small it was, "and yet it grows such apples," and with his cupped hands he indicated a volume about that of a small watermelon.

---

[33] The original is *Brattsy, ydarim palets o palets*, derived from the Russian idiom *ne udaril palets o palets* ("didn't hit one finger with another"), meaning essentially "didn't do anything much."

This episode shows him as he was in his role as our kindhearted and clever departmental head. When inspired he might have moved mountains—but only when inspired.

It was in his department that I was given my first independent course to teach: on the dynamics of guided missiles and rockets. I believe that this was in fact the first such course ever taught in Soviet tertiary institutions of learning. It was classified top secret, and I had to keep my notes in a safe at SRI-2. Dillon, my immediate superior there, looked over these notes more than once, and insisted that I submit them as my doctoral dissertation[34]—so I resolved to submit them in the very near future. Pobedonostsev also said encouraging things about this work, evaluated it favourably, and often had me accompany him to the seminars given at SRI-88, in Podlipki,[35] where at that time the foundations of rocket technology were being laid and projects to build future rocket systems were in the works.

Thus as far as my scientific career was concerned everything was as good as could be. Every day I learned something new and my perspectives seemed unlimited. I noticed, furthermore, that people were interested in my work, which was of prime importance to me since it made me feel that what I was doing was worthwhile. It gave me the feeling that I was doing my work not just to satisfy my own curiosity, but because it was needed—needed by my comrades and my country, which had just emerged from a terrible ordeal. Although I never spoke to anyone of these feelings, they afforded me crucial inner support. I don't know if everyone felt like that back then, but I would have found life insupportable otherwise. The darkest periods of my life have been those when I felt sure that my work would never be of any use to anyone. Although I have often had to work without being able to publicize my results, I have never learned to be satisfied with that. Which is why, when Gorbachev's perestroika started and the government's and country's interest in scientific research faded, I began spending time on publicizing science in various ways—although, having observed the vain efforts of dissident members of the intelligentsia, I understood very well how pointless such activity was. All the same, what I then wrote was published and people read it, which is more than one can say for most of the output of scientific research.

But that was much later. Back in 1949 I was in a state of great elation called forth by the work I was engaged in.

My research activity was going along swimmingly, and soon I was considered, if not a leading research theoretician of the field of rocket technology, then at least up-and-coming, which could not help but give me satisfaction. I was lecturing on a fresh and interesting topic in one of the most prestigious of Soviet higher engineering schools and my lectures were appreciated not just by the students but also by the colleagues from various SRIs who came to listen to them.

---

[34] The Soviet "doctor of science" was a degree beyond that of "candidate" (≡ Ph.D.) requiring proof of substantial further research capability.

[35] A village outside Moscow, since renamed Korolev.

And then—bang! It all came crashing down. My stepmother was arrested. At first I didn't correctly gauge the scale of my personal calamity. I was infinitely sorry for that innocent elderly lady, who had endured a difficult and bitter life without seeing anything much in the way of the bright side, but I failed to connect what had happened to her with my own fate, naively thinking myself shielded sufficiently by my qualifications and my former active service in the war, for which I had a very respectable collection of military decorations.... However, I was very soon to feel the weight of this event bearing down also on me.

On coming to work at SRI-2 one day, at the entrance control point I was told that my pass had been cancelled, and in the personnel department they informed me that I had been dismissed in the interests of staff reduction. General Zalessky refused to see me. Something similar occurred when I went to the MHIT, although it's true that there they were more polite, explaining that I had been stripped of authority to access classified work so no longer had the right to fulfill a dozent's duties in a classified department. And they offered me the post of assistant in either the Department of Mathematics or Physics, but only at an hourly rate—meaning essentially gratis. My leave-taking from Yury Alexandrovich was sad. He was sincerely upset by what had happened, accompanied me to the Metro, and gave me several pieces of absurd advice. I fully understood, of course, that he had no other recourse. I met him only once more after that, in 1960 at a conference in Baku. He was then retired. We drank a bottle of red wine in his hotel room, munched on grapes, and talked about old days. That meeting, following an interval of ten years, gave us both much pleasure.

But in 1949 I was not only cast out onto the street but deprived of the right to work in my specialty; there no longer seemed to be any possibility whatsoever of my being able to work as a scientist. The manuscript of my doctoral dissertation remained locked up in the safe, and I was never to see it again. However, I was told later on that it had seen use—but that was already in another life, when it had ceased to be of any interest to me.

For a month or maybe more I went around like someone under water. No one would hire me. At first they would address me kindly, but as soon as they saw the stamp in my workbook,[36] all conversation ceased. I managed somehow as long as I still had money left. Most of my friends avoided me. Gradually a real panic began to come over me: it was no longer a question of a scientific career but of life itself. All this was much more terrifying than anything I had experienced at the front.

Once again I was saved by sheer chance, by an unlikely combination of favourable circumstances.

One of my climbing friends and a colleague at the Zhukovsky Academy, one of the very few who then, in the winter of 1950, did not seek to avoid me, was Alexander Alexandrovich Kulikovsky. At that time he had the rank of major and was teaching radio technology at the academy.

[36] Every Soviet citizen who had at some time been employed had an official workbook, where the history of his or her employment was recorded.

On the night of the arrest of my stepmother, Sasha[37] and his wife Nina were visiting me in Skhodnya, and after the arrest they stayed on with me. During that whole winter the three of us lived together in the old dacha in Skhodnya. And then one evening when I had returned from Moscow in a thoroughly depressed state of mind after yet another fruitless day of looking for work, Sasha said to me: "Look, Nikita, why don't you go somewhere else nice and healthy, somewhere further off. While there's still time, you should just send Moscow to the devil." That's exactly what he said.

But where to go? Who am I? What could I do? I was an unrealized mathematician, an engineer trained to arm airplanes, who had been dismissed from work as an unreliable element. Maybe they would indeed hire me in some provincial university or college or high school. Mathematics teachers are after all in demand everywhere, aren't they?

So next morning I went to the Office for Universities in the Ministry of Higher Education, not imagining that my future was in the balance, that fate was indeed preparing a fluke encounter. In the corridor I ran into Professor Grigory Ivanovich Dvusherstov, a former subdean of the Mechanico-Mathematical Faculty of MSU. He recognized me and stopped me.

"Moiseev? So you're alive?" This was the standard question in those immediate postwar years, when every person returning from the front was greeted joyfully.

"As you can see, I am."

"So you fought," he continued, respectfully touching the military decorations on my epaulet-less tunic. In those days all those who had been at the front continued wearing their officer's uniform until it wore out, since in 1950 suits cost the earth. And it was also customary for military decorations to be worn along with the uniform.

"Well then, let's go and have a yarn."

It turned out that he himself was the head of the Office for Universities, that is, the very person with whom I had intended to make an appointment at reception. Our conversation got off to a good start.

"I'm glad that you remember me, Grigory Ivanovich."

"How could I forget you. The winter term came, and Moiseev had disappeared: gone from the games, gone from the ski team. So tell me how you fought, how your service ended up.

"In unemployment...."

And on impulse, feeling as if I was making a confession, I related all that had befallen me to Grigory Ivanovich.

He was a kind and compassionate fellow, and his students loved him. In this he had contrasted strongly with the other subdean, Ledyaevy, who had been dry and standoffish. Except that Grigory Ivanovich drank—was in fact a considerable tippler. Some years later, when I was already a professor at the MIPT, I ran into him by chance near the Pushkin Monument. He was evidently hung over.

[37] That is, Alexander Kulikovsky.

"Moiseev, good to see you!"
"Grigory Ivanovich, hello."
"Let's go and have a drink."
"I can't, Grigory Ivanovich, Alexey Andreevich Lyapunov is expecting me. He's flying to Novosibirsk tomorrow, and we have a lot to discuss."
"Not important. Your Lyapunov can wait. There's a place here, just around the corner."
At that time in a building at the bottom of Tver′skoy Boulevard—the building has been torn down since—there was the cinema "The Great Mute"[38] and next door a tacky eatery where, standing, one could eat a little and drink a lot.
We went up to the counter, and Grigory Ivanovich ordered:
"Two hundred-gram shots,[39] two mugs of beer, and cut that sandwich over there in two."
That was Grigory Ivanovich....
To return to 1950: After listening to my story, he began to reflect. Apart from asking me a couple of questions, he was silent for a considerable time. Then he examined me carefully as if assessing me:
"Go to Rostov, old chap. There they've jailed the whole of the Department of Mechanics, including Professor Korobov, its chair. There's no one left there to give lectures. You can teach hydrodynamics and classical mechanics."
"But I've never learnt classical mechanics; I specialized in functional analysis under Men′shov."
"Well, you know, if it's the head that's of concern, don't worry about the neck. Tomorrow I'll be seeing Belozerov, the rector in Rostov, and I'll tell him about you. Come back tomorrow at 11:30 and we'll settle the matter with him. And in a week's time let there be neither hide nor hair of you in Moscow!"
Thus it was that I went to Rostov-on-the-Don as acting dozent in the Department of Theoretical Mechanics of the university there. Dvusherstov also sent Iosef Izrail′evich Vorovich there to fulfill much the same duties. Like me, he had defended his candidate's dissertation[40] at the Zhukovsky Academy, and had also been unemployed, although for different reasons. But these were not the only ways in which our fates were similar: he, like me, would subsequently be elected to full membership in the Academy of Sciences of the USSR.
My "flight" from Moscow was to play a decisive role in my life.
This was, however, not just because my living conditions in Rostov and my teaching duties at the university there afforded me a few years of peace to do my work, to ponder matters, and to acquire the knowledge that would later form the basis of my professional research activity. As I appreciated only much later, of far greater importance was the circumstance that I was thus for some time out of sight of the state security organs. If I had remained in Moscow, then at any moment, in

[38] Showing silent films, or dating from that era.
[39] Of vodka.
[40] Roughly equivalent to a Ph.D. thesis.

the course of a routine "filling of the spy quota," as Volodya Kravchenko had put it, I might have been caught on the hook.

And sure enough, a year or year and a half following my departure for Rostov, the district security organs[41] began to take an interest in me. I later found out that it was they who had organized the denunciation and trial of my stepmother. According to my neighbours, they came to the dacha more than once, but it was locked up, and the neighbours genuinely didn't know what had happened to me since I hadn't told anyone in Skhodnya where I was going. It wouldn't have been very difficult to locate me, of course, but I was saved by bureaucratic disorderliness and the usual lack of initiative of security underlings.

All the same the security people did one day find me, but not till the end of 1952.

I now know for a fact that a dossier on me had begun to be compiled in Rostov. I even know who had been summoned to bear witness and where and what they had been asked. And I am happy to report, with absolute certainty that this is the truth, that none of them could be persuaded to write anything incriminating against me, not even amongst those outside my circle of friends. As a result, at that time, the beginning of 1953, the denunciation failed to gain traction. It must be remembered that then, just before the curtain was to be rung down on an epoch, it was an especially harrowing time, when those who had defended their Motherland were being slapped down. I, however, had escaped the blow. And then in March of 1953, Joseph went to meet his Maker, and by that autumn my stepmother had come back home from the Taishet camp, and a page in my life had been turned.

So fate, in the guise of favourable chance occurrences, saw me through those difficult years. Shielded by youthful obliviousness, I lived without especially reflecting on the fact that for many years an executioner's axe had been suspended over me. I neither knew nor intuited this—happily for me!

[41] That is, those of the district where the author had lived outside Moscow.

Chapter 2
A Few Really Happy Years

The Year 1921 and the Return to Moscow

Happiness is a concept with a large subjective component. Of course, everyone experiences minutes or even hours when a special lightness, a particular delight at being alive, comes over them. This happens, for instance, when one feels especially well or suddenly senses the beauty of surrounding nature, or when one's strivings are crowned with unexpected success.... A joyful feeling of this sort takes hold of me whenever I am deeply involved in my work—even now, when I'm no longer young and cannot boast of my health.

Sad to say, such moments of elation arise more and more seldom with the years. They still crop up, however, and even now sometimes when going to bed I am ready to repeat the words of the children's song: "There'll be another day tomorrow." Then, certain of its coming, I experience a joyful anticipation of the morrow—an anticipation consonant with the childlike optimism so aptly conveyed by that simple line from a lullaby.

Now, however, I want to dwell for a little on something else. We have all had periods in our lives that we distinguish from the rest for being generally happier and to which we return most often in memory—especially when we are alone or going through difficult times, when we look for present support in past memories. There were two such periods in my own life, periods free of shadows, without sicknesses, grief or arrests.

The first such period was that embracing some of my childhood years, when my family, still in its full complement, was living in Skhodnya. It was precisely during those years that I really did live through what they call a happy childhood—and felt what it means for someone, especially a child, to have a real family. The associated memories are sacred to me.

The second period is that following my demobilization, my stepmother's arrest, and the collapse of all my Moscow beginnings (which I have already related), when, suddenly and unexpectedly, everything came together for me: I obtained a real,

N. N. Moiseev, *How Far It Is to Tomorrow...*,
https://doi.org/10.1007/978-3-030-96651-5_2

and absorbing, job at Rostov University and at the same time acquired a family of my own and experienced the birth of my eldest daughter, around whom there immediately materialized a completely new life, full of enchantment.

Although these two periods were very different, there were things they had in common: a peaceful, rhythmical unfolding of life, a quiet domestic benevolence, the opportunity to busy oneself with whatever one wished, and the chance to spend a great deal of time out and about in nature. However, of greatest importance back then was the warmth of my relationships.

But to begin from the beginning.

I was born on August 23, 1917[1] in the mansard of the house at No. 7 Afanas′evsky Lane, now Myaskovsky Street. The street numbers have also been changed, although the house remains. It even bears a memorial plaque, though having nothing to do with my family. I was christened at St. Nicholas's church in Khamovniki,[2] where my father had been both christened and married.

Those years were very difficult ones for my parents. In 1918 my father was dismissed from his position at the university, and the family was left without means of support. A relative came to our aid, offering my father work in the countryside. Thus it was that we—Papa, Mama and I, not yet one year old—travelled to the village Gorodok in Tver′ Province,[3] situated on the bank of the Mologa River, 7 km from the large, and formerly rich, town of Sunduki, and not far from Maksatikha Station. It was generally reckoned that my father had been lucky: he was put in charge of a small office overseeing the preparation and floating of timber downriver to Moscow.

According to my father, we lived poorly there although we didn't go hungry. My father was officially assigned a horse, which he learnt to handle well. The horse loomed large in our life and even I have vague memories of it. We had a garden and were provided with milk by the prosperous peasant from whom the office rented the house we lived in. It was different with bread: Tver′ Province had never been self-sufficient in bread. The noise of revolution was somewhere off in the distance. On the banks of the Mologa people continued working and trying to survive.

We lived out the three most difficult and famished years of our revolution in Gorodok. We may have continued living there somewhat longer if not for the imminent arrival of my brother, which determined my parents on returning to Moscow.

Although life on the Mologa has left only hazy images in my memory, I recall very clearly the return trip to Moscow.

The journey from Maksatikha to Moscow took a whole week. We travelled in an overloaded goods wagon, for some reason called a "veal wagon." We were lucky in having an upper bunk. The train stopped often because the engine kept running

---

[1] So about two and a half months before the October Revolution in Petrograd and the installation of Lenin's Bolshevik government.

[2] A suburb of Moscow.

[3] Tver′ is a city to the northwest of Moscow on the Moscow-St. Petersburg railway line.

out of wood, and then men went into the woods with axes and saws. The engine stimulated the liveliest interest in me; even now before my eyes I see its tall funnel. It was evidently some sort of antediluvian locomotive, preserved by a miracle on some auxiliary track. Boys are always attracted by technology. I recalled that locomotive when I was told that the first word spoken by my eldest grandson was not "mama" or "papa," but "crane," which caused his parents some consternation. This was easily explained, however: construction had been going on outside his window, and the crane in action there had evidently made quite an impression on him. No less than that primordial locomotive had made on me.

But at last the train drew up safely at Moscow's Nikolaevsky Station, which was what Leningrad Station of the October Line was called then. I don't know if it was in the dead of night or late evening or early morning; I only remember that it was dark. Even now I can see the vast empty expanse of Kalanchevsky Square and the snow falling from above out of the nighttime blackness.

My father went off somewhere and was away for a long time. We were left alone. Mama was having a very hard time of it. My brother was due in two months. I pressed myself up against her legs and sensed how she wept. I think that she may actually have been moaning rather than weeping. She was cold and unwell. In earlier days, whenever things had not been going well for her, she liked to press me to her, and recite quietly: "Oh-oh-oh, life's hard for little Alyona in an alien clime."[4] My mother's name was Elena.[5]

But here at last was my father with a sleigh. Our scanty worldly belongings were loaded onto the sleigh, I was planted on top, and we set off on a long ride through the nocturnal Moscow of the year 1921. I see before me now that nighttime Moscow wilderness where not a single light showed. The sidewalks were heaped with snow, and a beaten track wended its way down the middle of the streets.

At last we arrived at the house on Afanas′evsky Lane in whose mansard I had been born. The house belonged to Nikolai Karlovich von Meck, son of the famous Nadezhda Filaretovna von Meck, who had done so much to ensure that Tchaikovsky was relieved of material cares and could devote his life solely to music. Nadezhda Filaretovna never actually met the great composer, but their published correspondence became a sort of classic.

More than ten years earlier Nikolai Karlovich had adopted my Mama after she had become fully orphaned overnight. He never made any distinction between her and his other daughters; in fact it seems she was his favourite daughter.

No one was expecting us. The letter that Mama had written had not reached "Granddaddy," as Nikolai Karlovich was called by his family. The whole house was

---

[4] In the original Russian the folk-ditty is as follows:

Okh-okh-onyushki,
Trudno zhit' Alyonushke
Na chuzhoy storonushke.

[5] So not Alyona, as in the folk-ditty.

roused, they began to ooh and aah, talk about how dangerous it was to go about Moscow at night, and say various other things that people say in such situations. They heated water on the *burzhuika*,[6] sat me in the bathtub and began to wash off the dirt that had accumulated over the week I'd spent travelling in the "veal" wagon—to be followed by a clean bed and blissful sleep!

Giving birth was beyond difficult for Mama. She caught childbed fever and a few months later died from a general infection of the blood.[7] During her illness her friend, or rather colleague—they had worked together as sisters of mercy in a hospital train on the Galician front[8]—visited us. Following on Mama's death she stayed on in the family, and after a short interval married my father. Thus it was that my brother Sergey and I acquired a stepmother.

My brother used to call her "Mama." She was in fact like a real mother to him: she had had the care of him, after all, since he was but a few months old. As for me, however, I could never forget how I used to cling to Mama's leg, and how she stroked my head while murmuring "Oh, Nikita, my Nikita." Throughout my life I was never to know greater love or tenderer caresses than were contained in those words. And I could never forget how she used to sing softly in my ear the words "... the dashing coachman slept but half the night." So it was that, despite all her love for my father and brother, and her capacity for self-sacrifice, my stepmother and I were never to become close. Something always separated us. Although this was a source of grief for me, I was unable to do anything with myself on that score.

Skhodnya

And now we come to Skhodnya.

Skhodnya remains the dearest place on Earth for me, and the time I spent there the happiest of my life—even though there were more than enough hardships and sorrows to go round. But perhaps it was precisely that combination of good and bad that was precious to me for having been a part of my life from the very beginning.

The Civil War[9] was behind us. The Far East had been restored to Russia, as a result of which my grandfather was able to return to Moscow. In 1915, Sergey Vasilievich Moiseev had been appointed head of the Far Eastern Railway

[6] A sort of little wood stove.

[7] According to a family archive in possession of the author's elder daughter Irina, Elena Nikolaevna Moiseeva died on May 7, 1926, so several years after the return to Moscow.

[8] In the early stages of World War I a front was established in Galicia, then a province of Austria-Hungary, with capital L′vov. In the Battle of Galicia of 1914 the Austro-Hungarian armies were forced out of Galicia by the Russians.

[9] The Russian Civil War lasted from November 1917 to October 1922, although peripheral skirmishes continued for another two years. The two largest groups of combatants were the Bolsheviks' Red Army and the loosely allied opposing forces forming the White Army.

District, comprised of all Russian railway lines running eastwards from Chita,[10] including the famous Chinese Eastern Railway.[11] For some time during the period of existence of the Far Eastern Republic,[12] Grandfather had been Minister of Railway Transportation or Communication Routes—I don't recall his exact title. During the occupation of the Russian Far East,[13] he had lived near Khabarovsk in temporary quarters set up in a wagon salvaged from the wreckage of an armoured train. My grandmother and great-grandmother were there with him. How they managed to survive those difficult times of occupation I don't know. They must have endured a great deal of hardship. To put it briefly, Grandfather had decided to stay, and the Japanese and Whites hadn't gone out of their way to harm them. Then when the war ended in 1922, Sergey Vasilievich Moiseev travelled to Moscow in the very carriage he had lived in for the previous two years.

In that same year we had moved to Skhodnya. In those days it was a very congenial exurban settlement. It dated from the time of the construction of the Nikolaevsky (later October) Railway Line, and was populated largely by qualified railway workers and employees of various ranks. Before the revolution a few well-built dachas had been erected there, and it was in one of these that we were renting a few rooms. Guchkov's[14] former dacha was there, converted under Soviet rule to a school for the local children.

Our village was remarkable in many ways. Most of its streets were paved, a rare thing for suburban villages in those days. The straight streets, then called avenues, ended at the ever so clean and cold Skhodnya River, a source of pleasure for the local kids. And the village was largely non-drinking—by contrast with the big fearsome village of Dzhunkovka whose precincts began on the other side of a ravine from Skhodnya and which was always in a general state of intoxification. The main asset of our village, however, was the railway workers' cooperative, dating from the 1880s. Many of the railway workers living in Skhodnya kept a few cows or other livestock. These formed the basis of the cooperative, which rented land for haymaking from the parish and had its own store, called, fittingly, the railway store. The cooperative produced and sold not only milk, but also sour cream and cottage cheese, as well as meat and vegetables. The enterprise flourished and contributed

[10] A city east of Lake Baikal.

[11] Built by Tsarist Russia across Manchuria from Chita to Vladivostok. The Chinese government had granted the Russian government a concession for the construction, completed in 1901.

[12] Also called the Chita Republic, a nominally independent state existing from April 1920 to November 1922 in the easternmost part of the Russian Far East, tolerated by the Soviets as a temporary buffer state between Soviet Russia and former Russian territories occupied mainly by Japan. The republic came to an end when the Red Army finally took it over in November 1922.

[13] Mainly by Japanese forces, but for some time also by White Army forces and an American contingent.

[14] Alexander Ivanovich Guchkov (1862–1936) was a member of the Tsarist government, and Minister of War for a short time in the interim Provisional Government, which, following the abdication of Tsar Nicholas II, ruled Russia from March to October 1917. Sometime following the Bolshevik Revolution he emigrated to Germany.

significantly to the village's prosperity. The cooperative survived both World War I and the Civil War. It survived collectivization. It even weathered the terrible years of the Patriotic War,[15] even though the front approached to within 3 km of Shkodnya. In the 1950s I was still going to the cooperative store to buy dairy products for my children. But it couldn't survive Khrushchov's reforms. They got rid of the cows, and the entire village, then numbering several thousand, was saddled on the government. The provision of facilities for the inhabitants worsened immediately, everything became more expensive....

Thus one day on a buffered sidetrack of Skhodnya Station there appeared an armoured train carriage bearing my grandfather and his family. The interior of the carriage was like an actual apartment. His abode for the previous two years, the carriage had simply been hitched to a train headed to Moscow. My child's imagination was impressed not only by the furnishings of that apartment with its fine writing table, beds, plush armchairs, and pictures on the walls, but also and especially by the tub of red caviar that had also made the long trip. I liked Grandfather very much: big and strong, bald and mustachioed. (At one time when at the front, I also grew a ginger moustache, which hung down on either side like those of my favourite Zaporozhians,[16] and in which noodles got caught just as they had in my grandfather's.) He and I at once became firm friends.

With the arrival of my grandfather, the happiest and most peaceful period of my life began—the six or seven years prior to the execution by shooting of Granddaddy Nikolai Karlovich.

Sergey Vasilievich had been brought from the Far East to work in the PCCR—the People's Commissariat for Communication Routes. He was given an imposing title: Member of the Collegium of the People's Commissariat and Director of the Financial Accounting Committee (or F-A-C, as Sergey Vasilievich called his committee), with good pay (or salary as he called it) and various other perquisites. My father worked in the same building of the People's Commissariat—the one near the Metro station Red Gates—as a senior economist in the Central Administration of Internal Waterways. Following a memorable conversation with Lunacharsky,[17] to be described below, my father understood that a university career, and in fact a scientific career of any kind, was closed to him once and for all. He fretted a great deal over the shattering of his scientific ambitions and the impossibility of publishing his dissertation. As I was to discover in the 1960s, however, his dissertation was published in an English translation by Yale University at the beginning of the 1920s. This my father was never to know. The Russian version was seized during a search[18]

[15] That is, World War II as experienced in the Soviet Union.

[16] Cossacks living in the district "above the rapids" of the Dnieper River called Zaporozhia. It was at one time fashionable for the men of the district to sport large downward-hanging moustaches of the "Fu Manchu" type.

[17] Anatoly Vasilievich Lunacharsky (1875–1933) was the first Soviet People's Commissar of Education, responsible for culture and education, an office he held from 1917 to 1929.

[18] By the NKVD.

and perhaps filed away somewhere. My subsequent attempts to locate it proved unsuccessful.

Evidently my father gradually became resigned to his fate and set to work conscientiously in what was for him a new sphere of operation. Many books on statistics appeared on his desk together with a variety of annual reports. He began to work seriously on a statistical analysis of river freight transport. It's difficult for me to estimate just how successful he was in the field of economics, but from time to time he had articles published in a specialist journal, for which he was paid well—a not unimportant consideration at the time. Furthermore, Professor Osadchy, one of the best known specialists of the time, wrote him a letter of fulsome praise proposing collaboration. (Osadchy was later to become a member of the "Industrial Party,"[19] and it was very likely because of his acquaintance with Osadchy that my father would be arrested.)

In a word, things were going well as far as the official duties of my grandfather and father were concerned. The family had sufficient material means—in fact greater than I was ever to experience again, even after I became a full member of the Academy. We had a house built for our own use, where I passed the years of my childhood and youth.

Ivan Bunin[20] once said: "Today it is difficult to imagine how replete with intelligence and substance our lives had been."[21] I can say something similar: today I recall with amazement how life in the bosom of my family was then so measured and meaningful and the application of intelligence so predominant. In the midst of the present hustle and bustle it's impossible to imagine a peaceful life, working without hassle or stress. In those days, by contrast, the whole of life's routine seemed somehow soul-refreshing. Every morning Father and Grandfather would leave home at eight, walk unhurriedly to the station, and travel to work (or, as Grandfather called it, "service") always by the same train, numbered V-15. Only steam trains plied the October Line then, yet nonetheless the ride to Moscow took only 40 minutes, which is faster than the electric trains now in use.

Some minutes before the arrival of the train a few engineers working at the PCCR would be gathered waiting at the spot on the platform where the first carriage generally halted. They were all acquainted and greeted one another by first name and patronymic.[22] The stationmaster would without fail come to see everyone off. On such occasions he always wore a red peaked cap, but at other times, when not on duty at the station, he wore an ordinary railwayman's cap with an engineer's badge.

---

[19] The Industrial Party Trial (November 25–December 7, 1930) was a show trial in which several Soviet scientists and economists were convicted of plotting a coup against the Soviet government. The defendants were ultimately given various prison terms or executed.

[20] Ivan Alexeevich Bunin (1870–1953) was a notable Russian novelist and poet. Fled Russia for France in 1920. Nobel laureate for literature in 1933.

[21] That is, before the revolution.

[22] All Russians have a patronymic as middle name. Using the first name and patronymic as a form of address is considered very polite.

This had been the custom since the time of Nicholas I,[23] when the railways were in the process of being built; those in charge were made to wear red caps so as to be visible from afar.

The stationmaster was very respectful of my grandfather, addressing him as "your excellency" without the slightest shade of irony. In fact, since Grandfather had served at the rank of general, "excellency" would indeed have been the appropriate form of address for him according to Peter the Great's table of ranks.[24] Those working on the railways clung to the old order long after the revolution—and for good reason: Russians had always taken great pride in their railway system. That system had been very quickly restored after the Civil War, so that during the era of the NEP[25] it ran as well as in "times of peace." In any case this is what my grandfather proudly affirmed. The stationmaster was one of our most constant guests. He would drop in whenever he saw a light at our window to drink tea and play a little cards—although, generally speaking, we did not go in for card-playing. The menfolk would occasionally play *vint*,[26] and Grandmother Ol′ga Ivanovna liked *rams*.[27] What sort of game that might be I have no idea. I think it's something like *preferans*,[28] though even more primitive than *vint*, which was considered a men's game as opposed to the "women's" *preferans*.

The train always came right on time. The October Line was proud of its punctuality, and its workers were zealous in ensuring that the timetable was adhered to. The first carriage would already contain a few railway employees on their way to work, these having boarded the train at Firsanovka or Kryukovo. Everyone would exchange greetings and sit down in their customary "appointed" places. If anyone should happen to occupy the wrong seat, he would immediately be told: "Sorry, but that's Ivan Ivanovich's or Peter Petrovich's seat." Grandfather always sat in the second *coupé* at the window, facing in the direction of travel, while Father unfailingly sat on his left.

From the station in Moscow, the whole group of PCCR employees would walk in leisurely fashion to their place of work. Work began at 9:30 and it wasn't far to the Red Gates.

Regularity being considered fundamental to the work of public transport and its ancillary services, they usually returned by a fixed train, and the rhythm of the return trip was not to be broken. Grandfather liked me to meet him, and this I did

---

[23] Tsar from 1825 to 1855.

[24] Introduced by Peter I in 1722 as part of his reform of the Russian hereditary nobility. It was abolished by the Bolsheviks in November 1917.

[25] Acronym for the New Economic Policy, allowing the operation of small private businesses, introduced by Lenin in 1921 to revive the collapsing Soviet economy. It was ended by Stalin in 1928.

[26] A Russian version of whist.

[27] A certain French trick-taking card game.

[28] A 10-card plain-trick game with bidding, played by three or four players with a 32-card Piquet deck. Popular in Russia since about the 1830s.

with pleasure. He and I would then walk ahead, with me telling him Skhodnya's news, and Father a step behind. We would be greeted by all passers-by as we went along.

There were many who liked to visit us.

Some of them dropped in just like that, on seeing a light at a window. Somehow or other Saturday was established as the day when we were "at home." After work on that day local railwaymen would visit us. I remember in particular the railway doctor N. A. Shalyakin, an amiable man who looked after the medical needs of the whole family. And, as already mentioned, the stationmaster used to visit us, as well as others whom I can now barely recall. We often had visitors come from Moscow. My father was a quite good amateur artist. In his student days he had studied painting and sculpture and he had a wide acquaintance in the artistic world. In particular, he was on friendly terms with Korin,[29] who was our guest on more than one occasion. Once the great Russian painter Nesterov[30] visited. He was my father's idol.

On such occasions no special food was prepared. The evening meal was always, in any case, quite simple—more like teatime than dinner. Grandmother usually baked pies. Her mushroom pies were especially delicious. As is well known, in those days there was no food cheaper than mushrooms! Grandfather always had a carafe of vodka in reserve, distilled from a brew made with wood-shavings. He served this up only on rare occasions, however: a single shot all round on the occasion of a guest's birthday or sometimes on one of the twelve post-Easter holy days,[31] although the family was not especially religious. Grandfather and Father went to church very rarely, but even though she was of the Lutheran persuasion my grandmother used to attend the Skhodnya church every Sunday. All official holidays were observed unswervingly, however.

When we had guests, by contrast with the custom nowadays in most families I was not sent out of the room. On the contrary, it was felt that it was a good thing for me to be present at the adults' conversations. I wasn't given a place at the dining table with the adults, however, but sat at a small table placed nearby. I liked very much listening to the grown-ups' talk. And they conversed about all and everything, in particular politics, without worrying about whether I was listening or not. They talked for the most part about history and literature, but sometimes veered onto unexpected topics. They discussed the theosophy of Rerikh,[32] whom my father considered not only a great artist but also a remarkable thinker. They argued about

---

[29] Pavel Dmitrievich Korin (1892–1967) was an accomplished Russian painter and art restorer. He is famous for the painting *Farewell to Rus'*, left unfinished, lamenting the loss to Russia caused by the Bolshevik revolution.

[30] Mikhail Vasilievich Nesterov (1862–1942) was a Russian and Soviet painter. He was one of the first exponents of Symbolist Art in Russia.

[31] These marked, in accordance with Russian orthodox doctrine, twelve events in the earthly lives of Jesus and Mary.

[32] Nikolai Konstantinovich Rerikh (1874–1947) was a famous Russian painter, writer, archaeologist and theosophist, with a consuming interest in spiritual practices. He left Russia for Finland in 1918, and subsequently lived in various places around the world.

Madame Blavatsky,[33] whose writings were later confiscated during another search of our home. I also recall them discussing the illness of the painter Kustodiev,[34] much loved in our family.

I listened attentively, although of course I understood far from everything and was not allowed to participate in the conversation or ask questions.

Sometimes they read poems out loud, and such evenings I remember very well. Until December 1942, when I received a minor contusion during a bombardment, I had a very good memory: I had by heart all sorts of things. At university, for a bet I learned by rote the whole of the second volume of Bukhgolts's textbook[35] on theoretical mechanics, a terribly boring book, and could quote verbatim from any given page of it. So I was able to memorize and later declaim at will almost all the poems that were read out over our Saturday dinner table. They recited the verse of many poets, in particular, Pushkin,[36] Tyutchev[37] and A. K. Tolstoy.[38] They loved the verse of Esenin[39] and Gumilev,[40] then considered seditious. Even now I can recite Gumilev's poem *The Captains* from memory. Around the dinner table they also tried out newcomers— Mandel′shtam[41] and Mayakovsky,[42] for example—but these didn't "take." As a result throughout my life I have been unable to enjoy such, as it were "unreal," poetry. Not long ago when Brodsky[43] was awarded the Nobel prize I tried to read his poetry, so-called. It seemed to me that, although written in

[33] Helena Petrovna Blavatsky (1831–1891) was a famous Russian occultist, spirit medium, and author. She founded the Theosophical Society in 1875.

[34] Boris Mikhailovich Kustodiev (1878–1927) was a noted Russian painter and stage designer. In 1916 he became paraplegic from tuberculosis of the vertebrae.

[35] N. N. Buchholz, *Fundamentals of Theoretical Mechanics*. Technical and Theoretical Publishers, Moscow, 1939.

[36] Alexander Sergeevich Pushkin (1799–1837) was a Russian poet, playwright, and novelist. Considered Russia's greatest poet and the founder of modern Russian literature.

[37] Fedor Ivanovich Tyutchev (1803–1873), was a Russian poet and statesman.

[38] Alexey Konstantinovich Tolstoy (1817–1875) was a Russian poet, novelist and playwright. He was a second cousin of Leo Tolstoy.

[39] Sergey Alexandrovich Esenin (1895–1925) was a Russian lyric poet of peasant stock. He was married to Isadora Duncan from 1921 to 1923. He hanged himself in a hotel room in St. Petersburg a day after penning a farewell poem in his own blood.

[40] Nikolai Stepanovich Gumilev (1886–1921) was an influential Russian poet. He cofounded the Acmeist movement in poetry. He was married for a time to the poet Anna Akhmatova. He was arrested and executed by the Cheka, the Soviet secret police, in 1921.

[41] Osip Emileevich Mandel′shtam (1891–1938) was a Polish-Jewish poet and essayist who lived in Russia during and after the revolution. He was one of the foremost members of the Acmeist school of poetry. Ultimately died in the Gulag in 1938.

[42] Vladimir Vladimirovich Mayakovsky (1893–1930), Russian poet and playwright of the Futurist school. Suicided in 1930 after becoming disillusioned with the Bolshevik regime.

[43] Iosif Alexandrovich Brodsky (1940–1996) was a Russian and American poet and essayist of Jewish background. Expelled from the USSR in 1972, he settled in the US. Nobel laureate for literature in 1987.

Russian, it had very little to do with Russian culture, or with our spiritual world, or, most of all, with poetry.

At these Saturday meetings there was much discussion of history and Russia's fate, a traditional topic for the Russian intelligentsia.

These evenings left an indelible trace in my memory and formed my worldview much more effectively than could any propaganda or short course in the history of the Party.[44] To them I owe my early reading of "grown-up" books. I was but eight years old when I read the whole of Merezhkovsky's trilogy *Christ and Anti-Christ*.[45]

Nowadays the culture of leisurely conversation, so widespread in past times amongst the Russian intelligentsia, has completely disappeared. It was simply a matter of people finding it interesting to talk over tea. Now, however, when guests come we drink a lot and "exchange information" about life's problems rather than engaging in reasoned discussion. Thinking out loud is almost unheard of. Our modern get-togethers remind one more of the American variety than traditional Russian gatherings.

The constant communication I had with adults would turn out to be of the greatest importance for my future. Much that I overheard during those discussions became embedded in my soul and remained with me my whole life. And whatever I failed to understand served later as a source of questions to be put to Father or Grandfather while out walking with them. I loved to go for strolls with the adults, loved the feeling these engendered of a united team. Perhaps I was something like a puppy which, when out walking with people, continually looks up at them, as if to reinforce its sense of itself as part of the company!

The life and work of my father and grandfather were of consuming interest to me. I learnt about what was happening in the world from their conversations with one another, on this basis forming an idea of the world, of my country, and of our situation within it. When alone with them I asked about many things, and they always willingly answered my childish questions. They also told me a great deal about the history of our family and the fate of our numerous relatives.

From what my grandfather and father told me I understood that in those blessed late-NEP years the country's situation was gradually stabilizing, that Russia was again becoming a power to reckon with, and that everyone was glad of this. Except that the Bolsheviks in their side-buttoned jackets kept on doing stupid things. They were catching on quickly, however. And my wise grandfather opined that in ten years or so everything would return to normal. Although my father was more realistic, even he overestimated the likelihood of common sense prevailing; up at the top a struggle for power was under way, and the scoundrels would win—in fact the basest of them would rise to the top. Of course a modern government would

---

[44] A veiled reference to the textbook entitled *History of the All-Union Communist Party (of the Bolsheviks): a Short Course*, commissioned by Stalin in 1935 and published in 1938.

[45] Dmitry Sergeevich Merezhkovsky (1866–1941) was a Russian novelist, poet, religious thinker and literary critic. A cofounder of the Russian Symbolist movement. Lived in France from 1906 to 1912 and from 1918 to 1941. The first volume of *Christ and Antichrist* explores, in the form of a novel, the theme of the "two truths" of Christianity and Paganism.

eventually emerge again, only not like the French or German ones, but our own, Russian, sort, and furthermore not soon but in the next generation. So averred my father, but he also was far too optimistic.

The opinions of my grandfather and father diverged in many ways. Sergey Vasilievich considered the February revolution to have been the main one, believing that at its core it contained the seeds of all the misfortune our people were enduring. If it hadn't occurred, if the democrats hadn't come to power—the democrats who subsequently fled Russia leaving us to fix the mess—the war[46] would have ended at the beginning of 1918. He considered the October Revolution to have been merely a coup, but one preserving the nation's integrity, which he regarded as the most important goal of any government. For this reason he was inclined to be much more tolerant of the Bolsheviks than my father. Father could not forgive them the Civil War with its millions of dead and its destructiveness. He was sure that Russia had been on the verge of a new flowering—of its economy but more especially of its culture. Her Silver Age[47] would very likely have developed into a new Golden one. He reckoned the basest of the actions of the Bolsheviks to have been the blow they delivered to Russian culture and traditions, and the inoculation of Russia with European ways of thinking with their atheism and hypertrophied economics.

My father's take on the various events comprising the revolution was perhaps similar to that of Churchill, believed to have said at the time: "The Russian dreadnought has sunk at the entrance to the harbour."[48] He was adamant in his belief that there were no arguments, including that of national integrity, capable of justifying the October revolution and the Civil War. He also thought that the February revolution had been inevitable, and that the basic cause of the evil that had befallen Russia had been its entry into the German war, as he and my grandfather called it. My father was a man of our country's "Silver Age." The cultural blossoming and the rapid progress of that age in every field had made a deeper impression on him than on my grandfather, and he greatly esteemed the originality both of our culture and our approach to life, bitterly regretting its loss. He talked a great deal of the heroism of our troops on the German front, but was of the opinion that nothing could have been done about that tragedy: it had begun to unfold earlier—with the murder of Stolypin,[49] he said. Whenever the subject of Stolypin arose, he would ask: "Why did the security people think this necessary?" I might now be able to answer that question for him.

---

[46] World War I.

[47] Term traditionally applied to the period of Russian poetry embracing the last decade of the nineteenth century and the first two or three decades of the twentieth, considered to be comparable with the Golden Age of Pushkin and others of a century before.

[48] The original source of this quotation has proved elusive.

[49] Pyotr Arkadievich Stolypin (1862–1911) was Prime Minister and Minister of Internal Affairs of Russia from 1906 to 1911. His tenure was marked by efforts to oppose revolutionary groups and introduce agrarian reform. He was shot by an agent of the secret police while attending a performance of Rimsky-Korsakov's *The Tale of Tsar' Saltan* at the Kiev opera.

However, there was one thing in which my grandfather and father were in accord: they were sincere Russian patriots in the most civilized sense of the word. In the 1920s opposition to Russian chauvinism was part of official doctrine. To declare oneself Russian and disclose an interest in or even sympathy towards Russian culture, especially as regards its traditions and history, was considered as close to being anti-Soviet. At school, we were taught absolutely no Russian history. Of Peter the Great or the victory at the Battle of Kulikovo[50] or other pages from history, we learned only from the members of our families—and even then in secret. At home there were many books on Russian history as well as historical novels by authors such as Zagoskin,[51] Tolstoy, and Merezhkovsky. My father gave me these to read—which I did with pleasure—and we would then engage in long discussions of the subject-matter. It was then our custom for someone to read aloud to the whole family. Sometimes I did the reading and sometimes my stepmother. In this way we got through *War and Peace*, for instance. The death-bed scene of Prince Bolkonsky affected me so powerfully that I afterwards hardly slept all night.

Thus it was that, somewhere between the ages of ten and twelve, I emerged into the world.

I recall my father relating the history of the Punic Wars to me during one of our walks, following which for a long time Hannibal was my favourite hero. At one time, in 1927 or 1928, he took me to a village situated in the watershed of the Western Dvina,[52] where we stayed almost a month. We had at our disposal a boat, a fishing rod, and all thirty years of the three musketeers' adventures described in French. We would row into some quiet inlet, anchor the boat (that is, throw in a rock tied with a rope), and spend engrossing hours, taking turns at checking the fishing line and reading the book out loud. I don't remember if we caught any fish, but even now I can reproduce the adventures of the brave Gascognard in detail. And my father enjoyed them no less than I.

This compatibility between the generations has completely disappeared since the war. The lack of close contact between my own children and myself grieves me exceedingly. When growing up they seemed to find me uninteresting. Perhaps this merely reflected a general trend. Or was I so caught up in my work, in sport, in my own life, that I failed to invest the requisite part of myself in them? Did I show them insufficient warmth? And apart from that, my attempts to "organize spiritual continuity" from me to them seemed doomed to failure; even my simple efforts to have them share with me some of the details of their lives were rebuffed in no uncertain manner. This detachment from my children is perhaps the greatest regret I

[50] Fought between the armies of the Golden Horde and those of various Russian principalities united under Prince Dmitry of Moscow in 1380. Considered the turning point between the wane of Mongol influence in Russia and the rise of Muscovite power and eventual formation of the modern Russian state.

[51] Mikhail Nikolaevich Zagoskin (1789–1852), was a Russian writer of social theatrical comedies and historical novels.

[52] A river rising in the Valday Hills in Russia, and flowing through Belarus and Latvia to the Gulf of Riga.

feel in my declining years. I reassure myself with the thought that this estrangement was merely endemic to the age we live in; after all, something similar is the case in nearly all families these days. And I have observed the same phenomenon abroad: children seem to start leading independent lives at a very early age. The situation is not thereby made any easier for me, however: a spiritual vacuum remains. And I think that this alienation of the generations is not just a personal sorrow, but represents a national danger. We have lost a great deal with the disappearance of intergenerational community, formerly so characteristic of the whole of Russian society, especially among the intelligentsia and the peasants....

Our Saturday gatherings continued for a good while longer. People felt an affinity for us—good people, as I now understand them to have been. However, the character of the discussions began slowly to change. The feeling of security originally fostered by the NEP was yielding to a sense of foreboding. The "purges" began. People began to be dismissed from their jobs, and many left the country. The government put no particular obstacles in the way of members of the intelligentsia wishing to emigrate, and many of these prepared to leave for distant places sure in their hearts that they would soon be returning. All the same, they experienced pain and grief, having a clear appreciation that there, beyond the cordon, lay alien lands, none promised. How different was the intelligentsia represented by the emigrants of the twenties from our current semi-intelligentsia, with their talk of Russia as "that country." Sometimes I feel like saying to them: Well then, clear off, and leave us be while we try to preserve this nation as ours!

Every time the conversation turned to emigration, I would hear the name of one or another family mentioned. I remember especially well the leave-taking of the Petrunkeviches. The head of that family was a colleague of my father, perhaps even a supervisor of some sort. Then one day they "purged" him, although he was offered, it's true, a job as accountant in a moderately sized office of the Department of Railways. They didn't even bother to conceal the reason for his dismissal: the Petrunkeviches came from an old land-owning family, and had a close relative—an uncle, I believe—who was well known to have been a "cadet."[53] My father had joined the Cadets' party as a student, but had soon become disillusioned with it. He was very careful to conceal the fact that he had at one time been a member, however. In those days this was considered a serious crime.

I remember how Madame Petrunkevich hugged my stepmother and cried on her shoulder. We tried to reassure her as best we could. Grandfather said that they'd be back within two or three years. Industrialization was beginning in Russia and good engineers would be needed. My dear, kind grandfather, so like Taras Bul′ba:[54] ever the optimist, just like his grandson.

---

[53] A member of the liberal Constitutional Democratic Party, formed in Moscow in 1905 during the Russian Revolution of that year. The party's deputies in the Duma (Russian parliament) formed the core of the new Russian Provisional Government formed following the revolution of February 1917.

[54] The eponymous hero of a historically based novella by Nikolai Gogol′, published in 1835, recounting the adventures of Zaporozhian Cossacks warring against the Poles in the seventeenth

I also remember the departure of the Shlippenbakhs. They were an amusing family: papa, two sons and a daughter, all about 2 m tall. As my father used to say, "four fathoms of Shlippenbakhs." They were descended from a Swedish prisoner-of-war at the time of the Northern War[55] who had chosen to remain in Russia. The Shlippenbakhs were also engineers, all working in the same plant. (I think it was the Guzhon plant, as the present "Hammer and Sickle" factory used to be called.) Their guilt hinged on the fact that a prisoner-of-war from the time of Peter the Great had been able to retain not only his name but also the title of baron.

My grandfather actually received an invitation to work abroad. The famous firm of *Westinghouse* invited him to work for them as consultant at some fantastic salary. One day Grandfather returned home from work very late, his face dark as a thundercloud. It turned out that there had been a meeting of the collegium of the People's Commissariat at which they had examined the letter from *Westinghouse*, which had been sent through official channels. The collegium had decided to recommend that Sergey Vasilievich go to America, urging him further to be sure to take his whole family with him, including his children and grandchildren.

Usually we dined around seven in the evening when Father and Grandfather arrived home from work, but on that day we ate much later. There was an atmosphere of gloom hovering over the table. Grandfather maintained a grim silence until at last he said: "I didn't quit Khabarovsk back then even though it was dangerous, and now…." He stood up removed the napkin tucked in behind his tie and left the room. He retired from his job not long afterwards.

The atmosphere had thickened; everyone felt it. On Saturdays politics were discussed less and less. And the number of participants at the gatherings dwindled, ultimately to zero.

Towards the end of 1928 Nikolai Karlovich von Meck, who held a fairly lofty post in the Supreme Council of the People's Economy[56] (SCPE), was unexpectedly arrested, and shortly thereafter shot. Our family understood that we were in their sights. A year later my father was arrested allegedly for his involvement in the Industrial Party. At the end of 1930, in the hospital of Butyrskaya Prison, he died of a heart attack—or so, at least, my stepmother was told. I have never been able to verify what exactly happened.

A few months later my grandfather also died, unable to get over the death of his son. Grief crushed the family. And the means for living were gone. Thus ended my happy childhood, and a page of my life was turned to a new, very difficult one.

century. The author's grandfather's moustache and bald pate would have made him resemble the hero of the novella.

[55] The Great Northern War, lasting from 1700 to 1721, between the Swedish Empire and a coalition, ultimately victorious, led by the Tsardom of Russia under Peter the Great.

[56] The ultimate institution then managing the economy of the Russian Soviet Federative Socialist Republic and later the Soviet Union.

Rostov-on-the-Don

The second very happy period of my life began after my arrival in Rostov-on-the-Don. It began unexpectedly following the terrifying predicaments, troubles and dangers that descended on me in the Winter of 1949–1950. Fortune suddenly smiled on me just when I was least expecting her to.

At the onset of the Winter of 1949–1950, I was dismissed from my workplace, and the threat of arrest hung over me. Then it was that I had to leave everything behind—my home and Moscow—and flee into the unknown. And moreover not only I, for I had just married so no longer had only myself to worry about. That Winter was one of the hardest of my life. My prospects collapsed suddenly and completely right at the very moment when it had appeared that nothing but success lay ahead. The war was over, and seemingly all doors were open to me. Life had settled down, a draft of my doctoral dissertation lay in a safe in my office at SRI-2,[57] and I had started a family.... How was it that I didn't break down under the strain? Incredible!

Certain circumstances—as related above—saved me, of course, and perhaps above all my friends Sasha and Nina Kulikovsky. They, and they alone, were genuinely on my side.

My stepmother, who had retired after a quarter-century of teaching at the school in Skhodnya, had been arrested earlier. That was a catastrophe, but, as it turned out, not only for her but also for me. Soon afterwards I was suddenly deprived of access to classified work, and also, therefore, to my dissertation, which was as a result consigned to oblivion. (I would defend a completely different doctoral dissertation five years later; the first one had had to do with the theory of guided rocket missiles, an important and highly classified subject.)

My scientific work in my speciality was thus at an end, and I had to look for some other kind of employment or means of living.

No sooner had I been stripped of the right to do classified work, than a stamp appeared in my workbook announcing that I had been dismissed through staff-reduction. However, in those days when there weren't enough people to fill available positions, such a stamp meant only one thing: that I had been dismissed as someone untrustworthy, or, in other words, as a relative of someone who'd already been repressed, and as a future arrestee. I applied to various places for work. At that time those staffing personnel departments were usually veterans of the front. On seeing my jacket without shoulder straps, decorated with my three orders and various medals—as was customary in those days for anyone still wearing his old army tunic—they would begin talking amiably to me, clearly disposed to be of help. But then as soon as they saw the stamp in my workbook, their faces would drop, and the standard response issue from their lips: "Sorry, but...."

[57] Scientific Research Institute No. 2; see Chap. 1.

Money was fast melting away. There remained the little that I had been able to put aside for my first postwar civilian suit. I had been saving up for it since hostilities ended. But now this money also was fast vanishing.... (I would buy a suit some years later when working in Rostov, on the eve of my doctoral defence.)

I was faced with an urgent question: how to survive? And now as a couple. My wife was then still a student at the Energy Institute.[58]

We were saved by the circumstance already described. I was offered the position of acting dozent in the Department of Theoretical Mechanics of Rostov University. This in spite of the stamp in my workbook, which I had, of course, pointed out to the rector of that university, Professor Belozerov. I will always be grateful to him! It was, after all, a terrible time, so he was taking a risk.

That is how I began a new life in Rostov-on-the-Don, lasting almost five very happy years. Despite many technical difficulties, as mathematicians call them, life very soon settled into a groove. The business of finding an apartment was soon dealt with: by the time of my elder daughter's birth we already had two large, fine rooms in a six-room professorial apartment in one of the best buildings of Engels St., right in the centre of town. The apartment's other rooms were occupied by the families of two university colleagues.

Nowadays it is customary to disparage communal apartments.[59] These are not fully fledged apartments, of course, and so all the more inferior to detached houses. However, we lived very amicably with our neighbours, and I recall with pleasure the frequent get-togethers in the communal kitchen. Strange to say, those most inclined to friendship were the womenfolk.

I shall never forget arriving home on a certain memorable March day of 1953[60] and finding the three adult female inhabitants of our apartment howling in unison in the kitchen. While walking home, my mood greatly elevated, I had been thinking to myself: Well, now, maybe, they'll return my stepmother to Moscow, and my ostracism will cease. In the blink of an eye they'll be letting us all return to Moscow. So when I observed the enthusiasm of the three kitchen blubberers, I said something for which they were long unable to forgive me—and the possible consequences of which I subsequently very much feared: "What are you bawling about, you stupid women? Perhaps now at last we shall have a new life without terror and having to continually look over our shoulders."

During those years hardly anyone talked politics. Bitter experience had taught us to consider it a forbidden topic. But in fact politics had ceased to interest us. We knew that no questions should be asked about anything, that we would be told everything we needed to know, that showing initiative in any regard was futile, and that any reference, no matter how oblique, to the competence of "the competent organs" was to be avoided. Mind your own business and don't stick your nose in elsewhere! That's how we lived, worked, and raised the children.

---

[58] The Moscow Power Engineering Institute.

[59] Apartments with shared bathroom, toilet and kitchen facilities.

[60] March 5, 1953, the day of Stalin's death.

At Rostov University I was for some unknown reason given the task of leading a philosophical circle on the theme of methodological questions of physics. One of the topics was "Criticism of the Copenhagen interpretation," of which I was hearing for the first time. I tried conscientiously to sort out what Bohr and Heisenberg and their students were asserting. From the library I got hold of various articles of Bohr's and of other major physicists who had debated such questions with him. The problem was for me a very interesting and genuinely scientific one, and I was pleased to have been chosen for this assignment by the Party. It was precisely at that time that I began to regard Bohr as one of the greatest thinkers of the twentieth century and as my first real philosophy teacher.

However, my engagement with methodological questions of physics almost ended tragically for me. Word got around about the discussion group, and I was called to the Science Section of the Regional Party Committee and taken to task: "What are you spouting all that made-up stuff for? Instead of busying yourself with creative work and using the recommended materials, you're popularizing Heisenberg (we had been examining a paper of his the day before). Well, you can hand over your Party card!" And I was relieved of my assignment as leader of the discussion group.

It was for reasons such as this that we avoided talking about anything at all to do with politics. Even Stalin's death was not discussed. The more intelligent shrugged their shoulders as if to say "Let's just wait and see," while the less intelligent repeated what was printed in the newspapers. There is only one conversation that I can recall from that time. In the building next door there lived a well-known professor of ichthyology by the name of Alexander Mikhailovich Probatov. One day I was out walking on Pushkin St.—a quiet street with a promenade down its middle—and ran into him. We exchanged greetings and sat down on a park bench. We both began to think—the same thought, as it turned out: "One would like to be hopeful, Alexander Mikhailovich." "Quite so, Nikita Nikolaevich. It can't get worse; that's impossible. There's a limit, after all." And that was the whole conversation.

My attitude to Stalin was unequivocal. It had been worked out in childhood, in the bosom of my family, on the basis of their misfortunes. My father ascribed these to Stalin's ambition to establish himself as autocratic, monarchistic master of the country. He had considered that the revolution was bound to end in absolute autocracy and that only Stalin was fitted for the role of tyrant. "The scoundrel must in this situation be an absolute one," he said to Grandfather. Thus from the beginning I apprehended Stalin and everything that occurred during his rule through the prism of conversations between my father and grandfather. But despite my utter rejection of Stalin as a political figure, during the war I was ready like everyone else to shout: "For the Motherland, for Stalin." However, in this there was not the slightest hint of that adulation of Stalin evident in some of Simonov's[61] poems. During those years I accepted Stalin as a necessary evil, even as an historical boon. Russia's territorial integrity had been preserved once again, this time under Stalin.

[61] Konstantin Mikhailovich Simonov (1915–1979) was a Soviet novelist and war poet.

In this I was following in my grandfather's footsteps. He detested the liberals of the provisional government, and forgave the Bolsheviks much because they had preserved the nation's territorial integrity. He liked to say: "The Bolsheviks have come and will go, but Russia will remain." I thought about Stalin in much the same way, although I never revealed this to anyone. It had seemed to me that, following the war, which had shown how undeniably united the people were, his personal aims to become "absolute overlord" and conqueror of fascism having been essentially realized, Stalin would begin to behave differently. It was my feeling that Yagoda,[62] Yezhov[63] and Beria[64] were no more than his creatures. I thought that postwar he would dispense with such people. What I had not yet understood then was that it wasn't just a matter of Stalin, that he was just the face or personification of the SYSTEM—a system that had achieved in him as its figurehead its "optimal" realization.

Thus my illusions, or rather hopes, gradually began to fade. I observed that my worst premonitions were being realized: all those "special types" lurking in the shadows would soon emerge once again. They would be considered necessary once more, and would again sneak up on us, the people outside the system—in particular me. I had escaped from Moscow thanks to the advice of my most wise friend Sasha Kulikovsky! And now Stalin was gone. An interval in which to catch my breath was now essential. And then? Well, history does not repeat itself, tragedy degenerates into comedy, just as democracy reduces to chaos, as Cicero said long ago—if I remember correctly. Life, even if miserable, would go on. It might be less dangerous, at least. And one day in the distant future all that scum would quit Russia, reprieving her at long last from torment at their hands. So said my grandfather and father. They were mistaken about the time it would take, however; they were far too optimistic, not imagining that one rogue would yield to another.

Still, one would like to think that leaders will arise in the future who are better than those of the past. And if this is so, then we must work harder and harder for the benefit of Russia. Such were my thoughts forty years ago, during the March days of 1953.

I occasionally aired such ideas to my wife. She had been trained as a nurse, and had been assigned to a medical battalion stationed near the Volkhov front in

[62] Genrikh Grigorievich Yagoda (1891–1938) was director of the NKVD, the Soviet security and intelligence agency, from 1934 to 1936. He supervised the show trials against the "Old Bolsheviks" and the construction of the White Sea-Baltic Canal using slave labour from the Gulag. He himself became a victim of the purge, being arrested in 1937 and shot in 1938.

[63] Nikolai Ivanovich Yezhov (1895–1940) was head of the NKVD from 1936 to 1938, when the "great purge" was at its height. Like Yagoda, he himself fell victim to the purge. He was arrested and, after forced confessions, shot in 1940.

[64] Lavrenty Pavlovich Beria (1899–1953) was head of the NKVD under Stalin from 1938 to 1945 and a Marshal of the Soviet Union from 1945 to 1953. Stalin introduced him to Churchill at the Yalta meeting of the "Big Three" as "Our Himmler." Like Stalin he was originally from Georgia. He was arrested in a coup by Khrushchov and others in June 1953 and executed in December of that year.

the Spring of 1942, so at that time we must surely have been at no great distance from one another. She had joined the Party under approximately the same conditions as I. And her medical battalion had likewise been surrounded. She listened to my reasoning with alarm, however, and without trying to argue with me simply begged me not to say such things to anyone else. And indeed I never spoke to anyone again about such matters—not even to Iosef Vorovich.

A New Life, New Work, and New Friends

So there we were in our two luxurious rooms right in the centre of Rostov. At first our funds were exceedingly limited since my pay was that of an assistant. I would say that our standard of living was about the same as mine is now, namely that of a scientific research worker living on a 1993 salary, with no allowance made for qualifications or rank. There is a difference, however: then I was forty years younger and wasn't an Academician.

But our narrow financial straits didn't really bother us. The main thing was that we were full of hope and certainty about the future, which nowadays, alas, is everywhere in short supply. I tried my best to earn a little extra on the side. I spent the summers in the mountains working as a climbing instructor. Not only did I thereby supplement my regular pay, but my family could live with me there in the Climbing Camp practically *gratis*. In any case, our financial situation soon improved. I was confirmed as dozent, my salary doubled, my wife began to work, and all this together with money earned on the side meant that our standard of living was fairly soon at the level of that of an Academician of the pre-perestroika period. At last I could afford a suit.

I gradually recovered from the state of shock I'd been in. At home it was pleasant and cosy despite a signal lack of furniture. But what is that to people who are young, healthy, and happy to be together!

I especially loved September in Rostov. During that month, the first of the academic year, my workload was moderate and I was usually home from the university by 4 pm. And in Rostov in September it's still warm, although not oppressively so as in July or August. We would often walk to the Don, rent a rowboat, and the three of us would row upstream, where there were several sandy islets frequented by very few on workdays. My little daughter was an enchanting creature; my wife and I called her "Slavnyushechka."[65] She was very winning, stamping her little feet at the edge of the water and laughing without cease.

We sometimes went with company to the Don, picking up a keg of beer and a bucket of crabs on the way, and sometimes treating our guests to black caviar bought in the bazaar from fishermen poachers.

65 "Marvellous little one."

Being a northerner, in Rostov I missed the Winter. However, Spring in Rostov came early and was enchanting. Once we were entertaining a visitor from Moscow, and the evening was already late when we—my wife and I—set off with our guest to accompany him back to his hotel. Returning along Pushkin Boulevard in the midst of a Spring frenzy, holding hands with my wife and negotiating puddles, I composed the following lines, inspired by what was going on all around us:

The wind ripped, playing with the spray,
Gusting a wet melting warmth.
The night breathed, moist, alive,
And one was reluctant to go back inside.

Many years later I would repeat these lines, though in a completely different context. But I leave that to later.

Arriving home, we opened the window wide. Spring rushed into our apartment. Next morning the weather was clear and sunny: the southern Spring had begun.

We soon became part of a friendly group comprised of the families of those I worked with at the university. The whole company would often of a Sunday spend the whole day at the Don, and we would frequently visit one another at home. Communication within the group was relaxed; there was neither squabbling nor gossiping. Rostov took us to its bosom, quickly counting us as of its own. Since our apartment was spacious, we often played hosts. The group also quite frequently foregathered at the Probatovs' home, since Alexander Mikhailovich liked to have us join in singing Russian songs. Although I have a tin ear—even on parade I was forbidden to sing in case the column got out of step—I very much enjoyed Probatov's singing. He and I. I. Vorovich sang duets rather well; although untrained as a musician, the latter had a fine ear.

The atmosphere in the Faculty of Physics and Mathematics also turned out unexpectedly business-like and agreeable. A highly qualified group of dozents had been assembled there by Professor Morukhay-Boltovsky, when he arrived in 1914 from Warsaw. They may not have been really first-class scientists, but they knew their stuff, and were good teachers, well up to university standard. I now feel I can say that all the dozents of that faculty demonstrated a high degree of professionalism. It was they who determined the atmosphere in the faculty, which then compared better than just favourably with those of other provincial universities. And furthermore, there was the particularly gratifying circumstance that all of the faculty's teachers were what one might call very non-Party. This was very different from the situation I had encountered in the Mechanics and Mathematics Faculty of my *alma mater* MSU,[66] where a group of Party Komsomol activists assumed the right to decide the fate of both individual people and the faculty as a whole.

Of the Rostov university faculty members, I remember with especial warmth Mikhail Grigorievich Khaplanov, head of the Department of Mathematical Analysis. He helped me a great deal, in particular with his criticism of my first articles, which he read in manuscript. Among the other instructors in the faculty there were a few

[66] Moscow State University.

"cutting" ladies. As a rule, once they have obtained a secure position, such scientific ladies become very "concretely educated," by which I mean that, although their knowledge is not especially wide, on the other hand what they do know they know so well that they immediately latch on to any inaccuracy. At scientific seminars, for instance, they carry on as if examining students.[67] In their presence one always needs to keep on one's toes. However that also has its usefulness!

The atmosphere at work was relaxed and cheerful. Following meetings of the department or the scientific council, we customarily went to the "second hand bookstore," which was what we called a small wineshop on Engels St. near an actual second-hand bookstore. The wineshop sold draught wine from the region of the Don. The wine was of good quality and cheap, but there was no food sold there, so that sometimes we went elsewhere to dine—usually to the restaurant "Don" on the same street. (Dining in a restaurant on a lecturer's salary!? But this sort of thing did occur; those times were so much more easygoing, if you can believe it.)

My colleagues were always telling jokes or playing practical jokes. Once someone of the company at the "Don" sent the rector of the university a page taken from the Complaints Book containing the following grievance: "When I asked for a third half-litre, I was impolitely refused!," signed "University Dozent Vorovich." It should be explained that the future full member of the Russian Academy of Science I. I. Vorovich almost never touched spirits, especially at that time. The message composed in the restaurant was exhibited at a general Party meeting of the faculty, but the signature was never subjected to expert verification.

I once again took up various kinds of sport. I was a member of the faculty volleyball team and president of the municipal branch of the Alpinism Society, which used to begin preparations for the forthcoming season in early Spring. I shall describe my climbing activities below.

I stopped writing verse altogether; real preoccupations took up all my time.

I. I. Vorovich

Right from the first my workload in Rostov was very heavy, especially the lecturing load. I had absolutely no lecturing experience, and now, looking back over forty years, I am astounded at my chutzpah and levity. How had I been able to take on so much responsibility so lightly? In my very first semester I was given five(!) completely different courses to teach—and made no objection. One of the courses was on theoretical mechanics, compulsory for all students of the faculty. My knowledge of the subject was but so-so. I also had to teach the theory of relativity and Riemannian geometry to theoretical physics students. I had attended such a

[67] Examinations in the Soviet Union always had an oral component.

course taught by Academician Tamm,[68] and had the notes of his lectures. However, of the subject-matter of the other three courses I was totally ignorant.

I had only two or three months to prepare to lecture on the most difficult of those courses, namely the one on fluid mechanics, a subject I had never studied. I ended up teaching it always just one step ahead of the students: I lectured today on what I had learned just yesterday. I had no idea then that four years later I would be defending my doctoral dissertation in, of all things, fluid mechanics! And moreover at the Steklov Institute.[69] Those early days of my academic activity in Rostov now seem to me almost fantastic. All the same, I understood even back then that it's much easier to give lectures than pass examinations on their content!

Of course, I had youth on my side—and health. And there was also that extraordinary postwar atmosphere of general uplift.

The nation was on the rise. Everyone had a positive attitude to their work. There was hardly any talk of life's difficulties, even though life was very hard indeed in the early 1950s. I don't know another era to compare with it. Certainly not the early 1990s! Then we waited every day for something good to happen—and amazingly enough it did!

The working environment in the Department of Mechanics, my home department, was positive. The department was completely new, having been reconstituted following the period of arrests and general havoc.[70] The rector of the university, Professor Belozerov, had hired three Moscovites: I. I. Vorovich, N. N. Moiseev, and L. A. Tolokonnikov. We were all Candidates of Science[71] fresh from the defence of our dissertations and without the slightest teaching experience. But we all set to work forthwith, grappling seriously with our teaching assignments, and this set the tone of the department. We worked under the dozent A. K. Nikitin, who seemed to us elderly although he was only about forty years old. He had been lecturing for many years, and, although he had scarcely any publications, was a very competent instructor.

The department was not only new but young. Apart from Nikitin, we had all served in the army, and this was one further circumstance uniting us. The spirit of the "brotherhood of the front" could be felt long after the war ended.

As I said, our departmental head Nikitin was the only member of our department with any experience of teaching at a university. At first he was the lone dozent, we others all having the rank of assistants. Although he largely refrained from interfering, he did at least keep a check on the quality of our teaching. He would

[68] Igor′ Evgenievich Tamm (1895–1971), Soviet physicist. Together with Andrey Sakharov proposed the "tokomak" system as a possibility for a fusion reactor. Nobel laureate for physics in 1958.

[69] The premier Soviet institute for mathematical research, affiliated with the Soviet Academy of Sciences. Named after Vladimir Andreevich Steklov, founder of the Institute of Physics and Mathematics in Petrograd in 1919.

[70] Of Stalinist repression?

[71] That is, science Ph.Ds.

drop into class and make comments. On one occasion he actually taught me a lesson, thus making a lasting impression on me. In preparing a lecture, I had formed the habit of writing out detailed notes to take to class with me, and would frequently have recourse to them while lecturing in order to check a calculation or the final formulation of a result. Following one such lecture, Nikitin reprimanded me: "Can't you prepare well enough in advance so as not to have to refer to your notes?" I turned lobster-red in shame. Thus it was that I learned to lecture without notes. Thenceforth, while I still wasted a lot of paper making detailed notes, I went into class without them. Only now, in the second half of my eighth decade, when I am required to give lectures on themes having to do more with the humanities, so lacking the logic of mathematical proof, do I take a list of questions with me for fear of forgetting something important.

Iosef Izrail′evich Vorovich, whom I had known earlier in Moscow, arrived in Rostov at the same time as me. His presence turned out to be a great boon; his help was crucial, especially at the beginning of our sojourn. As I now understand it, during my undergraduate years sport had occupied a larger place in my life than was appropriate—to put it mildly. My study habits were slip-shod, science had then been of only secondary interest to me, and I had studied only during the examination period. And now in Rostov the considerable gaps in my education were catching up with me. I felt them very keenly and was ashamed of my ignorance. When preparing my lectures and, especially, my seminar talks, I often found myself in urgent need of help. Since Vorovich was like family to me, I felt able to ask him questions without feeling compromised by my ignorance. He had been a student of a rather different type from me, and he never once declined to help me. I have remained grateful to him ever since.

My relations with Vorovich were, in fact, of a special sort. Iosef Izrail′evich was two years younger than me, and fate had brought us together at the student residence on Stromynka St. when I was already a "mature" student in my third year and he was just beginning. That was, I think, in September 1937. Our room in the residence, then occupied by five third-year students, happened to have a bed free. And thither, to that cloister of "mature" students, they sent a freshman—none other than the future full member of the Russian Academy of Sciences I. I. Vorovich.

We would often try to recall the circumstances of our first meeting. It has to be said that my recollections of it differed markedly from his. As Iosef Izrail′evich remembers it, on entering the room he saw several half-naked lads completely caught up in card-playing, who received the new tenant unenthusiastically, but then immediately showed enthusiasm in dispatching him to buy beer, which was at that time widely available and affordable even to students!—fantastic as that may seem now.

I remembered it differently. Into the room came a shortish, skinny boy with big sad eyes registering all the world's woes. I recall especially clearly his big suitcase or rather bag, bound about with straps under which felt boots had been thrust—boots which would have to serve little Iosik in the cold Moscow Winter. I didn't remember the episode with the beer, and he didn't recall that concerning the felt boots.

But are the details of our first meeting as we remember them so significant? Of much greater importance is the basic fact that we have lived our whole lives more or less in parallel. Only I did everything a little earlier than he. I was born two years earlier and began my university studies two years sooner. We both entered the Zhukovsky Academy, although I, having completed a full program at the university, was at the Academy only a year, whereas Vorovich was there for the full three. Similarly, I defended my Candidate's dissertation and was awarded the degree of Candidate of Technological Science two years before him. And then two years separated the defences of our doctoral dissertations. And we were both elected to full membership of the Academy of Sciences, only once again in my case some years earlier.

At the beginning of our time at Rostov University we took the shared initiative of organizing a seminar devoted to problems of the theory of elasticity and fluid mechanics. Our seminar quite soon became very popular with the students, and from it there eventually emerged quite a few first-class mathematicians. I can say now that that seminar played a significant role in the strengthening of the Mathematics Faculty in Rostov, and at one time even defined its character.

The reason for this had to do with the fact that prior to our arrival on the scene the instructors had all been specialists in classical mathematics, which was what they taught both the undergraduate and postgraduate students. Our seminar did not fall within the standard scheme of things.

In the first place, we were doing "new" mathematics—new at that time, of course—such as operator theory, nonlinear analysis, etc. And paramount for us, the cornerstone, as it were, of our approach, were applications to specific problems of physics and applied mathematics. We considered that mathematics, even of the most advanced sort, was nothing more than a means to the solution of such problems. Not for nothing had Vorovich and I studied with D. A. Venttsel′![72]

Our seminar was attractive to the young; even those running it were but thirty-year-old dozents. And it should be mentioned that the success of our seminar prompted in some of our faculty colleagues the usual feeling of envy. This was especially the case with the venerable mathematician Professor F. D. Gakhov, a superb specialist in the theory of boundary problems for functions of a complex variable. He regarded his speciality as that holding out the greatest promise for the future of the "Rostov mathematics" of the period. I called his research activity "boundary-ism," thereby angering him. In fact he seemed to like getting angry, especially with any young folk showing too much independence.

I. I. Vorovich has always been one of the closest of my friends; I trusted him absolutely, just as I did Andrey Nesmeyanov, Yura Germeier, and Volodya Kravchenko. Vorovich was one of a very few to whom I could turn for advice in difficult situations.

[72] Dmitry Alexandrovich Venttsel′ (1898–1955) was a Soviet scientist working in exterior and interior ballistics. Stalin Prize laureate in 1951.

We worked hard and systematically. We often went to Moscow. I was soon giving talks at the seminars of M. V. Keldysh, S. L. Sobolev, and L. I. Sedov, thereby entering into a scientific world new to me, and I began publishing articles in serious scientific journals. I gradually ceased fretting over the loss of my doctoral dissertation. New horizons were looming, but of these I shall write later on.

On Mountaineering and Igor′ Evgenievich Tamm

In relating my life story, in particular the good parts of it and the episodes that most readily spring to memory, I feel a compulsion to include the tale of my mountaineering activity. I didn't achieve great heights, so to speak, in this sport, and in the list of my professional climbs there are no peaks of the sixth category of difficulty of which every climber dreams. However, I went on climbs with real first-class climbers. Observing how they went about the sport, I lost any illusions concerning my own ability in that sphere. Over a period of several years I climbed on a single rope with Valentin Mikhailovich Kolomensky, tackling various climbs of the fourth or fifth degree of difficulty, and I understood that things he did with ease would never fall within my own capability. But I wasn't overly concerned about it.

But although a rock-climber of very moderate ability, I did have one quality that was appreciated and by virtue of which I was made welcome in many a climbing team: I was a good Sherpa, in the sense that my endurance on long climbs at high elevations was very good. In skiing I likewise preferred long distance trails—especially 50 km races, where I performed better than at sprints. My ability as a "stayer" was also of great help to me in wartime. It is likely that, had I dedicated myself more to excelling in sport, I may have been up to qualifying as a Master of Sport. But ... science got in the way.

Following on my demobilization I fell in with mountaineers working at the MHIT.[73] The team was led by Slava Lubenets, an excellent climber and decent fellow, with whom I am to this day still friends. They were then making preparations for a record-breaking traverse Dykh-Tau—Mizhirgi—Kashtan-Tau.[74] I was told unambiguously that I stood a definite chance of being included in the final team of climbers, but that I would have to begin training hard and without let. So what did I do? I went off to work as a climbing instructor at the Alibek[75] mountaineering camp. I had made my choice.

Every climb at or above the fifth degree of difficulty requires not just physical preparation and good technique, but also considerable psychological preparation

[73] Moscow Higher Institute of Technology; see Chap. 1.

[74] Peaks over 5000 m of the central Caucasian Mountains in the Kabardin-Balkaria Republic of Russia.

[75] Now a resort town in the Karachay-Cherkess Republic of Russia, situated in the Caucasian Mountains.

in the form of a demanding mental workout. In mountaineering there is no encouragement from the gallery: it's just you and the rock! And this was a matter of 18 days on a crest of the fifth degree of difficulty. Preparation for such a climb would take up the whole Winter and maybe more—in fact one would be occupied with that and that alone. It is possible that a year earlier I would have agreed to prepare myself to participate in such an attempt at a record-breaking traverse. Just at that time, however, my mind happened to be on other things. Following one of my talks at S. L. Sobolev's seminar, Sobolev himself told me that the results I had obtained might suffice for a doctoral dissertation, in the defence of which he would be prepared to act as opponent. He even informed the council of the Steklov Institute of this, and I was given leave from work to write up the dissertation. In other words, "science took off," as Gorbachov might have said, and I was thenceforth unable to dedicate myself to anything else. Notwithstanding my love of the mountains, from that time on mountaineering would be of secondary interest.

That was when I began working as climbing instructor. As an occupation to tide me over the academic summer break, this was entirely to my liking. I worked in the company of mountaineers who had already acquired official ranking, and enjoyed doing climbs of moderate difficulty with them. This satisfied my appetite for the sport and opened up unlimited possibilities for interesting hikes and new, not very difficult, but still engaging, climbing routes. For the most part I worked at the camp at Alibek in Dombay, but I often worked also in Altay[76] as Director of Rescue Services of the First Mountaineering Camp, located in the Aktru gorge. Once I was on Tien Shan,[77] where I worked in the Talgar[78] camp, again as Director of Rescue Services.

It was a pleasant incidental feature of my work as climbing instructor that I was afforded the opportunity of meeting many interesting people. One such was Igor′ Evgenievich Tamm, one of the greatest of our physicists and a person of extraordinary charm and kindness. He was to play a very important part in my life.

At the end of the 1930s I spent a month at a "school for climbing instructors," as, with some exaggeration, we called it. The Dombay clearing was then in its original, splendid state. There was just one building, which we named "Ksu house" since it had been put up by the Commission for the Promotion of Scientists (initials KSU in Russian). It was an attractive two-storey wooden building. And on the opposite bank of the river, right near the start of the climb onto the Ishachy Pass, as we called the beginning of the track over the Ptysh Pass, our University Sporting Society "Science," as we had proudly named it, set up a moderately sized camp of ten tents. It was here that prospective climbing instructors were trained. My main teacher was the Austrian Franz Berger, expelled from Austria as an active member

[76] The Altay Mountains lie in Central and East Asia, where Russia, China, Mongolia and Kazakhstan come together.

[77] System of ranges in Central Asia.

[78] A settlement near a peak of the same name in the Tian Shan range.

of the *Schutzbund*,[79] a communist workers' organization. He was a professional climber and through him we acquired a reasonably good understanding of modern mountaineering technique, of which we had hitherto but a dim idea.

Upon completion of this course I was invited to work in the Alibek camp as trainee instructor. They entrusted me with a small group of scientists. I was to be their shepherd: taking a rope and an ice axe just in case, I was to accompany them on walks and not interrupt the high-flown scientific chitchat they carried on amongst themselves. And that's how I came to meet Igor′ Evgenievich. But first a little background.

At MSU our course in the theory of electricity had been taught by Professor Belikov. I don't know what sort of physicist he was, but his lectures were incredibly tedious. As a means for preparing for the examination, he recommended the book by Eikhenval′d,[80] noting that it was "real physics, without any mathematics." For my part, the "Eichenwald barrier" proved unbreachable: the book consisted of a dense collection of disparate examples unconnected by any general guiding principle. And I had failed the exam—after which I left for the mountains with the failure needing to be made good, taking with me the book *The Theory of Electricity* written by the rising star of Soviet physics Professor Igor′ Evgenievich Tamm. And by coincidence that same Igor′ Evgenievich turned up in the group I had been charged with "shepherding." However, initially I had no idea that the author of the very textbook that I was studying from was a member of my group.

My duties were minimal; those I was in charge of walked by themselves without paying much attention to me. I began preparing for my make-up exam. One day when I was sitting on a rock near my tent reading Tamm's textbook and making notes, I suddenly heard a quiet voice behind me say: "It's amusing, after all, when my instructor reads me." Jumping up and turning around, I saw standing before me, smiling as he smoked, the shortish man who while hiking had rattled me with his liveliness and fearlessness, or, more accurately, apparent ignorance of the dangers involved. "I'm Igor′ Evgenievich, the author of that book," he continued. "What are you doing reading that rubbish here in the mountains?"

I repented of my sins, and he responded very indulgently. He spoke to me on two or three further occasions, asking how the book was going, but I was too shy to hold much of a conversation with him.

Then at the beginning of September I received a directive from the Dean's office to take the make-up exam with none other than Professor Tamm! On arriving at the Department of Physics, I began by explaining to him that his appointment as my examiner was a complete accident. "I swear before God it's a pure coincidence"—

[79] The *Republikanischer Schutzbund* (Republican Security League) was an Austrian paramilitary organization established in 1923 by the Austrian Social Democratic Party in order to secure power in the face of rising political radicalization following World War I. It was banned by Chancellor Dolfuss in 1933 and ended in 1934 in the brief conflict called the "Austrian Civil War" between conservative and socialist factions.

[80] Alexander Alexandrovich Eichenwald (1863–1944) was a Russian and Soviet physicist. The 8th edition of his textbook *Electricity* appeared in 1933.

these were my concluding words. "Well, now we'll check," said Igor′ Evgenievich and, after asking a bespectacled young man, whom he addressed as Misha, to examine me, immediately absented himself—for a considerable time, as it turned out. My dealings with Misha were over quite quickly, and we then sat around waiting for the professor. It was two hours before he returned. My examiner told him that I had acquitted myself well. Igor′ Evgenievich put a couple of questions of a general nature to me and then asked: "Well, Misha, shall we give this mountaineer a five?"[81]

Misha expressed agreement, and my failure was redeemed. And what's more, Tamm suggested I attend certain of his courses and his seminar.

I made an effort to follow this advice. I attended his course on the theory of relativity, at least, and this made a great impression on me. I took detailed and complete notes; this was possibly the only one of my university courses from which I had notes. Twenty years later they would come in very handy.

The following year I met Tamm again in the vicinity of Tiberda.[82] He was with his children, a boy and a girl. His son Zhenya would become a famous mountaineer; he was the leader of our first Himalayan expedition to Everest. And even back then he was not so much Zhenya but Evgeny Igorevich Tamm.

Later, in the 1950s I, in the company of others, met up with Tamm several times in the mountains and had real scientific discussions with him. At Rostov University I conceived the idea of giving a single unified course covering every topic of mechanics considered standard for a university program, so including those required for a general course in physics, classical mechanics, and the special theory of relativity. I believed that such a course should be taught by a single professor, who would then have the task of uniting into a whole the worldview on the one hand and the experimental and mathematical components on the other of what is understood as essential to mechanics. I gave such a course twice, obtaining enormous satisfaction from the work involved. It was important for me to discuss my experience in teaching this course with others. Such a course was of interest to Tamm, in particular, and we talked about it many times.

Two or three years later, when I was already working at the Moscow Institute of Physics and Technology (MIPT), I tried teaching an analogous course on the mechanics of continuous media, embracing fluid mechanics, the theory of elasticity and magneto-fluid mechanics, and once again I consulted with Igor′ Evgenievich. He warmly approved the idea, and, with his blessing, I taught such a course at the MIPT for several years. Again I felt it important for the course to be taught by a single professor for the sake of systematization and also in order that a single point of view of the subject-matter be maintained throughout. When I eventually ceased giving this course, there was, unfortunately, no one at the MIPT prepared to teach it

[81] The traditional marking scheme throughout the Russian educational system awards students grades of from one to five, where three represents a pass and two or less a fail.

[82] Now a ski resort in Dombay.

all by himself. Corresponding Member[83] Sokolovsky and Professor Voit, who were next assigned to teach the course, subdivided it once again into three parts.

Thus it was that mountaineering brought me into close contact with a person who would greatly influence the formation of my worldview. First there were his lectures, which, in their attitude and approach, were so unlike any other physics professor's lectures. What he lectured on and how he lectured struck a chord in tune with my idea of mathematics and, if I may so express it, I listened to him "awestruck." And then when I myself became a professor, the advice of I. E. Tamm helped me reach my own understanding of the fundamentals of teaching and learning.

Following an address I gave to a meeting of the MIPT Methodological Committee I found myself rebuked by Professor Rytov: "You don't teach physics but only physical models." I agreed with him, adding that this is a principle with me: physics education—indeed education in any field—has to be based on a thoroughly worked-out system. There is nothing else in the works of humankind comparable in its wholeness and internal logic with the system of models of physics. Once having assimilated this system so that it has become second nature, one more easily absorbs concrete facts than is possible by means of physics teaching of the traditional sort. Thus not only theoreticians but experimentalists also are well-advised to master the system of "physics models." Igor′ Evgenievich confirmed me in this view—as also in my estimate of Niels Bohr as the greatest thinker of the twentieth century. The 1960s were fundamental to my subsequent enquiries of a methodological character, to which I ascribe great importance, and I. E. Tamm was one of just two people discussions with whom led me to the definition of my own "paradigm."

And that's why I have devoted so much space to the story of my mountaineering career.

In 1960 that career came to an end, and for good reason. I almost fell doing a relatively easy climb. This occurred during a climb up a rock face of Karatash, a rocky peak of modest height in the Aktru gorge in Altay. The degree of difficulty was not high: 4-A, and then mainly because of the first 200 m of a rather steep wall. This I managed without especial difficulty. From there one had to climb across flattish rocks like sheeps' heads, where the degree of difficulty was at most three. At a certain moment when I happened to be leading, my climbing partner shouted to me from below: "Drive in a piton!" But I didn't do as he asked, thinking that I had enough strength for another 2 or 3 m. I did in fact have enough left, but just barely. I felt the blood drain out of my face and took a good while to recuperate my strength.

When relating what had happened back at the camp, I felt acutely that the sentence spoken by Ktorov[84] in the marvellous movie *St. Jorgen's Holiday*[85] was relevant to me. His character says: "In the profession of scoundrel the main thing is knowing when to disappear!" This applies just as much to the profession of mountaineer. One's eyes continue to see just as well but one's strength, alas,

[83] Of the Academy of Sciences of the USSR.

[84] Anatoly Petrovich Ktorov (1898–1980) was a famous Soviet film and stage actor.

[85] A Soviet comedy first shown as a silent film in 1930, and dubbed into sound in 1935.

diminishes. A mismatch of that sort is very dangerous. I came up against this and resolved never to repeat the experiment.

Throughout my life I have adhered to this "scoundrel's principle." Thus, when the time came, I withdrew from the faculty, then from the chairmanship of the department, and then, some years later, taking advantage of a new regulation concerning council members, I think, became the first member of the Academy to retire completely. So that now, alone with my computer, I am in a position to do something both useful and of interest to myself, without needing to meet demands requiring excessive effort and the sacrifice of my health.

In 1961 there began a new but no less appealing stage in my life in the mountains, which I had no intention of retiring from. I no longer recall whose idea it was, but we organized in lighthearted fashion a club with the facetious name "Warm belly lovers." I have an idea that this name was suggested by the late Professor Vadim Borisovich Ustinov of Leningrad. The club admitted only people forty years old or more who were certified senior mountaineering instructors. The club had a "Führer," and the distinguished master of sport Vasily Pavlovich Sasorov was unanimously elected to fill this position. It was decided that the club should also have a president, however, and the person who agreed to fill this post was none other than Igor′ Evgenievich Tamm.

The goal of this "club" was the very simple one of having a group of mountain lovers of long acquaintanceship and mutual sympathies gather together somewhere in the Caucasus. They came in their private cars with their families and set up tents on a small campsite where they lived for a few weeks enjoying themselves. We always chose a campsite near some regular mountaineering camp, from which, being well known in the climbing community and having friends there, we could expect a little aid.

Our "Führer" made sure that the bellies of the club's members did not expand. Every three or four days we set off on a moderately strenuous hike, and as a result we were kept in good shape. For the remainder of the time we thought up pastimes no less pleasant. I remember especially well the evenings spent around the campfire. Since my fellow campers were an interesting lot, the talk was absorbing. We drank tea, but not because of any prohibition on alcohol; it was simply that no one thought of strong drink. We were often joined around the campfire by climbing instructors of an older generation, and sometimes also by acquaintances from Moscow, Leningrad, Sverdlovsk, etc.

On such occasions Igor′ Evgenievich revealed yet another remarkable skill: he was a wonderful *raconteur*. And since he knew all of the world's top physicists and many interesting details concerning them, the stories he related around the campfire and his running commentaries were something of a cultural phenomenon. For me they were like an echo from the past: in spirit those campfire conversations reminded me of those Saturday evenings in Skhodnya of 1925 or thereabouts: a similar circle of people, the same disposition to attend to one another, and the wish—or rather need—to simply communicate with each other.

At some point two Leningrad physicists joined the group: Nikita Alexeevich Tolstoy and Alexey Mikhailovich Bonch-Bruevich. Knowing that these two came from an old noble line, I suggested as a subject for discussion the determination as to which of their lineages went back further. As Vadim Ustinov later said to me: "My Leningraders didn't let me down. They really knew their genealogy." And their knowledge did in fact extend well beyond their own family trees. Witty and good-humoured, they transformed the evening into a marvellous show, persuading us that the Bonches certainly go back further even than Rurik[86] and all his ancestors, and that the Tolstoys even date back to Cicero!

A few days later, with my Moskvich[87] taking on the additional load of sturdy Nikita Tolstoy, I headed for the Crimea. However, the extra one and a half quintals[88] of Count Tolstoy turned out to be too much for my "antelope-gnu."[89] The car kept stopping; clearly it was protesting. And I was surprised—and secretly gratified—to discover that when it comes to automobiles, a mathematician and an experimental physicist hardly differ in terms of knowledge and resourcefulness. We arrived jointly at the conclusion that my Moskvich simply did not want to carry two Nikitas! Thus were we reconciled to the situation. And all the while my younger daughter whined: "I want to go floating on the sea on an air mattress." Nikita Tolstoy very charmingly urged her to be patient, promising that she would one day soon certainly find herself floating on an air mattress in Koktebel.[90] And so, to our astonishment, she did.

[86] The Viking Prince Rurik established himself in Novgorod in the year 862. The dynasty founded by him ruled various principalities in what is now Russia and Ukraine until 1610, when it was succeeded by the Romanov dynasty.

[87] A small family car of Soviet make.

[88] One quintal = 100 kg.

[89] Name given to a jalopy by the hero of the popular satirical novel *The Golden Calf*, written by Ilf and Petrov as a sequel to their popular novel *The Twelve Chairs* and published in the USSR starting in 1931.

[90] A resort town on the shore of the Black Sea in south-eastern Crimea.

Chapter 3
Outcast

The Moiseev Family

I have already related something of my childhood, of the few happy childhood years passed in the bosom of a family still secure—until overtaken by the catastrophe triggered by events at the end of the 1920s, resulting in its complete and utter destruction. My childhood years during the NEP era were definitive of much in my life. They gave me an understanding of basic humanity, of the goodness that unites people, and equipped me to endure the many difficult and dangerous times I would experience on my path through life. But my family life is far from being the whole story. "That's the truth but not the whole truth," as they say. There was also society, hostile and cruel. Even during that happy time I became aware of the existence of something supremely evil and unquiet, originating from somewhere outside, from society at large. Its malevolence insinuated itself into my life, and over the course of many years overcoming a feeling of ostracism was one of the defining motives of my behaviour. I now feel a compulsion to describe this situation in detail.

The first intimation that I somehow stood outside society came to me when I was still a schoolboy. This was one of the sharpest and most painful sensations of my childhood and youth. The feeling receded to some extent when I started demonstrating sporting ability. But even in the company of my fellow sportsmen I felt distanced from them; I was the only one who was not a *komsomolets*;[1] I was from an alien world, as it were. There were, of course, people like Andrey Nesmeyanov and Yury Germeier, whose sincere friendship eased this feeling somewhat, but all the same .... I told no one of this and no one guessed at it—except possibly Andrey. It sometimes seemed to me that perhaps he, even though a *komsomolets*, had the same feeling. I tried hard to conform: I twice applied to join the *Komsomol* and was twice rejected—publicly and derisively! It was as if

[1] Male member of the *Komsomol* (Communist Youth League).

N. N. Moiseev, *How Far It Is to Tomorrow...*,
https://doi.org/10.1007/978-3-030-96651-5_3

my inadequacy and inferiority were being proclaimed without my being able to do anything about it. I was made to understand that society merely tolerates me and that I had no claim on it.

It was only during the war that at last I began to sense my full worth to society. The ability to see myself as a useful citizen needed by society was absolutely essential to me, and without it life simply lost all meaning. I continually tried to reinforce a sense of being socially worthwhile. In this respect sport was a great help: in that sphere no one asked where or who one's father was. This striving was doubtless the main reason for my rejection of the flattering offers I received following my graduation from the Zhukovsky Academy. The front and only the front! At the front I joined the Party, and moreover in the uneasy situation where some of its members were getting ready to bury their Party cards. It wasn't "belief in Leninism-Stalinism" that ruled my conduct, however, but the idea that I was Russian and wanted to be with those fighting on the front lines. And furthermore, just as in the sporting arena, there at the front it never occurred to anyone to ask which social class my father had belonged to or whether there were members of my family who'd been repressed.

I had already begun to be cured of my affliction when, following on my stepmother's unexpected arrest everything came full circle back to the beginning. It took till 1955, when I was finally given top-level access to classified material, before I was able to do work of interest to myself without having various scumbags looking over my shoulder. That would seem to be the time when at last I began to acquire a real sense of social acceptance. But even later I refrained from telling anyone, even my very closest friends, that my mother had been the adopted daughter of Nikolai Karlovich von Meck, shot in the Winter of 1929, and that my father had died in the Butyrka just prior to 1931 because he had been a colleague of Professor Osadchy, a member of the Industrial Party.

My family had belonged to that large—perhaps the largest—part of the Russian intelligentsia that over many generations lived only by work of a technical nature using their hands. The Moiseevs never possessed the means to an income not deriving from manual work of some kind. The family had always been very Russian and very devoted to Russia. They had been nudged towards emigration but had preferred to stay home and work for the good of their own (not "that," as they say now) country. This attitude was typical of that stratum of the intelligentsia to which my family belonged; in fact, the majority of the intelligentsia, especially those of its technological stratum, were genuinely disposed to be patriotic, and never identified Russia with Bolshevism. But, despite their rejection of Bolshevik ideology, they were ready under any conditions to work for their country, not out of fear but in good conscience. (I would subsequently convince myself that the *émigré* Russian technological intelligentsia abroad continued to have concern for their country foremost in their minds—their country remaining as always Russia.) Yet in spite of all this, in the 1930s I found myself alone in the wilderness: a general liquidation of my relatives had taken place around me. The occasional family member surviving by chance, together with a few distant relatives, were then finished off at the front. I alone miraculously survived.

My father, Nikolai Sergeevich Moiseev, graduated from the Faculty of Law of Moscow University, where he specialized in economics and statistics. He was kept on at the university to be "prepared for the calling of professor," and was sent to the Russian legation in Nagasaki to write a doctoral dissertation on Far Eastern economics, with emphasis on the history of economic relations between Japan and China.

In 1915, when the war was well under way, my father was called back to Russia to do military service. He was sent to work as a volunteer male nurse, that is, health worker, in an ambulance train serving the south-western front. There he met my mother, who was working as a nurse in the same train. His army service didn't last long. A few months afterwards he was taken out of the army and once more sent to Japan, this time not to Nagasaki but Tokyo, and not to do research for his dissertation but as an official in the Russian embassy, where his knowledge of the Japanese language and economy might be put to good use.

The few months in the ambulance train followed by a month spent at Voskresensky-on-the-Desna,[2] the family estate of N. K. von Meck, were enough to ensure that my father was accompanied by a young wife on the trip back to Japan. Mama was then only 18 years old. My parents returned to Moscow in July 1917, one month before my birth. My father was appointed acting professor (or, as it was called back then, extraordinary professor or *Privatdozent*) at Moscow University. This position gave him the right to give lectures and came with a salary, which, although very modest for those times, was enough to live on—all the more so in that the von Mecks put the two-room mansard of their house at the disposal of the young family. And that is where I was born.

At that time my grandfather, Sergey Vasilievich Moiseev, was in the Far East, where he occupied a high position in the Department of Railways: he was the head of the Far Eastern Branch of that department. Grandfather was of an old noble family—not of the landed but the administrative nobility. He wasn't a landowner. There were, in fact, no familial memories of family estates or ownership of any lands whatsoever, while on the other hand recollections of service to the emperor were plentiful. Grandfather loved to tell stories of the great variety of adventures undergone by his ancestors and of their service to the nation, chiefly in the military arena.

The noble line of the Moiseevs was a venerable one. At the very least it hails from a time before Peter the Great. Tradition held that some clerk from Roslavl[3] by the name of Ivan Moiseev travelled with a certain ataman to the lower reaches of the river Ob′,[4] or maybe somewhere else, and later wrote down something about the journey. Since the Moiseev line did in fact originate in Roslavl, Grandfather wanted to believe that this Ivan was our direct ancestor. In any case, he liked to preface his lectures on morality, to which he used to subject me quite often, with

[2] A tributary of the Dnieper River, flowing through Russia and Ukraine.

[3] A town in Smolensk District.

[4] A major river of western Siberia flowing into the Arctic Ocean. The world's seventh longest river.

the injunction: "Remember, Nikita, that in your veins there flows the blood of an explorer." I suspect that the clerk from Roslavl was a product of his imagination, which tended to run away with him. And even if that mythological clerk existed, then an indefinite number of inhabitants of that town might claim him as an ancestor since in those remote times everyone in government service in the pleasant town of Roslavl had last name Moiseev or Naumov or Ilyin![5] Even now Roslavl boasts of a great many citizens named for "prophets."

But there is one thing known for certain: my grandfather's father was the last stationmaster, and then postmaster, of Roslavl, which lies on the Main Smolensk Highway. Grandfather was the eldest of the many sons of Vasily Vasilievich, who was married to a daughter of Captain First Class Belavents. (It would seem that pre-revolution all Belavents men were first-class captains.) The Moiseevs were related to many well known Smolensk families: the Buzhinskys, the Belaventses, and the Engel′gardts.

My grandfather and his younger brother Uncle Vasya became engineers, while all the other brothers graduated from the Cadet Corps[6] as officers and disappeared into the vast Russian military. One of the brothers perished in Manchuria during the Russo-Japanese war,[7] and another, apparently already at the rank of general, in the German war.[8]

When close to forty my grandfather married Ol′ga Ivanovna, a daughter of a mathematics professor at St. Vladimir University in Kiev by the name of Ivan Ivanovich von Shperling. This great-grandfather of mine came from a Russianized German family, preserving, however, its Lutheran religion and certain other customs of Baltic origin peculiar to Russified Germans. Thus my grandmother Ol′ga Ivanovna adhered to the Lutheran persuasion, although she attended the Russian orthodox church, and evinced a strong dislike for Latvians even though she had apparently never had anything to do with a single one of them.

Everyone in the extended family adored and respected Grandmother, and whenever any of them turned up in Moscow they would be sure to call on us to see her rather than Grandfather. Despite the apparent ease with which she related to people she was a very solitary person; she much preferred listening to others than speaking of her own intimate life.

Despite the almost twenty-year difference in their ages, Grandfather and Grandmother lived a long and, it seemed to me, happy life together. Ol′ga Ivanovna was in many ways a remarkable person. It is not too much to say that she was the cement that held the large extended family together, scattered as it was across Russia and

5 Derived from Moses, Nahum and Elijah.

6 An admissions-based military school, which prepared boys to become army officers in Imperial Russia. Many of the cadets and the graduates of the Corps, called "Junkers," supported the White movement opposing the Bolshevik takeover.

7 Waged 1904–1905 over rival imperial ambitions in Manchuria and Korea, and ending in a decisive victory for Japan.

8 That is, World War I; see Chap. 2.

indeed throughout the world. A certain Germanic pedantry in her character did not prevent her from being very kind and sensitive to others' misfortunes. And, what was exceedingly important during our harsh epoch, she was a person of enormous innate courage. When, following the death of my father and, in short order, that of my grandfather, the family was left practically without any means for living, Grandmother, although elderly, began giving lessons in German. A sort of rigorous resoluteness arose from within her to ensure her grandsons got a good start.

Grandmother was highly educated. She spoke and read in three European languages, was well-read not only in Russian but also in German and French literature, and could recite at will a great deal of verse: in German, mostly Goethe and in Russian Tyutchev and Alexey Tolstoy. She astounded everyone with her ability to concentrate. Everything she put her hand to she did well. She was a splendid cook, never disdained any work needing to be done, and kept the house in a state of ideal cleanliness. She was never seen in a dressing gown or with her hair uncombed. With me she was strict, and used to check my homework carefully. I owe her a great deal.

Although I understood this, alas, too late.

School and the Demise of My Family

In Skhodnya there was only one school, a "School for Peasant Children" or SPC, to which I was sent on turning seven in 1924. At that time I was already reading for pleasure; for my seventh birthday I was given an illustrated edition of *Tom Sawyer*, and I read it at one sitting. I had no time for arithmetic, thinking that, since I intended to become an astronomer, it would be of no use to me. I knew the constellations and used to explain the peculiarities of the calendar to the adults. I spoke French and German quite freely. I subsequently lost my German completely, but found it easy to retrieve my French when much later I had to lecture in France.

The first day of September 1924 was for me memorable for being a very bad day. Grandmother took me to school to begin Grade One. I returned home blubbering: I had been beaten and smeared with dirt, but the worst of it was that I had been called a bourgeois. And they had told me I was one of those who should be finished off. At school I turned out to be a real outsider and felt my situation keenly. I didn't understand the reasons for the general spitefulness towards me, why they were beating me, what it was about me that my classmates didn't like—and more generally, why people fight and whence their animosity towards others.

Later I myself learned to fight and gave as good as I got. When my younger brother started school, they didn't touch him, knowing that they wouldn't get away with it since Sergey Moiseev had a brother called Nikita.

But for the first few years I detested school and feared going. My father obtained permission for me not to have to attend every day. My stepmother, who was a teacher at the school, taught me lessons at home, and Grandmother checked my homework. My actual teacher at the school, Zinaida Alexeevna, would verify my progress now and then and, as I remember it, was generally satisfied with how I was doing. Grades were not handed out then, and I was moved up from class to class without any fuss.

In Grade Five I was transferred to a "second level school," so-called, for Grades Five, Six and Seven. This school was small, consisting of just three classes each of twenty or thirty pupils, and the teachers were good, and furthermore I had by this time managed to adapt and even began to like going to school. The school occupied a most beautiful *dacha* situated high above the river. I mentioned this dacha earlier: before the revolution it had belonged to the famous Guchkov.[9] A little while after I began Grade Six, our "Guchkovka," as we called the school, burned down. At first we walked about with a kind of joyful bewilderment among the ashes, but then, there being no other schools in Skhodnya, we were of necessity transferred to a school in Moscow. I was enrolled in School No. 7, located in Skornyazhny Lane in Domnikovka.[10] I was then twelve years old.

The times were rapidly changing. The era of five-year plans and collectivization were upon us. First of all our railway line, that Nikolaevsky, or October Line whose honour had been so stoutly upheld by all those old railwaymen, was transformed. The number of the latter gradually dwindled until soon they were gone altogether. The patriarchal system with its unhurried tempo, of which I wrote above, disappeared. The trains began to go more slowly, and it became usual for them to be late. Suburban trains were often cancelled, as now happens with the electric trains. Waiting times were often considerable, and we could never be sure of getting to school on time. The trains were now overcrowded, there appeared a great number of profiteers,[11] and there was thievery, fighting and general hooliganism.

The country began to experience hunger. Food coupons were introduced, for which one could obtain a daily allowance of 200 g of moist, undercooked bread. Life became a struggle against hardship, especially hunger. Our vegetable garden was of some help in the struggle to survive. And we also gathered mushrooms in large quantities. Mushrooms then grew in the woods around Skhodnya and I knew exactly where to find them. We both dried and salted them. Following my grandfather's death I became the "man of the house." It fell to me to fetch water and cut and saw enough firewood for the whole winter. There were difficulties with the supply of kerosene, which was needed since Skhodnya was then without electricity. It had to be brought from Moscow by stealth since it was forbidden to transport flammable liquids by train. We began to look shabby. There was no money for clothes. My grandmother and stepmother were always altering old garments for my brother and me, who were becoming taller in apparent indifference to our situation. I continued attending the school in Domnikovka. In those days coupons for cheap clothing were distributed to schoolchildren in need, but, although I was among the most needy,

---

[9] See Chap. 1.

[10] *Domnikovkaya* was the name of a street in Moscow that has since been renamed *Masha Poryvaeva*.

[11] In the original, *meshochniki*, from *meshok* ≡ sack, meaning people reselling marked-up scarce goods out of sacks.

being a bourgeois and the son of someone who'd been repressed, I was never given any.

In 1932 I turned 15, and applied to join the *Komsomol*, only to be rejected at a meeting of my peers. I was shocked and traumatized, though not so much by the mere fact of my rejection, which I had in any case been prepared for, as by the behaviour of my classmates at the meeting. I had thought they were all my pals and that I was well respected by them. I had made cheat sheets for many of them, had helped those lagging behind, and had played on the school's volleyball team—and yet here they were unanimous in protesting against me and hurling insulting words at me. Especially zealous in this regard was Rakhil′ Sklyanskaya, the niece of a well known Bolshevik, one of Lenin's former comrades-in-arms, who then occupied a very high position in the Party. A few years later Sklyansky was shot, but the fate of Rakhil′ is unknown to me. Back then, however, to the applause of the meeting, she said things about me and my family that were so offensive and untrue that I was unable to contain myself and at the conclusion of the meeting, despite my fifteen years and my sense of myself as having achieved manhood, I burst into tears. Mishka Lisenkov, the son of a mathematics teacher at a Moscow institute of higher education, took me to his home, where his father plied me with tea, listened attentively to our story, and then placed his hand on my shoulder and said: "Hang in there, Nikita. Today we need to be patient. God willing, times will change."

In my class there was another outcast: Prince Shakhovskoy. Lanky, awkward and very quiet, he was a less than average pupil. I visited his home once, and drank tea with his family. His father, a quiet God-fearing old man—at least that's the impression I had of him—worked as an accountant somewhere. He was one of the "disenfranchised," that is, officially stripped of all voting rights. It was said that before the revolution the father of my fellow pupil Shakhovskoy had been a dashing guards officer. Somehow I had difficulty believing it.

Shakhovskoy was a year older than I and his application to join the Komsomol had been rejected at a meeting the year before. He was an outcast and behaved like one, avoiding everyone. Since I myself was unable to comport myself that way, I suspected him of having some sort of a mental failing. One day just before the war, when I was about to finish my university studies, I was returning from a concert at the Conservatory when I bumped into him near the Nikitsky Gate.[12] He was walking along with a ladder over his shoulder. It turned out that he was working as a night-shift electrician. Fate works in strange ways.

Several years later I again tried to become a member of the *Komsomol*. I was then in my second or third year at the university. The meeting seemed to be positively disposed, and doubtless I would have been accepted if not for Vice-dean Ledyaev. He put just one question to me: "But is it true that your father, Professor Moiseev, belonged to the nobility?" What answer could I give to this question? I could only confirm his suspicions. He then shrugged his shoulders and said, addressing the meeting at large: "This is, of course, your business. Let Moiseev continue his studies

[12] A square located at the site of a former gate in the outer wall around Moscow.

now that we have already permitted him to do so, but why accept him into the *Komsomol*?" And that put paid to my chances. So I never became a *komsomolets*.

Gel′fand's Circle

Viewed from the side, it might appear that in my attempts to join the *Komsomol* I was trying to penetrate to some forbidden place, to break through to certain people for the sake of my career, but that some force working in the interests of justice always repulsed me. Such a force did indeed exist, and did indeed refuse to admit me: it was the system of Soviet power, expressed through Soviet society, that in actuality rebuffed me. Back then, however, I gave no thought to such a force. It was fortunate that I didn't think in terms of the position I occupied relative to that society. I simply did what seemed necessary at each given moment. I felt like a normal person, and wanted to be recognized as such, to be like everyone else, to merge with the surrounding society. Everyone else was a member of the *Komsomol*, why was I an ugly duckling!? So I tried to "break into the *Komsomol*." I had no thought for the rhyme or reason of that organization, I was innocent of ideology. I simply did not want to be a second-class citizen. That's all there was to it! Every outcast is consumed by the desire to be like everyone else, not to differ from others, to be lost in the crowd, as Dostoevsky put it.

It's likely that this desire was a big factor in determining my behaviour. I was just a boy and wanted to be at one with the people but was excluded. It even came to the point that I almost resigned myself to the situation and began to become used to regarding myself as a second-class person. Two or three years previous to the above-described episode, I was told directly by that same Ledyaev that this was indeed the case, that I really didn't have the same rights as others. (I shall give the details of this below.) I wished fervently to continue with my education, and was very much afraid that I wouldn't be allowed to. Although I had been a good pupil at school, I had no certainty as to my future.

Although in 1924 I had disliked arithmetic, in 1935 I decided to apply to enrol in Mechmat,[13] moreover in the mathematics section, rather than apply to study astronomy, as I had wanted to do in childhood. This alteration in my priorities was essentially random, as with much that befalls us.

The process whereby I was ultimately admitted to university is the most extreme example from my youthful experience of the malevolent attitude of society towards people of my background. It may easily have ended in personal catastrophe—may indeed have turned my life completely upside down. It was only through the goodwill of two people prepared to break the rules of admission to MSU, plus frenetic work over a period of several months, that I was admitted into the students' world. The story is worth telling in greater detail.

[13] Short for the Mechanics and Mathematics Faculty of Moscow State University; see Chap. 1.

When I was in Grade Ten,[14] the Academy of Sciences and Moscow University jointly organized the first mathematical olympiad to be held in our country. In order to train prospective participants in the olympiad, a mathematical circle for high school students was organized in the Steklov Institute—the celebrated "Steklovka" of those days. The circle was to be run by Izrail′ Moiseevich Gel′fand, a first-class mathematician and future academician, but then only a dozent in the Mechmat Faculty. He would play a decisive role in my life, changing its course in the space of a single hour. But I shall come to that in good time.

The mathematics teacher at our School No. 7 was Ul′yana Ivanovna Loginova, considerably gifted mathematically and an attentive and kind teacher. She taught us well, but, even better, around her there crystallized a coterie of pupils of more than average mathematical ability keen to study the subject more deeply. In this group Monya Birger was a star of the first order of brightness; I believe he would have gone on to a first-class scientific career if he hadn't been killed at the front in the very first days of the war. Other gifted pupils in the group included that same Rakhil′ Sklyanskaya, as well as Yashka Varshavsky, Zhenya Shokin,.... And these all attended Gel′fand's circle.

Ul′yana Ivanovna advised me also to join the circle, but I felt mathematically somewhat insecure so not ready for such a step. I considered myself much less able than our acknowledged top pupils, at least. And furthermore the ski season was almost upon us, and I had been included in the Moscow Junior Team. When I told our teacher this, she scolded me, ending with: "You could do as well as them if only you ski-ed less and studied more." She insisted that I attend the lessons at the Steklovka and leave sport to a more appropriate time. "And what's more," she more than once upbraided me, "it's high time you thought seriously about your future, especially since you have no one supporting you."

The Steklov circle turned out to be very interesting. I can now make a professional assessment: it was brilliantly organized. Gel′fand wasn't the only one to thank for this: several other talented young mathematicians worked with the pupils in the circle. They solved nonstandard problems with us, illustrating by means of the solutions the extraordinary possibilities of mathematical inventiveness, and gave lectures to us. Among the pupils there were several highly gifted young people. It was here that, among many others, I met my friends Yura Germeier and Boris Shabat, both destined to become professors at Moscow State University, Volodya Rokhlin, later a professor at Leningrad State University, and Oleg Sorokin, a marvellously talented young man who died on the frontline in 1941. The circle met on Sundays, so I had to sacrifice my Sunday ski training in order to attend; for that Winter at least I had definitely decided to follow Ul′yana Ivanovna's advice.

In the Spring of 1935 the olympiad was held. Only two pupils from our school survived the first round: Monya Birger and I. I was the only one to get through the second round. Monya Birger was upset at himself for not having solved one relatively easy problem, but a competition is a competition. I very nearly fluffed the

[14] The final year of the Soviet school system as it was at that time.

third round, but managed to hold on. In the end Germeier, Shabat and I emerged as the laureates of the olympiad, thereby gaining the right to have the mathematics entrance examination waived should we apply for admission to the mathematics section of the Mechmat Faculty of MSU. This outcome decided me: I would become an MSU student in that section of Mechmat. I began to study for the other entrance exams and already saw myself as such a student when my hopes were dashed by a terrible blow which for some time reduced me to a state of stupefaction and hopelessness.

I passed all the exams, not brilliantly but without threes.[15] However, according to my rough calculations there shouldn't have been any problem about my being admitted: on average the level of the examinees was not very high, and only a very few did well on the exams. Germeier and Shabat did about the same as me. Only Oleg Sorokin got fives in everything. Most of the examinees did far worse than I. Yet I was not admitted!

During the exams I had befriended another examinee by the name of Semyon Shapiro. This was his first time in Moscow; he had come from some small town in Belorussia to try for admission to MSU. He was a decent, quiet young fellow. His preparation for the exams left something to be desired, and Germeier and I made a considerable effort to help him. He obtained many threes in the seven or eight examinations, including a three in mathematics. But all the same he figured among those admitted.

When I was sure that my name was neither in the list of those admitted not in the list of candidates—a secondary category existing then—I was overcome with despair. I didn't know what I should do next or even how to go on living. Again some cruel hand was barring the way. Semyon empathized with my misfortune, consoled me as much as possible, and dragged me off to see Vice-dean Ledyaev, the official chiefly responsible for admissions.

Where had Semyon Shapiro's quiet self-control disappeared to? He began to explain loudly and angrily that an injustice had occurred, that in his opinion there had been a mistake, and that it should be corrected before it was too late. Ledyaev interrupted him. Turning towards me he asked dryly: "What do you want, Moiseev? Look at yourself and at him." He pointed at Semyon. "Think about the kind of person a worker-peasant government should admit to university, on whom it should spend money. Is it possible you don't understand this?"

My fate was decided.

And Grandmother also despaired.

[15] Recall that in the Russian tradition, grades are awarded out of five, with three a pass and less than three a fail; see Chap. 2.

Nevertheless, I Become a Student

The Autumn of 1935 and the Winter of 1936 together comprised the most critical period of my life. I am not talking here so much of my shattered morale, but the problem of what to do and where to go. I couldn't continue living off my stepmother and grandmother, whose combined income was paltry. Society had rejected me, had thrown me on the dust-heap, and I felt this with all my being. I went into a sort of trance. I was overcome by desperation and a sense of helplessness; there was no one to come to my aid or even to ask for reasonable advice. I was ready to do anything: contract to go somewhere or other in the Far North or work as a fisherman in the Sea of Okhotsk.[16] Yet somewhere inside me common sense prevailed and enough courage remained to stop me from doing anything stupid, and the upshot was that, as I now see it, I managed somehow to make the right decision.

I enrolled at the Moscow Pedagogical Institute and left Skhodnya to live in the student residence there. The main thing was that I received a stipendium, a less than miserable sum but just livable. And with it I acquired a certain independence and a limited "time-out." I could use this time to look about me and ponder my situation.

The Pedagogical Institute itself made a very poor, I would even say dismal, impression on me. The mass of students reminded me of my classmates in the School for Peasant Children I had attended in Skhodnya, who were a far cry from those clever and educated young people whose company I had shared the previous year in the mathematics circle at the Steklov Institute and with whom I had hoped to study at the university. As for the instructors at the Institute, well, doubtless they were experienced teachers and knew how to train teachers for the schools of the period, but how little they resembled those young mathematicians who trained us, the members of Gelfand's circle, and to whose lectures we listened so enthusiastically every Sunday!

In a word, I didn't feel like studying at that institute, so I did no work. I attended lectures only once in a while. That Winter I was only 18, so still had the right to compete in junior skiing competitions. I entered the competition for admittance to the Moscow Junior Ski Team and was successful. Incidentally, as it were, this sporting interest provided me with sustenance: as a member of the collective of the Junior Ski Team I was entitled to free lunch coupons, which was for me no small matter. That year I went as a member of the team to Kavgolovo[17] to compete for best team in the Soviet Union.

The team as a whole performed extremely well: it came first in every category. My own performance, however, was merely average. It was only through my having been a member of the relay team that I was included among the champions of the Soviet Union "in the junior category," as they say these days. Having absolutely zero interest in studying at the Pedagogical Institute, I didn't even bother taking the

---

[16] The sea or gulf separating the Kamchatka Peninsula from far eastern Siberia.

[17] A winter sports resort in the Vsevoloisky District of Leningrad Region, where world-class competitions in cross-country skiing were held.

exams at the end of the Winter term. It looked as though I would be quitting that institute and taking up professional sport. I might enrol in the Institute for Physical Culture, for instance, whither they were trying to attract me and where there were no entrance examinations. But fortune was disposed to deal with me otherwise. She occasionally smiled upon me—or at least allowed unexpected alternatives to turn up.

In Spring, the skiing season being at an end, I went wandering to Mechmat to look up my more fortunate friends. In the corridor on the third floor of the old Mechmat building on Volkhonka St., by chance I bumped into Gel′fand. Izrail′ Moiseevich examined me from under a furrowed brow and asked:

"Moiseev, how is it I haven't see you about? Why aren't you attending seminars? How did you fare on the Winter term exams?"

"I'm not a student. I wasn't admitted."

"How can that be? Did you fail the entrance exams?"

"No, I passed them."

He was silent for a moment and then continued:

"And what are you doing with yourself?"

"I'm skiing!"

Again he fell silent, and then grabbed me forcefully by a button and said:

"Come with me."

He took me to the office of the faculty's dean. At that time the post was held by Lev Abramovich Tumarkin. When we entered his office he was alone; to my great good fortune, Ledyaev wasn't there. Gel′fand said the following, word for word: "Lev Abramovich, I am requesting that you allow this person (that's what he called me: "this person") to try for all necessary first year credits. He was a member of my mathematics circle. If he passes all tests and exams, then I guarantee that he will make a no worse than average student." That's exactly what he said: "no worse than average." Tumarkin gave his assent, flying in the face of all regulations.

I was given the necessary forms concerning the tests and exams, which I had to pass outside normal regulations and deadlines. I began working madly. Oleg Sorokin was of great help to me; without him the task would have been almost beyond me. It's one thing to attend lectures and learn how to solve problems at seminars, and quite another to master the material alone using others' notes and moreover in a very short time. And there is the further consideration that the advanced mathematics, albeit introductory, taught in first year university courses is significantly more sophisticated than school mathematics.

It proved possible to overcome these difficulties, however. In fact, I obtained excellent grades in every subject except higher algebra. In that subject the dozent Litsman, a harsh and pedantic German, gave me a three. But that was already neither here nor there: I was admitted to the mathematics section of the Mechanico-Mathematical Faculty and began studying in the same group as Germeier and Sorokin. Being excused from military service, Boris Shabat was assigned to a different group. At that time the program of study of those liable for military service was of six years duration, a year longer than normal.

Thus it was that in spite of everything I became a student at Moscow University, the same university where my father had studied. My grandmother was utterly overjoyed!

More About Gel′fand

Many, many years later I was elected to full membership of the Academy of Sciences of the USSR in the same year as Izrail′ Moiseevich. In those safe years the president of the Academy used to host magnificent receptions *à la fourchette* in honour of the newly elected academicians. In that memorable year the reception was held in the restaurant of the Hotel Rossia, and we both attended. Gel′fand came up to me with a goblet of champagne in his hand and, after congratulating me, said: "I knew, Nikita, that you would be a not less than average student!"—especially pleasant words for me to hear.

In return I congratulated him on his election, which should have occurred at least twenty years earlier, and once again thanked him for his support throughout my student years. Without that support, and the dean's willingness to waive the regulations governing admissions, it is highly probable that I'd never have studied at the university. The only real option then left to me would have been enrolment in the Institute of Physical Culture. I would have become a professional sportsman of average ability, and then, in the best case scenario, a physical education instructor!

Thus I had once again been the beneficiary of someone's goodwill, clearing the way to an occupation congenial to me. Is there anything that needs to be added to this?

So I became a student at the university. My status as an outsider, however, remained: Soviet society would continue demonstrating my unworthiness for a considerable time to come. I have already described how I was rejected by the *Komsomol* membership. Later on there would occur another even unhappier episode, which may have ended tragically indeed.

Like all of those liable for military service, at the university I went through training for the army reserve, at the end of which I was to be given the rank of sublieutenant in the reserve. I was assigned to a group of airmen, where I acquitted myself quite well. But then it was suddenly discovered that I wasn't a *komsomolets* and shortly thereafter the reasons for my non-acceptance emerged. A kerfuffle ensued: those in charge got it in the neck for assigning me to pilot training, and I was ultimately ejected: aviation was not for the likes of me. As a result I wasn't after all given officer's rank, and in the event of war I was to be called to the front lines as an ordinary soldier. It was as such that I was called up to serve in the Finnish

War[18]—not, it is true, as a soldier but as a skier. My interest in sport has many times in my life caused a lifeline to be thrown to me.

When The Patriotic War[19] did start, they ceased examining people's biographies, and sent me to study for a year at the Zhukovsky Airforce Engineering Academy, from which I graduated in May 1942 as senior technician in the arming of airplanes. I was given the rank of lieutenant and dispatched to the Volkhov front.

The End of Ostracism and Adventures of My Army Cap

A person tends to forget the gloomy times of his life, but remember favourable episodes well. One fine day the painful memories recede into forgetfulness, leaving only the joyful and, even more vividly, the humorous ones. The above rather dismal tale of my ostracism, which affected my whole life, but especially the youthful part of it, certainly begs to be told, but I would like to conclude it by relating a humorous episode. This also has left its trace in my life and in a way capped the years of my exclusion.

The air-arm of the regiment to which I had been posted after graduating from the academy was staffed by pilots from civilian aviation. These were excellent pilots and navigators but they were uniformed according to the wartime standard. On the other hand, since I had come to the regiment from the academy and was considered a career officer, my uniform was appropriate to that calling. The main difference was that my peaked officer's cap was decorated, in standard prewar aviation officer fashion, with a "crab"[20] and was almost the only such cap in the regiment. The rest wore the cotton pilot's caps of previous usage. Thus my cap made me stand out from the other officers—in much the same way as the red cap of the stationmaster used to distinguish him from other railway workers on the platform. The stars on the shoulder straps of the officers' tunics could not be used to distinguish us since we all went around in overalls. When my sergeant-major was asked where the engineer was, he would answer laconically: "At the aerodrome with his cap and moustache." This defined me uniquely since, although the regimental commander had the same style cap, he was moustache-less.

Well, that invidious cap of mine was not only an object of envy but a stimulus to desire. Try to picture to yourself a war pilot with a row of orders and medals jingling

---

[18] Also called the Winter War, this conflict began on November 30, 1939 with an invasion of Finland by the Soviet Union, and ended with the signing of the Moscow Peace Treaty on March 13, 1940. The Finns showed determined resistance against the much more numerous and well-equipped Soviet army, which, weakened by Stalin's purge of Red Army officers of 1936–1938, endured unexpectedly heavy losses. However, in the end Finland was forced to cede 11% of its territory.

[19] World War II.

[20] That is, the crest at the front of the cap above the peak featured a crab as well as other items such as oak leaves grouped around it.

on the breast of his tunic—it was then the custom to wear them—getting ready for a tryst with a maiden but having only that old cotton pilot's cap for headwear. It may be possible to imagine this, although only just. For a damsel, however, it's an impossibility; for her such a sight constitutes unbearable torture. So there comes up to me a certain hero—a real hero, that is, reckoning his heroism, his daily duel with death, to be merely a natural everyday affair—and says: "Captain, could you lend me your cap for the evening?" Is it possible to refuse him? But I don't feel like lending my cap for nothing. "Take it," I say, "but tell me everything that happens!"—to which the response is in the affirmative.

So that is how my cap began going on dates. One young man would wear it pushed to the back of his head, another would have it slipping down onto his nose, but in any case the cap went on assignations, and, as a rule, successful ones: in this sphere also my friends were heroes. Then came all their tales, which were always entertaining though doubtless somewhat embroidered since a hero must be a hero in all things!

Once many years later, during a summer break, I conceived the idea of writing a book entitled "Tales of My Cap." I recalled a sufficient number of the pilots' stories to fill a substantial volume. I doubted there was a publisher willing to print such tales, however. But that was then: now it's no longer a problem. They publish even Miller![21] If only I had the money!

I tell this story for the following reason. Whenever I loaned my cap to one or other of my friends, and heard the tale of its adventures, my sense of being an outcast abated to zero; I had become a member of a special brotherhood. Here, among these my fellows, I felt totally cured of my inferiority feeling. I never felt myself so worthy a scion of my people as back then at the front amongst healthy young Russian and Ukrainian lads all living the same life.

That's what they were like, those friends of mine, going on dates wearing my cap.

Pashka Anokhin, who shared his surname with the famous test pilot,[22] once flew on a sortie to photograph the port Pillau[23] without the benefit of cover by fighters. The navigator and gunner were both hit. Anokhin himself was spared but not his plane. However, he managed to fly it back safely to the aerodrome and deliver the necessary photographs. Here is a description of him:

> The aircraft flew, its rudder gone,
> The engine smoked, the field was cleared.
> Overhead, amid tongues of flame,
> One last time it roared enticingly.
> Heeling over on its right wing,

[21] Presumably this is Henry Miller (1891–1980), an American author whose writings include many sexually explicit passages.

[22] Sergey Nikolaevich Anokhin (1910–1986) was a famous Soviet test pilot. Among other exploits, he test-flew the first Soviet jet-propelled airplane.

[23] A seaport formerly in East Prussia, now in Russia's Kaliningrad Region. A U-boat training facility was located there during World War II. It has since been renamed Baltiisk.

At last the many-tonned mass,
Obscuring the pale window of sunset,
Buried itself in the snow a hundred metres off the tarmac.
And one hour later, playing with his pistol,
Just now smashed by a bullet in battle,
Over a shotglass he jokingly tells the story,
As if he'd known beforehand what fate had in store.

And he was one of those who wore my cap! Therefore am I one of them!

Chapter 4
War's End and the Search for Myself

Victory Euphoria

Following World War I, the phrase "the lost generation" entered literature. It referred to those people who, when that world war—the first one—ended, were unable to find themselves, whose lives ran downhill, so to speak. I myself knew strong, brave people who, after earning a good name for themselves and many decorations at the front, found themselves unable to get used to life in postwar[1] peacetime. For that life different qualities were needed in order to cope with a hard and cheerless daily round, often lacking hopeful perspectives.

One such was Karelin. I had known Dimka Karelin as a first-class navigator, a wonderful companion, and a subtle and sensitive person. When I happened to meet him a year and a half or so after leaving the regiment, he had become transformed from a bold, strong, healthy fellow—despite a limp caused by a bullet wound to a ligament in one of his legs—into a wreck with hands trembling from alcoholism. After being discharged from the army, he had been unable to find work that suited him and lived on a paltry pension, or, more precisely, on the goodwill of his wife.

But in this country those who had "lost themselves" were few and far between; we were spared the scourge on the scale that postwar Germany, for example, endured. The end of the war and the first postwar years were almost intolerably hard. It was not just that people lived very poorly; there was a greater difficulty. During the war everyone was fully occupied, and then suddenly their occupations vanished. People had to think about what to do with themselves, to search for a new line of work and then get used to it. And it was natural that not everyone was able to cope with the never-ending difficulties that arose when one went back to being a

[1] That is, post-World War II.

N. N. Moiseev, *How Far It Is to Tomorrow...*,
https://doi.org/10.1007/978-3-030-96651-5_4

civilian. But all the same our reality wasn't as bad as that of the Germans post-World War I as described by Remarque,[2] for instance.

We were, after all, the conquerors. The end of the war marked our victory! *My* victory! For those who had fought on the front lines, the feeling of jubilation at what had come to pass long remained with us—jubilation mixed with astonishment, perhaps, but none the less for that. Victory generated optimism, belief in the future, and our horizons had become unlimited, our energy boundless.

Very few recall that postwar atmosphere—and even fewer speak of it or understand it and fewer yet there are who desire to understand it. Yet this is Russia, and that mood was characteristic of her, and in order to live properly in Russia, it must be remembered. I began to understand Russia's essence only when I was at the front. I reached a full understanding of my country only in those first famished and impoverished postwar years. Today it's customary for emigrants and so-called democrats to facilely paint a black picture of that period, without mentioning the deep springs of vigour enabling the recovery of our country.

In his two-volume work *Our Problems*, I. A. Ilyin[3] continually emphasizes that Russia must not be identified with Soviet power, with the Bolsheviks. One has to agree with this, but it's true only in principle. In the final months of the war and the first postwar years, the Party, the government, and Stalin himself enjoyed such unstinting support of the people as has throughout history never, perhaps, been granted to the leadership of any state! The scale of the victory, the unity of purpose, the general anticipation of the future and desire to work for the common good—all this opened up unprecedented prospects for our country.

We failed to take advantage of all these favourable opportunities, however. We understand now that it was in fact impossible for us to realize the great promise of our Victory. The system's priorities had little to do with the people's needs. The rulers didn't trust the people, they were afraid of them, so they were mainly concerned to rein us in. As a result the people gradually stopped believing in the system, their energy evaporated, there arose the contradistinction between "us and them," and then hatred for "them"—that is, those within the green fences.[4]

But in the first postwar years, we still had no thought for that, although many of us were shocked by and lamented the deportations of people from the Crimea and the Caucasus.[5] There was no concerted reaction, however. It was easier to believe

[2] In his novel *All Quiet on the Western Front*, first published in 1928, Erich Maria Remarque (1898–1970) describes the German soldiers' extreme physical and mental stress during World War I and the detachment from civilian life felt by many of them on returning home from the front.

[3] Ivan Alexandrovich Ilyin (1883–1954) was a Russian religious and political philosopher. He was one of many prominent members of the intelligentsia expelled from Russia by the Bolshevik regime in 1922. He was a conservative Russian monarchist in the Slavophile tradition, writing many books on political, social and spiritual topics pertaining to Russia's historical mission.

[4] Around the dachas of the *nomenklatura*, or ruling *élite*.

[5] Stalin had people deported *en masse* throughout his rule: kulaks from the early 1930s to 1953, and, from 1939 to 1953, Poles, Romanians, Lithuanians, Estonians, Volga Germans, Finns, Crimean Tartars and Greeks, various peoples from the Caucasus, Koreans, and others. (The latter

back then—and it was of course convenient to believe it—that they were deporting Hitler's accomplices rather than innocent peoples. Nevertheless, even in the army this action didn't pass unnoticed. Some in our division made a fuss when a pilot who was a Crimean tartar by nationality was discharged and sent to live in Kazakhstan. He had been wounded twice and awarded the order of the Red Banner of Battle four times. The head of the Political Section of the division, Colonel Fisun, himself a fighter pilot, threw his arms wide in incomprehension.

But despite the excesses that were beginning, we continued to believe: the Party, which in the most difficult of conditions had led us to victory, would surely be able to open the door to a "bright future." Of course, the question did arise as to precisely what kind of bright future. But that was not the one chiefly occupying us.

For the time being, most of those twenty-year-old boys simply returned home, removed the shoulder straps from their tunics—since they would be wearing these for some time to come—rolled up their sleeves and set to work—while keeping an eye out for girls! Life went on, and we were in eager anticipation of the morrow.

Doubts crept in on us somewhat later, when at the end of the 1940s there was talk of new arrests and of what was going on in Magadan, Kolyma[6] and other prison camps, and of incipient arrests of us, those who had fought at the front, and of our partisans! The involuntary thought occurred to all of us: is it possible that 1937 is beginning all over again? We wondered how it could possibly be that they didn't trust us, our generation, who had looked death in the face in Leningrad, Moscow, and Stalingrad, the generation that reached Berlin? And we began to discuss this with each other, moreover out loud!

But as a matter of fact even then, in the Spring of 1945, far from everyone shared in the euphoria of victory or looked to the future as optimistically as the present author.

Ivan and the Leningrad Medal

At the very end of the war, in early May 1945, while standing on one of our divisional airfields, I was struck by a bullet. Although we were very far in the rear—the front having moved all the way to Berlin—there was plenty of shooting going on all around us, especially by friendly Poles. In that Springtime everything that could happen did happen. It remains unknown who shot me. In any case, it was not serious: it meant merely a few days in the regimental medical unit and a scar on my forehead. And that's where Ivan Kashirovsky or Kashperovsky—I don't recall

deportations would now come under the head "ethnic cleansing.") It is estimated that almost half of all the resettled people, of whom there were over three million, perished from disease, malnutrition, or exposure as a result of being uprooted.

[6] Kolyma is a region in the far northeast of Siberia, and Magadan its principal town. The forced-labour camps of the Gulag situated there, operative 1932–1954, were notorious for the extreme harshness of their conditions.

the exact spelling—found me. In the Autumn of 1942, when my regiment left for Alatyr′, temporarily leaving me and my armourers behind in the 14th Air Force, I happened to intersect with Ivan in a squadron flying IL-2 attack planes. These planes were each armed with a 20 mm automatic ShVAK[7] cannon. This was an excellent rapid-fire gun, but at that time it was being manufactured in factories just evacuated to beyond the Volga! There the cannon were being assembled under the open sky by women, children, and cripples, as a result of which their quality left something to be desired. They tended to malfunction frequently—or refuse to fire, as we armourers used to say. This was the Achilles heel of these cannon. It was especially terrifying when they failed in flight, and many lives were lost on this account. However, I managed to work out how to quickly locate the cause of the failure and fix the problem.

There were usually not enough gunners for the IL-2s. Their turret was, unlike the pilot's cockpit, not armour-plated, and since gunners were a prime target of enemy fighters, their casualty rate was higher than that of pilots. For this reason, when demand was high, armourers, being familiar with the cannon's workings, often filled in as gunners. In particular, I myself, the leader of our team of armourers, occasionally had to fill the role of air gunner. I usually flew with Ivan. We were hit twice and on those occasions limped home with great difficulty. And that is why the Leningrad medal—for "the defence of Leningrad"—is dearer to me than all the more prestigious decorations I was later awarded.

The situations in which I found myself with Ivan are never to be forgotten; the comrade with whom one shares such experiences becomes more kindred than one's closest kin. And that is why when Ivan heard that I was close by—in a neighbouring division—he searched me out in the infirmary of the regimental medical unit.

It's early May. Through the wide-open window a blossoming cherry tree is visible. The sky is cloudless. The war has retreated over the horizon and everyone is hoping it's forever, and my mood is commensurate. I got off lightly: a minor concussion, and no damage to the bone. It's true that the bullet has done a good job of ploughing up my forehead; there was a lot of bleeding and now my head is swathed in white. But this doesn't spoil my mood. I am being cared for by a very kind and amusing Ukrainian girl, a lieutenant in the medical service. In the course of her work—and sometimes outside it—she often comes to check up on me, and my hands involuntarily move towards forbidden places. She brushes them aside, whispering: "Comrade Captain, what are you doing? You mustn't do that sort of thing now. You'll feel bad again."

And that was what I was doing when Ivan at last found me. He had a flask with him, a flat German-style flask containing pure alcohol, and I began begging my sympathetic boss lady to bring us something to eat. For a good while she resisted,

[7] Acronym for *Shpital′nyi-Vladimirov Aviatsionnyi Krupnokalibernyi* (Shpital′nyi-Vladimirov Aviation Large-calibre). The cannon was designed by Boris Shpital′ny and Semyon Vladimirov and went into production in 1936.

trying to convince me not to drink. "For you," she said, "it's very dangerous." But then she went to the kitchen and brought us back some food.

I myself drank very little, but Ivan downed two large glassfuls. From the way he was drinking and not taking anything to eat as he drank, I sensed that something was wrong. But no: externally everything seemed in order. He looked well, he was already a lieutenant-colonel, was flying a new bomber, and had been awarded several new decorations. The devil-may-care attitude of that senior lieutenant I had met two and a half years previous was gone, however. I felt that something had given way inside him. The saying came to me unbidden: "The blue-grey horse has been ridden hard over steep mountains,"[8] and I was saddened that something had apparently snapped in him.

I, on the other hand, was in a completely different frame of mind. I talked to him of victory, described my plans, and painted the future in bright colours. I was proud that our country had become the strongest among European powers. The century-old struggle between the Slavs and the Germans had at last been decided in our favour. A marvellous life awaited us! And there was a great deal more twaddle I came out with on that May day.

In spite of having imbibed a large dose of almost yeast-free alcohol, Ivan wasn't showing any sign of drunkenness. He listened to me silently. His silence was mournful, like the monologue that soon followed: "You're a member of the intelligentsia," he said with a wry smile, "yet the war hasn't taught you a thing. Do you think that up there," pointing a finger at the ceiling, "anything has changed? The same scum, concerned only to feed themselves and hold on to their power. It will all be just as it was before. You'll see: they'll bestir themselves a little, and then relapse to looking after number one again, to jailing people once more. Without that they can't survive. And then all those "special types" can't be left without anything to do. But the main thing is that they can't endure without an enemy, since then everything becomes too transparent. How can they control people without having an enemy to blame? It was the Germans, next it'll be the Americans. What's the difference? You think they care about the people? They'll repress them without blinking an eye. They'll continue repressing them. Is it possible you really believe anything they say?" And so on in this vein, on and on.... And then, almost at the end of his harangue: "It's OK for you, an engineer. You'll always be able to find something to do, something suitable. But what about me? Once I'm done flying, they'll discharge me. And then what? Where will I go?"

And indeed, I later heard that he was discharged in 1947 as not suitable for further work as a pilot! So my lieutenant-colonel comrade, with his four Red Banner Battle orders, went back to the Ukraine, where it appears he worked as a tractor driver. I heard that he took to drink, and then either froze to death or was drowned. That's how the life of that brave war pilot, generous and full of heart and of infinite daring, ended.

---

[8] A folk metaphor for someone no longer their former robust self.

I listened to his gloomy words, so jarring with my upbeat mood, and doubt crept in: perhaps he's right, perhaps there'll be no new dawning. But even if there is one, how far off it may be! When Ivan had gone I looked through the window at the blossoming cherry tree with altogether different eyes.

And I ceased flirting with my female medic, who then, naturally, lost all interest in me.

Autumn 1945

Dreary prognoses and expectations of new woes were realized not only by my lieutenant-colonel. Some of the more general among his predictions began to come true. Just before being demobilized, my sergeant-major Eliseev also became dejected: he was worried by news from his village in Ryzan′ Province.

The Autumn of 1945 found us in Tunoshnoe, or Tunoshna as the locals called it, a village situated on the banks of the Volga between Yaroslavl′ and Kostroma. There was an old military airfield there to which our division, renamed the 4th Guards Bomber Division under General Sandalov, had been relocated. We were to be retrained: Our task was to learn—very quickly—how to operate the new Tu-2 bombers[9] and then fly them to the Far East. The first part went swimmingly since our flying and technical people were first-class, but the delivery of certain essential equipment was held up. It was still Autumn when the divisional regiments were at last fully equipped, but by that time, fortunately for us, we were no longer needed in The Far East. The war with Japan[10] was already history and had even begun to be forgotten.

With Eliseev I was quartered in one of the furthest huts, closer to the airfield. The village, once a thriving town, made a grim impression. It was clear that it lacked skillful male hands. Over the years of the war everything had fallen into ruin; the huts were aslant and there was hardly any livestock. We were billeted on an elderly, infirm woman whose husband had been killed at the front. She awaited the return of her two sons: they had been conscripted in 1944 and had apparently survived. Eliseev spent a great deal of time helping her around the house: during his off-time he would always be repairing something or other, or chopping firewood for the Winter.

---

[9] The Tupolev Tu-2 was a twin-engine Soviet high-speed bomber aircraft playing a key role in the final Soviet offensives of World War II.

[10] The Soviet-Japanese War began on August 9, 1945, quickly ending Japanese control of Manchuria and northern Korea, in particular. It is considered a significant factor in the Japanese decision to surrender unconditionally to the Americans on August 14, 1945.

One rainy Autumn day I wrote the following lines in memory of that village:

There it is, a village without streets
And with hovels for houses.
There an old church crouches
Over a pond.
There's no one about....
The huts are tight shut
Against who knows whom.
And in front of mine,
As if on purpose,
There's a puddle
Of depth about an arshin[11]
Causing grief to laden vehicles.
And one has already almost forgotten that there are homes
With warm toilet, with bathroom and light,
With a bookshelf where Dumas père[12]
Lies on top of Blok[13] like a 100-volume athlete.

Closing my eyes now, I once again see those scattered blackened huts, the half-ruined church over the pond, forgotten by the Almighty, and the poverty and dolour of the inhabitants. And to think that it was once a prosperous town, a trading Volga town on the road linking Yaroslavl′ and Kostroma. And in that town there lived independent peasants—people of the Volga, which says it all.

That day a fine rain was falling. When I returned from the command post, I found Eliseev seated by the window, glumly examining the drops of rain as they slid down the glass. A letter lay on the table. Eliseev was so sunk in thought that he appeared not to have noticed my entrance.

"Why so gloomy, Eliseich? Longing for home?"

"Oh, how I do so miss it, sonny," and then coming out of his reverie and jumping up, he said: "Oh, please excuse me, Comrade Captain."

He then told me about the contents of the letter from home. Akim, who had returned armless from the war in 1944, had been arrested after a year as chief administrator of a collective farm. He was an honest abstemious fellow, who thought more of others than himself. But he had refused to let them requisition the potato harvest, so they'd arrested him.

"What will happen now?"

Eliseev was silent for a moment, then sighed and said: "Now we have neither Akim nor potatoes. Once again they'll be cleaning us out. How will we live?"

A week later our regiment flew off to the Baltic littoral, and Eliseev, as a soldier of more advanced years, was discharged, returning to his home near Ryazan′. I gave him my stepmother's address and begged him to write and tell me how things turned out. I never had a letter from him, however. Perhaps he lost the address I had

[11] An old unit of length equal to 28 in.

[12] Alexandre Dumas père (1802–1870), the great—and prolific—French author.

[13] Alexander Alexandrovich Blok (1880–1921) was a major poet of the Russian symbolist school.

given him. Or the same fate had befallen him as Akim. He also was an honest and uncompromising person, after all.

The Volga

The pessimistic forebodings and cheerless conversations, slow in coming in the first postwar months but then more insistent, were incapable of darkening my joyful sense of incipient peace and a fresh beginning to life. We frontline warriors all looked around us in anticipation of a turn to something new.

You have become strange, unlike
The one I knew before.
And at the door of your house,
I wait like an unexpected passerby.
I shyly knock, uncertain,
At the window's clouded darkness.
Time measured in versts[14] has passed,
So will she smile at me again?
Today the night unfolded amiably,
The wind gently urges me onward.
Soon daylight will come; will its dawning
Show me the way to her?

That is how I pictured the way ahead, sitting in my hut or walking by the Volga. But at the same time there was a feeling of anxiety inside me: "How far it is to tomorrow...." These lines were the constant *leitmotif* of my thoughts.

That Autumn I had rather a lot of time on my hands. The wars were all over and done with, and my duties were no longer onerous. I often found myself alone with my thoughts—now not by Lake Ladoga but on the banks of the Volga. The Volga here was not very wide; she didn't have the grandeur she has further downstream at Saratov. Her 200 m wide stream, rushing eastwards with surprising obstinacy and energy, had a spellbinding effect on me. Framed with yellowing trees, the Volga was splendid that year.

The more I was alone with nature, with the Volga, the stronger became my faith in the morrow; there was born in my heart a conviction of my ability to overcome the difficulties that would inevitably arise on my way forward.

I had no thought for demobilization; it seemed to me that as a career officer with experience of the conditions of war and with an academic degree, I was bound to the army for life. What I failed to understand then was that a wartime army is one thing, while the routine of service in peacetime is something else altogether.

---

[14] A traditional Russian measurement of length just over a kilometre.

Kostroma

Peacetime entered into our lives, cocooning us, altering our psychology.

Army service was easy and sometimes even interesting. My regiment received around thirty new Tupolev[15] KB[16] bombers. At that time these were considered the most up-to-date short-range bombers. And their cannon were of a new design. The gunsights were especially interesting; even in the academy we had never heard of such as these. My immediate superior officer, Lieutenant-Colonel Ivan Timofeevich Tamara, the divisional weapons engineer and a zaporozhian Cossack by origin, dispatched me, as the only "academic" in the division, to Moscow to learn the technicalities of these cannon. With alacrity I travelled back to my Zhukovsky Academy to consult with teachers known to me from before in the department of Lieutenant-Colonel Sassaparel′, who had supervised my degree thesis. I remember that once when he had looked at my work, including the devil only knows what kinds of graphs I'd sketched, he had said in distaste: "We'll never make an engineer of Moiseev!" (So, to be frank, this time around I very much wanted to show off my three "engineer's decorations" to him!)

There at the academy I spent a week mastering the new technology, and the following week gave myself over to pleasure, both permitted and otherwise.

On returning to Tunoshna I was given the task of instructing all airmen and auxiliary staff in the new weapons technology, and I did this with pleasure. The instructing was to be done in Kostroma,[17] where one of the regiments of the division was stationed. In addition to mere instruction, practice firings were to be carried out there.

It would appear that I fell in love in Kostroma; there was a romance which almost ended in marriage. There's no real point in telling the story; taken altogether, it's a rather banal tale. Four years of life in the army, and more particularly at the front, transforms a healthy male twenty-something into something very soft and highly responsive to feminine endearments. He'll interpret a few kind words as who knows what! But his ardour tends to fade rather quickly.

I repeat here the lines I wrote early one morning in the celebrated city of Kostroma—I mean the one on the Volga, should there be another town of that name. The least I can say is that Kostroma, together with everything that happened to me there during that Autumn, seemed to me at the time marvellous and unrepeatable—though not for long, it's true! The following indicates what it was like:

In the chasm of the greying street
A weary visage and compressed lips.
Will tomorrow come? No answering words,

[15] Andrey Nikolaevich Tupolev (1888–1972) was a pioneering Soviet aircraft designer. From 1929 till his death he was a leading designer in the Central Aero- and Hydrodynamics Institute in Moscow.

[16] The initials of the Russian for "Design Bureau."

[17] An historic city located on the Volga about 80 km downstream from Yaroslavl′ and Tunoshna.

Just thunder from an overcast sky.
Daybreak is misting the room's windows,
And the cobbles are damp with dew.
And that morning—will you remember
The night's minutes and the day's hours?—
You left, subtly swaying,
In the raw and misty dawn,
For somewhere far away, without a backward glance,
Taking with you both happiness and the dark.

In Kostroma's annals there is the following entry which I reproduce here verbatim: "Female Volga pirates from Novgorod captured Kostroma [then still a mere town] and did all sorts of terrible things to its menfolk."

Well, during my time there nothing like this occurred. For one of the menfolk, at least, it all ended safely—and for the female pirate also, I imagine.

To put it briefly, life in peacetime, about which we had completely forgotten, is totally different from that of wartime. It brings us fresh pleasures and fresh woes. One has to get used to everything from scratch—which seems easy at first, but actually isn't.

Anticipating Tomorrow

In that memorable Autumn of 1945, morning frosts began early. The weather was splendid, a real golden Autumn: sun, a limitless blue sky, golden foliage. Life seemed to begin afresh every morning. Feeling in tune with the morning I would hurry towards my airplanes in the frosty early hours along a path on the bank of the Volga.

The clear freshness of an Autumn morning,
The bright flush on women's cheeks,
And underfoot the scrunching powder
Of hoar-frost on the grass and yellow leaves.
And with the spring in one's step, desires were born,
In tune with the wind, the frost, and the dawn.
I greet you, morning, and the Volga's reaches,
And the woods turning blue on the high mountain!

I decided to resume working at mathematics and to this end went to the public library in Yaroslavl′. I soon convinced myself that I had forgotten mathematics completely! Science seemed infinitely far removed from me—a distant part of that completely unreal past life. But I couldn't ignore its existence, and all the more in that I felt it coming back to me after all. Reality was little by little being re-established and I needed to be ready for it. I was trying to write verse, but quickly grasped that that's not for me: an occasional pastime, for my own pleasure when in the mood, but not as a serious pursuit.

I felt certain that it would be best for me to remain in the regiment, that the "how" and "what" of the future would be decided for me. In the meantime, the regiment was treating me well. Apparently I knew my stuff, and those in charge valued me, as did my comrades. And so what if I was rising but slowly in rank? Is that the main thing? I was a technician. There appeared to be no plans to discharge me. I was a graduate of the Zhukovsky Academy, and there were few with such a qualification; in fact, I was the only one in the division. That was how my thoughts tended back then.

I nevertheless understood that the situation in which I found myself was only temporary. I sensed the approach of changes and was waiting for them, although without being able to guess from which quarter they might come. Despite my post-victory euphoria, I lived with a feeling of anxiety.

My Last Military Parade

In the first days of November, my division flew to the Baltic region. Its regiments were dispersed among the airfields at Jakobstadt, as it is called in both Russian and German, or Jēkabpils in Latvian, and Krustpils, two towns situated on opposite sides of the Western Dvina. The divisional headquarters were established in the ancient German city Mitau, renamed Jelgava by the Latvians, and formerly the capital of Courland.[18] To accommodate the general staff, an old palace—or perhaps it was just a large house—was commandeered, which, so it was said, had once belonged to Biron.[19] I shared quarters with Volodya Kravchenko, also a captain; we rented a room from a certain teacher of Russian. My sergeant-major Eliseev was no longer with me. He was to be demobilized, and had remained in Tunoshna. Generally speaking, demobilization was proceeding slowly. In my division only a few older technicians who had voiced a wish to return to civilian life had been demobilized. The pilot complement was kept intact; medical commissions[20] were due to arrive only in the following Spring. Our superiors wished as far as possible to preserve the professional cadres of pilots and technical personnel, but all the same a few pilots were excused from flying on medical grounds. In peacetime the health requirements for pilots were more demanding, partly perhaps because the planes themselves had become more complicated.

In the Spring of 1946, Lieutenant-Colonel Tamara, the divisional engineer and my immediate superior, was demobilized. It was his lack of higher education that had done him in: he had attended a parish school, so had the equivalent in present

[18] Former name for an historical and cultural region of western Latvia. In 1918 it was for a period of a few months a German client state.

[19] Ernst Johann von Biron (1690–1772) was the great favourite of the Russian Empress Anna, who reigned from 1730 to 1740. In 1737 she appointed him Duke of Courland. On the death of Anna, he was for a very short time regent before being exiled to Siberia. He was restored to the Duchy of Courland by Catherine the Great in 1763, and died at his palace in Mitau, which had been designed by Rastrelli.

[20] Presumably tasked with assessing the health of the division's members.

terms of four years at a village school. He had been trained by simple gunsmiths, and had gone far ahead in his profession. During the war he showed himself well able to cope with his duties, and I had learned a great deal from him. He was especially expert in small arms, but understood gunsights less well, and calculations were completely beyond him. For instance, he said to me once: "Well, tell me why the sine of an angle is sometimes large and sometimes small." He could make neither head nor tail of firearms reference tables, especially those pertaining to rocket shells. On the other hand he was remarkably good at diagnosing and repairing any type of malfunctioning small arms, and gave us all instruction in this art. He was a decent, kind fellow, and I became a firm friend of his during the war. He liked to tipple, but at that time who didn't? All the more so in that alcohol was available by the bucketful.

So Ivan Timofeevich Tamara returned to his northland, and we were later told he'd obtained a job as mechanic in an MTS.[21] I was appointed to replace him, and before he left he showed me the ropes. Everyone congratulated me: for a captain to become divisional engineer was an honour, especially since the posting had naturally been expected to go to the most senior engineer of the division Major Alexeev, the regimental engineer of an adjoining regiment. But the top brass chose me instead.

The winter of 1945–1946 was a strange time. We were unaccustomed to everything, most of all the idleness. Over the Summer and Autumn of 1945 in Tunoshna we had at least been busy becoming proficient with the new bombers and the new unfamiliar firearms, and there had been training flights and practice firing the guns—in a word, we had had meaningful work to do. It wasn't like being at the front, of course: gone was the constant strain, the need to be always on duty. But we had been occupied, whereas in our new location in Latvia it was a matter of make-work. I gradually came to understand what military duty is like in peacetime. Leo Tolstoy called it legalized idleness, but I would add to that the need to continually search for or else fabricate something to do. They had stopped supplying us with gasoline since apparently it was needed somewhere else, and as a result flights practically ceased.

All this had the most unpleasant consequences: widespread drunkenness, debauchery, and resultant waves of venereal disease. And this was facilitated by Latvian morals as they were then: their women seemed to be extraordinarily available. I had never seen anything like it in Russia. Every booze-up became an orgy. Discipline fell by the wayside, and there was an infinite progression of emergencies and sortings out of personal affairs.

There were moments, however, when we once more felt like real military officers. I recall the date of May 9, 1946, when the first anniversary of Victory was celebrated. In Jakobstadt a military parade was organized, and I marched in it as a member of a battalion formed from the officers of our division. We marched past our commanders

[21] Possibly a *Mashinno-traktornaya Stantsia* (Machine-Tractor Station), a depot for farm equipment to be supplied to collective farms and *Sovkhozy* in Soviet times.

listlessly, any old how, almost out of step, and were on the point of leaving the square when someone began to sing the satirical military song heard in every educational establishment affiliated with the Air Force:

We learnt long ago
How to march,
But we still march
Like a herd. . . etc.

The battalion straightened up, its step became brisk—and suddenly it was a sight to inspire! The division commander caught up to our column in a jeep: "What's up with you, you scoundrels? You can't march past the tribunal properly, and then suddenly you remember your lives as cadets?" And, by way of an answer, the battalion's two hundred young voices spontaneously barked "Hurraaah," making it clear that there was still gunpowder in the powder flasks.[22]

For me personally that winter was quite hard: I continued looking about me for something real to occupy myself with. Service in the military gradually began to lose its lustre for me. We organized reviews of the squadrons and various practice "alarms," even engaged in parade-ground drills. Service in the regular airforce began to dampen my spirits.

However, nothing that ever happens is uniformly negative. I found an outlet of a kind for my energy. Amongst all the war-trophy trash, of which there was a vast quantity, I discovered a curious automobile. It was a Fiat "Western Desert."[23] It's hard to imagine how it came to be in Latvia, since it was designed to be driven on sand. It was equipped with balloon tyres like those of airplane undercarriages. It was wonderfully manoeuvrable. The steering wheel was on the right, and to its left a machine-gun was mounted. The first such vehicle was discovered by my mechanics when going through a war-trophy junk yard. We towed it onto the tarmac and repaired it. It turned out eminently drivable. Busying myself with this bizarre vehicle gave me something pleasurable to do. I began driving it further afield and ended up driving it all over Latvia. I had the machine-gun removed, but always took an automatic rifle with me since occasional shooting was still going on in the woods, although the roads—or at least the main roads—were already safe. All the same, once, not far from the town of Madona, it came in handy. I returned fire with abandon, despite which, as I later discovered, the body of the vehicle was punctured by several bullets.

I often went to Riga, sometimes twice a week. I found the city interesting, and even more so the people. There I met a few members of the Russian intelligentsia left over from pre-revolutionary times, as well as many Latvian intellectuals. Here I have intentionally not used the term "intelligentsia" for the latter, because in fact I didn't come across anyone who could be said to belong to a Latvian intelligentsia. At first I was surprised, but then I realized that no such class had had time to "mature" in Latvia. Before the revolution, the Riga intelligentsia had been

[22] A widely known quote from Gogol''s *Taras Bul'ba*.

[23] A kind of automobile manufactured for the Italian Army in World War II for use in desert terrain.

comprised of Russians and Germans—and in this connection it must be noted that Baltic Germans are completely different from Germans from Germany, even those from nearby Prussia. Not only did the Korffs, Rennenkampfs and Pleves[24] serve the Russian imperium with zeal and honour, they also made signal contributions to Russian spiritual life. They really did espouse our culture. Their absorption into Russia changed them in many ways, and demonstrated the feasibility and fitness of the symbiosis of Russian orthodoxy and Lutheranism. The Latvians were, unfortunately, much less receptive of our culture than the Germans[25]—which explains why Latvians so predominated in the revolution[26] and especially in the Cheka.[27]

I travelled to Dvinsk—in Latvian Daugavpils—a town not far from Jakobstadt. But, of course, distances are generally not very great in Latvia. Dvinsk was essentially an old Russian town, still populated mainly by Russians, and had preserved an older way of life of a kind dear to me. There I befriended a middle-aged mathematics teacher, and took to dining at his home and even staying overnight.

As I have already mentioned, although in the Spring I had been promoted to the position of divisional engineer, I had nonetheless not moved to the divisional headquarters in Mitau since I had plenty of work to keep me busy in Jakobstadt and Krustpils, where the regiments were stationed—not to mention that it suited me to be at some distance from headquarters. Furthermore, Kravchenko and I had found a decent room to rent in Jakobstadt, whereas in Mitau I would have had to make do sleeping on a couch in my "office," which is what I called the closet-sized space under the stairs assigned to me as service quarters in that old palace.

But then in July something extraordinary happened!

It was a very warm Sunday, and I was lounging on the bank of the river Dvina, when suddenly a soldier from the regimental headquarters ran up: "Comrade Captain, you're required immediately at headquarters!"

I was met by the duty officer: "The divisional personnel officer requests your presence. You are to be flown to Mitau in the commandant's U-2."[28]

An hour and a half later I was standing before the divisional personnel officer, a grumpy, aging major. "You have—what?—an aunt in Moscow? Read that!," says he, handing me a telegram reading: "Dispatch Captain Moiseev immediately to

---

[24] For example: Johann Albrecht von Korff (1697–1766), a Russian diplomat and president of the St. Petersburg Academy of Sciences from 1734 to 1740; Andrey Nikolaevich Korff (1831–1893), from 1884 to 1893 first viceroy of the Russian Far East; Paul von Rennenkampf (1854–1919) and Pavel Adamovich Pleve (1850–1916), Russian generals serving in the Imperial Russian Army up to and including World War I.

[25] Perhaps because Latvia had been unilaterally absorbed into the Russian imperium in the eighteenth century (and earlier ruled for various periods by the Polish-Lithuanian Commonwealth, Germany, Sweden, and Russia). Latvia was independent 1917–1939, and then again post-1991.

[26] The 1905 and 1917 revolutions both enjoyed especially strong support in Latvia.

[27] It is claimed that in 1918, 38 of the 70 members of the Cheka, the Soviet secret police agency, were Latvians. In 1937 those remaining were purged, along with many thousands of other Latvians.

[28] A basic Soviet military trainer biplane in use from 1927 to 1978.

report to the head of command personnel at the headquarters of the SAF,[29] signed Vershinin."[30] At that time Marshal Vershinin was the commander-in-chief of the Air Force. Never before had our division received a telegram signed by such a one. "You will hand over your affairs to Alexeev tomorrow. He's already been summoned. Acknowledge the order, and let there be no trace of you here two days from now. Clear?!"

Why was I suddenly needed by Moscow? I didn't know what was going on, but prepared to leave as ordered—and—why pretend?—gladly.

So what had happened? What force was propelling me, a mere regimental engineer, to the headquarters of the Air Force of the Soviet Union? To explain this sudden twist of fate, which I had had a premonition of, had even been expecting, yet was nonetheless completely surprised by when it came, I have to backtrack somewhat.

Professor Kranz's[31] Exterior Ballistics

The head of the Political Section of my division was Lieutenant-Colonel Fisun. Or maybe he was a full colonel; I don't now recall exactly. He was a big slow-moving Ukrainian. Earlier he had been deputy political commissar in our regiment. Fate had brought us together in 1942 when he had provided me with a recommendation for membership the Party. Though totally inadequate as a political operative, he was an excellent pilot. He had flown many sorties, willingly and successfully; he was one of those who had the happy knack of battle. But then his health began to fade, he was forbidden to fly, and he went over completely to political activity. After he was promoted out to the division, his relations with me remained good and he regularly showed one or another sign of regard for me.

He once summoned me to the Political Section and gave me a war-trophy book. It was a work devoted to the exterior ballistics of rocket shells written by the well known German ballistics expert Kranz. "Look at this, Moiseev. In your line, perhaps?" I had then not yet completely forgotten my German and began reading Kranz's work without encountering much difficulty. This turned out to be a pleasant and interesting occupation, providing me with diversion from the tedium of military service. I had my acquaintance in Dvinsk give me a course of sorts in the higher mathematics—that subject having quite thoroughly gone from my head—and I began to sort out for myself the subtleties of what was then a new science: the calculation of the trajectories of rockets shells.

---

[29] Soviet Air Force.

[30] Konstantin Andreevich Vershinin (1900–1973) was commander-in-chief of the Soviet Air Force 1946–1949 and 1957–1969.

[31] Possibly Bernhard Kranz, a highly decorated *Hauptmann der Reserve* in the Wehrmacht during World War II.

In fact, I rather rapidly recovered my mathematical knowledge, and Kranz's essay turned out not very difficult. Not only was I able to understand the work, but thought of various ways of perfecting it. Although the purpose of Kranz's work was purely practical, he had developed, with that pedantry and lack of a sense of humour characteristic of all Germans, a general theory of the motion of a rocket in the gravitational field of the rotating Earth, assumed spherical, and from this had deduced the rules for calculating those trajectories of rockets that we now classify as "Earth to Earth."

I myself, however, had studied ballistics at the academy under the supervision of D. A. Venttsel′, one of the most brilliant professors that I have ever heard lecture. He used to flavour everything he said with a large dose of humour, and in science preached the religion of his teacher, the famous admiral and academician A. N. Krylov, to wit, a false value of a numeral in a calculation is an error, and an excess of places after the decimal point represents a half-error. And unneeded complicating calculations not essential to the work at hand constitute a mortal sin! So there! An applied theory must go right to the point and be as simple as possible!

In those days, of course, rockets couldn't fly very far. Even the infamous V-2 flew only a little over 200 km. For this reason Kranz's theory, considered as an aid to solving the ballistics problems of that era, seemed to me "over-theorized," and I didn't like his book.

I set myself a simple problem in the manner of Krylov and Venttsel′: How might one calculate the trajectories of ballistic rockets over moderate distances in the simplest possible way, using methods already known to the science of artillery? I would still be disposed to claim that I really did solve this problem, producing simple formulae for corrections to entries in the ballistics tables already in existence back then. When written up, my essay occupied around ten pages. The question then arose as to what I should do with these pages.

During my time at the academy I had attended a few lectures on the ballistics of rocket-propelled shells. These had been delivered by Yu. A. Pobedonostsev—"civilian professor," so-called—and by Venttsel′ himself, "the father of Soviet reactive technology." These lectures had made a certain impression on me. I had spoken with Pobedonostsev a few times, and he entered my soul, as they say—so much so, in fact, that I had devoted my diploma thesis to a ballistic computation of a concrete-penetrating bomb with additional, that is, reactive, acceleration. As it turned out, Pobedonostsev remembered me.

E. Ya. Grigoriyev, one of my instructors at the academy and a very capable young lieutenant-colonel, had contacted Professor Pobedonostsev on my behalf; it was to him at the Zhukovsky Academy that I had sent my few written pages with the request that he pass them on to Yury Alexandrovich. I had subsequently learnt that they had indeed reached their addressee safely, but now it turned out that they were actually the reason for my being unexpectedly summoned to Moscow and for my final departure from my regiment. At the time I had no inkling of this; it didn't occur to me that my letter or my outline might be behind the summons.

Leaving My Regiment

The consequences of my exercise in German and ballistics became known to me only some days later. In the meantime I handed over my divisional responsibilities to others and returned to my regiment, where I quickly gathered together my few belongings. But there was a hitch. I had been hoping to take my Fiat "Western Desert" with me and make a triumphal entrance into Moscow. I can imagine what a furor—in those days they used to say "furage"—it would have caused! I regarded the vehicle as my own on the grounds that it had been my mechanics who had resurrected it from absolute nonexistence. Others thought differently, however. It turned out that the divisional quartermaster had long had his eye on it. As long as I was divisional engineer he had allowed me the use of my Fiat, but quietly, without informing anyone, he had early on had it registered as the property of the division, so now it belonged to the Soviet army. Thus it was that I travelled to Moscow, like all ordinary mortals, by train.

My departure from the regiment was marked by a drinking bout the likes of which had never been seen before, apparently—even on Victory Day.

It began early in the morning when the regimental commandant phoned requesting the presence of Kravchenko and myself—at once! Lieutenant-Colonel Andrianov was, as they say, born to the military: the son of a military man, he had from childhood been groomed to serve in the military. He always looked spic and span and youthful and held himself ramrod straight. He never got drunk, flew a great deal and with pleasure, and had been in some dreadful scrapes. Everyone in the regiment felt that he should long ago have been officially acknowledged a Hero of the Soviet Union. However, exceptional courage and good fortune in war together with independence of mind were not much in favour with the powers-that-be.

Five or six years later I bumped into him in Rostov—in, of all places, a bathhouse. I'm walking past a mirror and unexpectedly see a familiar face reflected in it: Andrianov in his military tunic with shoulder straps removed, is standing in front of the mirror sprucing himself up. He sees me in the mirror at the same time, recognizing me at once even though I was attired like Adam: "Engineer, you mother-. . . , where did you spring from?"

I dressed quickly and we went to my apartment. My wife brought to the table whatever God had provided—we were living close to the bone at the time—and we talked long and very agreeably. Not long after I left the regiment he had been promoted to colonel and made second in command of the division. He got on poorly with the divisional commandant, however, and was given a temporary leave and then, in the course of a routine reduction in manpower, demobilized, or, more precisely, retired. He was now working on the Regional Executive Committee of a Cossack village. However, a medical commission had evaluated him as still able-bodied, and he was getting ready to return to flying, only now as a civilian pilot. In that profession he would once again have real work to do since he could fly anything, even a broom.

But back in the summer of 1946 he commanded the regiment, and had the use of a decent house with a garden, in which he organized a farewell "breakfast" for his former engineer. It was attended by almost all members of the original officer staff still alive. These were some of the most picturesque personalities of my experience—and that quality essentially explains why they had survived! They all had the same attitude: mischievous boys is what I would now call them, never mind their iconostases of decorations and the childishly pompous titles associated with their ranks. And what boys they were! The pick of the Russian Air Force. I was tremendously proud that they had assembled for my sake.

But the commandant's "breakfast" was just a gentle introduction—the warm-up before the real "battle," if you like. Following that, the all-out spree began. In the evening a crowd of them went with me to the Krustpils railway station to see me off on a train for Moscow. There they continued drinking and carousing. At the station a major of artillery headed for somewhere or other joined up with us, and he was very shortly reduced to our general state.

At that time trains ran very irregularly, and the train by which I was supposed to travel to Moscow failed to arrive. Instead of my train a troop train with a few first-class carriages drew up. My friends knew how to ensure I got aboard. Even better, they managed to obtain a separate *coupé* for myself and the major: nothing is beyond the Air Force! The major had to be carried aboard but I was able to get on unaided.

I awoke late. The train had come to a standstill and the major was snoring on the divan opposite. On the table someone had obligingly prepared a spread consisting of a bottle of vodka, a chunk of black bread, two cucumbers and a lump of fat—a most enchanting *nature morte* worthy of the brush of Dutch artists, and very apposite after yesterday's farewell.

Evidently the train had been standing for some time. In our carriage complete silence reigned and the platform outside was deserted. I woke the major and said the first thing that came into my head: "Get up, Major, the vodka's getting cold. We've arrived at Velikie Luki."[32] He got up, looked at me without recognition, and said: "What Velikie Luki? I need to go to Vindava." He grabbed his rucksack and got out onto the empty platform. I have my whole life been tormented by the unanswered question: Did my major make it to Vindava—or Ventspils, as Latvians call it?

The troop train arrived in Moscow only in the very early morning of the following day. And instead of arriving at Riga Station, as a normal train from Riga should, it pulled up at Kiev Station, and moreover on a side track. But that was as nothing to me: I was disembarking in Moscow!

The station was grey.
The heavy dome of the cathedral
In the starless morning stillness
Was silent without encounters, without fuss or hubbub
In the dim space of the gloomy courtyard.

[32] A town in Pskov Province.

The final minutes of the final versts,
And through the dewy sieve of glass
There are already glimpsed, snug throughout,
Sleeping suburban Moscow buildings.
Somewhere not far from here, in Nara or Skhodnya,
Was my fate decided. Seems just yesterday. . . .

I remember the morning of my return in great detail. Although already July, it was still cold. The sun had just that moment risen, and below in the square it was murky. But the windows of the buildings' upper floors glowed in the rays of the rising sun. I was home, really home. I repeated these words to myself but couldn't believe them.

The Metro was still closed and trams were not yet running.

I had with me a heavy rucksack and two suitcases; over a single year of life in peacetime, junk had accumulated, and I had even acquired a few books. I went out onto the square and sat down on one of the suitcases to wait for the Metro to open. Taxis were then just as far beyond my means as they are now. Incidentally, this situation was easier to bear back then since there didn't appear to be any.

A man in an army tunic with shoulder straps torn off came up to me. "What's up, Captain? Service over?"—"Not yet."—"Mine is. I'm waiting for the Metro, in a hurry to get to work," he said with a certain pride. This chance fellow traveller helped me with my luggage in the Metro and even to board the electric train. I was headed for Skhodnya, where my stepmother was still living, now with my younger brother, returned from the war an invalid.

My Return to Moscow

There remained a few days till my appointment with the Air Force's personnel section, and I spent those days light-heartedly immersing myself in Moscow. I was completely stunned by that city, and by the sensation of having it once more my town. It was as if I was getting to know it anew. I wrote poetry, understanding that in all probability these would be the last verses I would compose since my life was about to take a completely different turn. Life would take up all my strength and time, and I would simply cease needing poetry as a pastime. With real interests to absorb me, I would have no time for it.

Meanwhile I visited acquaintances, who treated me to tea substitute as a rule made using carrots. Moscow life was not easy then! I didn't return to Skhodnya every day, sometimes spending the night with friends, and I went walking, walking, walking. Most of all I liked investigating the old lanes in the Arbat:[33] Afanas′evsky, Sivtsev Vrazhek, and others of the neighborhood where I was born and whither we returned in 1921 from Tver′ Province. Then I would go via Vozdvizhenka St.

[33] An old area of Moscow just west of the Kremlin.

towards the Kremlin, and drop by my old faculty MechMat at the university.[34] Since it was the Summer break, however, I didn't find anyone there. Only the Admissions Committee was in session, but I saw no familiar faces there.

I wandered around the city unable to slake my passion for it:

Moscow, Moscow, she's just as always:
The Metro, trams and busyness.
And everyone, in a state of agitation,
Rushes from morn till night.
The quiet of the Arbat's lanes,
Their pleasant, cosy slumber,
And the cordon of wide, resounding squares
And frenzied streets
Around the ancient Kremlin,
A place close to my heart.
And far down Moscow streets,
Drawn into the vortex,
Only in the evening, with difficulty, bending double,
I came upon an apartment's PFP.[35]
Then through drawn curtains
I heard the city's booming,
The tumult along the corridors of streets,
Always full of crowds. . . .

And here is something else I wrote under the influence of that mood, of which only the last lines remain in my memory:

And there, high above the rooftops,
Where the stars are already out,
I hear what I had long not heard:
The nighttime voice of Moscow.

I looked up acquaintances and friends. Many of them were no more, but a surprising number had survived. The demobilized ones were already working in full swing. They were glad to have survived and to be back in Moscow. I talked endlessly with them—but not about politics nor the hardships of postwar life. We mainly talked of work, the future of the country, its resuscitation, the problems of educating the young, and the situation in the schools and institutes of higher education. And, of course, of our domestic lives.

But these conversations always began with the same topic: the fates of mutual acquaintances and friends: who had fought where, who had survived, who was still single and who had married. My former girlfriends, all of whom were of the same age as me, no longer interested me; they seemed to have become ladies of a respectable age. But perhaps it wasn't in fact a question of age. At the front, notwithstanding its rigours, we had remained those same boys who joined up in

[34] Moscow State University was then located in the centre of Moscow. Most of its faculties were relocated to its present site on Sparrow Hills, a half-hour bus ride from the centre, in 1953.

[35] "Permanent Fire Position," a small-scale strongly fortified position for protracted defence of the apartment against enemy fire.

1941, while in the rear our female peers had had to bear an enormous burden of care: how to get enough food and clothing, how to help the family survive—something akin to our present situation, now in 1992. This kind of daily responsibility stresses and ages a person much more than exposure to immediate danger, which, once past, assumes the guise of vicarious recollection.

Back to the Academy

But at last the day came when I appeared before the piercing (not really) eyes of Lieutenant-General Orekhov himself, head of the Soviet Air Force personnel section, and, as I was later to find out, a hard man before whom all who had business with him trembled.

A vast gloomy room with austere expensive furniture in a huge building on Pirogovsky St. When I entered there was a colonel standing bent over the table. This turned out to be the "head of the administrative personnel section," to which section I had been assigned. He just happened to be in the middle of explaining the terms of my assignment. Contrary to standard procedure, these had not been given to me under seal when I left the division, but dispatched to Moscow by field post. This explained the delay in my meeting with my prospective high-level superiors: they needed time to familiarize themselves with my assignment. Next to my dossier there lay a sheet of paper which the colonel kept prodding with his finger. Having announced my arrival, I waited, standing at attention. The general was thumbing through papers and muttering under his breath while asking more or less pointless questions. Finally he said: "You'll be working in the department of the first secretary of the commander-in-chief. You come with good recommendations. You know and love rocket technology. That's what we need now." And that was all!

While I was standing quietly by the table, my eyes were not on the big boss but on the paper lying near my dossier which, clearly strange to him, the colonel kept poking with his finger. My eyesight was somewhat better then than it is now, and I could make out the letterhead "Ministry of Farm Machinery Production." This was a codename for the wartime ministry charged with planning and building rockets. Curiosity aroused, I tried hard to read the text below the letterhead, but this was, of course, next to impossible. I was nonetheless able to make out two things of interest: first that the letter was addressed to Air Marshal Vershinin, then commander-in-chief of the Air Force, and second that across the whole page a flourishing hand had scrawled in red pencil: "Use in the central apparatus," with Vershinin's signature.

So my fate was being decided by the commander-in-chief himself. This explained the appointment with Orekhov, who rarely vouchsafed anyone a meeting in person, and also the non-standard dispatch of my dossier. But beyond this I understood nothing—nor was I told anything.

The department whither I was sent to begin working consisted of a large room occupied by various colonels and lieutenant-colonels. There weren't even any majors, while I myself was a mere captain. My immediate superior, a colonel, told me that I had been very lucky: the work here was a bed of roses and admin service ranks were generally higher than those in a regular division! "How did you get hired? Clearly someone with influence on your side," he added without irony and even with a certain respect. His relations with me remained kindly, though a little apprehensive, since he really couldn't comprehend how a mere captain, moreover one from a regular fighting unit, had attained to such a position.

But the facts of the matter were completely other than what my coworkers thought. I myself discovered the circumstances of my return to Moscow only several years later. They are as follows.

My letter had reached Professor Pobedonostsev; I remain grateful to Grigoriyev for handing it to him in person. Furthermore, Yury Alexandrovich remembered me. At that time he filled several exalted posts. As one of the founders of the SRI-88[36] in Podlipki, he had been appointed its chief engineer, with responsibilities which at that time included those of the scientific and technical supervisor of the forges used in cosmological rocket technology, and he was also a member of the ministerial collegium. Unlikely though it may seem, certain pages of my calculations had proved very useful to him. It so happened that my note was historically the first critical commentary on German rocket research, before which everyone was then in awe. Furthermore, in my work I had suggested an alternative approach to solving problems relating to the ballistics of reactive shells of the type "Earth to Earth." Whether my approach was good or bad was another question altogether. In fact, I can now affirm that it was completely primitive and perhaps even plain bad. However, it differed from the German version, and, despite its shortcomings, was more convenient that Kranz's method insofar as it allowed the standard methods of ballistics computation to be brought to bear.

My letter had been stamped "top secret" and then Pobedonostsev had reported it to the minister, who had liked it: so we too are capable of such things! He had then written to the commander-in-chief: "There's an Air Force captain Moiseev who etc. etc." In effect, have him demobilized and hand him over to us. His former duties can be fulfilled by any of the regiment's engineers. Evidently Yury Alexandrovich had described me in rather too glowing terms, however, since Vershinin had evidently begun to feel that handing that Moiseev fellow over would be a pity, like giving away a prized personal possession. So on the ministerial letter he had scrawled : "To be kept in-house and used!"

All this I heard from dear Yury Alexandrovich—in fact twice: the first time following the defence of my candidate's degree[37] and the second time when, following my demobilization, he invited me to work in his department at the MHIT.[38]

---

36 Scientific Research Institute No. 88.

37 Roughly equivalent to the Ph.D.

38 Moscow Higher Institute of Technology; see Chap. 1.

But for the time being, without knowing how or why, I found myself in Air Force headquarters, in a section where the work was indeed a picnic. As lowest ranking officer my main duty was wangling soccer tickets. Apart from that, I occasionally had to examine war-trophy material concerning rocket technology and write something informative about it—which, as I soon understood, no one read. The only hard thing was the work regime, which was strange indeed. We arrived at work towards evening, and then, provided there was no soccer that day, sat there till early morning waiting for the commander-in-chief to leave. And he—Vershinin—was in turn waiting till Stalin himself went to bed. So there was a hierarchy of waiting: what if a question should issue from the top?!

I grew miserable: Moscow had not fulfilled my expectations; even returning to my regiment would be better. At least there in the division, and especially in the regiment, I had felt much more at ease. Even in idle times one had to check that the machine guns had not rusted and that one's subordinates weren't bingeing! While all I was doing here was sitting for hours waiting for promotion to a higher rank even though I had never set much store by rising in the ranks. I began to think seriously about extricating myself. I weighed up various means to demobilization, that is, returning to civvy street, as they used to say back then. I had an academic degree, however, and for that reason was considered a career officer, and such as those were not to be released into civvy street.

At that time, in the academy[39] they were creating a new department in the Faculty of Aircraft Weaponry No. 2, namely, one of Reactive Aircraft Weaponry. As head of department they appointed my former teacher E. Ya. Grigoriyev, with whom I was on the best of terms. We often saw each other on a friendly basis, outside working hours, and I asked him to take me on in his new department. Although I didn't have an appropriate scientific degree, I did have practical experience of the use in wartime conditions of contemporary reactive shells in attack-planes and bombers. Such experience had once been important! The faculty head General Solov′yev—or Solovey[40] in the vernacular—approved my application.

Appropriate letters were sent, essential phone calls made, and a week later there emerges an order with Vershinin's signature, appointing me junior instructor—assistant, that is—in department number such and such in the Zhukovsky AFEA.[41] Apart from Grigoriyev and myself there was only one other member of the department: Senior Lieutenant P. A. Agadzhanov, lieutenant-general-to-be and future corresponding member of the Academy of Sciences of the USSR. At that time he was designated departmental engineer, which is to say that he worked as a laboratory assistant.

My co-workers gasped and sympathized: that captain who had been so good at obtaining tickets to all interesting matches, has been transferred to a new position

[39] The Zhukovsky Air Force Engineering Academy.

[40] Nightingale.

[41] Air Force Engineering Academy.

involving demotion by at least two ranks. So there was after all no one pulling strings for him as we'd thought!

But that was the turning point I had been waiting for on my path through life. At least, that's how I see it now. And I bless the fates for it—and likewise for Solovey, destined to rescue me from a bad situation on another occasion.

Chapter 5
Climbing Mt. Olympus or Seven Very Strange Years in My Life

Yet One More Metamorphosis

The years from 1947 to 1955 were indeed the most astounding of my life. In the course of those seven years I went from being a regimental armourer at the rank of captain, first to candidate of technological science, then to doctor of physico-mathematical science, and finally to respected professor and dean of a prestigious faculty of the most prestigious institute of higher learning in the country. That which takes most men of science decades and sometimes their whole lives to realize, in my case took but a handful of years. And if one considers that it was during that same period that, following the arrest of my stepmother, I was fired from my job, forced to leave Moscow, and had to begin all over again, then the rapid succession of events becomes something beyond extraordinary that continues to this day to amaze me. How did it all happen? I still don't understand how I managed to survive this string of vicissitudes. There was, of course, my youthfulness as well as a perhaps unusual lust to live and work. But these were incidental to the improbable sequence of unexpected bits of luck and personal kindnesses shown to me, interspersed with blows capable of turning the brain to mush. And there were also, of course, my friends.

Perhaps it was just the way the country was back then, when the times were such that the unlikely seemed everyday.

Or as Ostap Bender[1] expressed it once: life's a soul-stirring racket.

---

[1] One of the main characters of the very popular classic satirical novel *The Twelve Chairs* by Ilf and Petrov, published in 1928. Ostap Bender is a charming, resourceful and street-wise con-man.

N. N. Moiseev, *How Far It Is to Tomorrow...*,
https://doi.org/10.1007/978-3-030-96651-5_5

The Starokonyushenny Academy and Professor D. A. Venttsel′

Being normally very difficult, my ascent of Mt. Olympus was completely unanticipated. It wouldn't have figured in my wildest dreams. Yet it took me literally only two or three years to rise to the level of specialists who were household names in engineering and scientific circles. No matter how my life should turn out subsequently, whatever sorrows might befall me, I was to become a specialist to be reckoned with who could count on a place in the sun.

My "ascent of Olympus" was in the first place linked to a particular problem discussed at a certain meeting of the Academy of Artillery Science.

This academy had been founded following the war but ran for only a few years. It was presided over by the academician Blagonravov, with, if I remember correctly, my former academic supervisor at the Zhukovsky Academy, Professor Dmitry Alexandrovich Venttsel′, as vice-president. It was located in Starokonyushenny Lane,[2] so it was nicknamed the Starokonyushenny Academy. I succeeded in finding a new approach to the analysis of the problem under discussion there of estimating dispersion of aircraft missiles, and, much to my surprise, obtained a solution that was at the time completely satisfactory from an engineering point of view. My solution served as a sort of launch pad allowing me to bypass the usual graduate work and advance directly to my first higher academic degree, and moreover securing for me a certain standing in the world of science. And doubtless also securing my future.

Yet once more a chance event and once more good luck. Fortune would continue to favour me from time to time and place me in situations impossible to anticipate. Once, when I happened to read Einstein's famous statement "God doesn't play dice," I thought that perhaps each individual has his or her personal God, with mine not just playing but even winning occasionally!

But I'll be discussing God seriously later on. For now, let me return to the Zhukovsky Academy, where at war's end I was appointed junior instructor in the Department of Reactive Aircraft Armaments, and where earlier, in 1942, I had obtained my second degree, this time in engineering, and to which I am infinitely obliged—not only for the degree but for the leg up in life it gave me.

I taught in the academy for only about a year and a half, but that short period had disproportionate significance for my life or, more particularly, for my development. I met people there whose work made a huge impression on me and I wished to further my studies under them.

I can now affirm with utter certainty: the Faculty of Aircraft Weaponry of the academy represented at the time in question a truly unique enterprise. This was primarily due to the high professionalism of its teaching staff, whose undisputed leader was the head of the Ballistics Department Professor Major-General Dmitry Alexandrovich Venttsel′, of prodigious authority and popularity. He was, so to

[2] Oldhorsestable Lane.

speak, a good head taller than the rest of the teaching staff of the faculty in terms of general learning, liveliness and keenness of intellect, and also kindliness to the young. Venttsel′ astonished not only by the independence and acuteness of his judgements, but most of all by a boldness in his pronouncements unusual in a career officer. When I was forced to quit Moscow shortly after the arrest of my stepmother, General Venttsel′ was the only one of my former academic teachers to maintain friendly relations with me.

I recall the last time I met with Dmitry Alexandrovich. This was most likely in 1954, when Stalin was already dead. He told me then how much his boldness had cost him, how throughout his life he had been afraid of arrest and how he considered it a miracle, an absolute miracle, that he'd never been imprisoned. Since I likewise considered it a great miracle that I had escaped incarceration, his words resonated with me.

But then, when teaching in the academy, I had already come to understand that our lives were so arranged that everyone not under arrest should silently thank Stalin for granting him or her the kindness of not having them languish in the camps.[3] This was how I interpreted the omnipresent slogan "Thank you, Comrade Stalin, for a happy life." For those on the outside it was like a prayer. This prayer was relevant to all of us, but to Venttsel′ especially. I said to my first wife once: "I thank Comrade Stalin for allowing me to leave Moscow after the arrest of my stepmother, and not sending me to work in the uranium mines"—which horrified her. But that's exactly how it was with us. Anyone who spoke out freely was suspected of being a snitch since that was the only explanation for their being free! More generally, the non-arrest of any individual seemed like a unusual circumstance requiring explanation. I do not exaggerate here: this is an exact characterization of the psychological state of mind of a significant portion of the intelligentsia.

Dmitry Alexandrovich came from a family tracing back to a certain Estland[4] noble who had sometime during the reign of the Empress Elizabeth gone into the service of the Russian Empire. He had received a splendid education in engineering and the military. He considered Alexey Nikolaevich Krylov to have been his main teacher, and professed adherence to the latter's principles and scientific outlook, which he tried to instill in us in turn.

Thanks to my contact with Dmitry Alexandrovich I gradually began to appreciate the appeal of applied science and of problems arising in engineering practice, which require no less sharpness of mind and inventiveness than more abstract ones. Furthermore, I progressively came to the realization that science constitutes a single whole—provided, of course, that what is being pursued is genuine SCIENCE. The distinction between primary and secondary sorts of science is nonsense; the division is made on completely different grounds: on the one hand real, profound science, and on the other, speculation about science.

---

[3] That is, the Gulag.

[4] What is now Estonia was from 1721 to 1917 the Province of Estland of the Russian Empire.

Although to some obvious, this view of scientific activity was new to me: it did not fit well with the mathematical snobbery that flourished—and, I imagine, still flourishes—in the mathematics section of the MechMat Faculty of Moscow University. Only pure science, only ideal constructions unrelated to anything too earthbound—that's the true calling of the true mathematician! This was the "Lusitania Law,"[5] and we had been indoctrinated with it, even though Academician Luzin himself had worked on applied problems to a considerable extent. I recall an entertaining episode in one of the seminars I attended during my student years. One of the best-loved professors of the faculty, Alexander Gennadievich Kurosh, was asked: "What are ideals of algebras over a field needed for?" Kurosh thought for a bit and then in all seriousness answered: "For the theory of ideals of algebras over a field!" But in the academy I began to understand that, despite its attractions, "pure science" by no means exhausts the world of science, other branches of which are no less splendid.

Dmitry Alexandrovich's irony completely cured me of any remaining mathematical snobbery, thereby saving me from the fate of the great number of unfortunate ones who, having obtained a first-class mathematical education, failed to find a place for themselves in life. Many of these felt that the only worthwhile occupation was proving theorems, failing to understand that this requires a special talent in much the same way as chess-playing. Mathematics, the most beautiful of all the sciences and arts, is all the more marvellous by virtue of the fact that it helps us understand relatively easily things that are otherwise difficult to grasp.

Perhaps I exaggerate somewhat here. What I have just said, however, reflects my immediate post-MechMat perception of science, and my subsequent disowning of that view of science and transferral of allegiance to concrete applied science was of very great importance, even crucial, to my future. And that is why I include D. A. Venttsel′ among my most influential teachers. There were essentially just two people who imbued me with the the view of science that I have had most of my life: Venttsel′ and Tamm.

The chief worth of the Faculty of Aircraft Weaponry in those first postwar years resided in the young men sent to the academy after the mobilization of June 1941. Under the most punishing of conditions, our government found the courage to preserve the country's university-trained young, sending a large proportion of graduates and higher-year students to the military engineering academies. Although I don't have any figures, I believe, nonetheless, that the rapid development of the defence industry and of our rocket and nuclear potential was due to that action. In the immediate postwar years, in scientific research institutes and design bureaux I was forever encountering graduates of various military academies who had been sent there in the first month of the war to further their studies.

---

[5] A partial pun based on the similarity of the name Lusitania, that of an ocean liner sunk by a German U-boat in 1915, and that of Nikolai Nikolaevich Luzin (1883–1950), a Russian and Soviet pure mathematician. The name "Lusitania" was applied to a loose group of young Moscow mathematicians of the first half of the 1920s strongly influenced by Luzin's set-theoretic approach to mathematics.

The faculty to which I had been assigned postwar was made up mostly of a few dozen genuinely talented senior lieutenants remaining there as adjuncts, junior instructors or engineers. The result was an exceptional, youthful team which, together with Venttsel′, Pugachev, Pokrovsky and several other very able scientists of an older generation, represented an enormous national asset. Unfortunately, the Air Force top brass did not prize this collective at its true value for solving problems relating to the development of aircraft weaponry. Instead, using a variety of justifications, they began gradually dispersing the faculty.

Of the considerations entering into the authorities' decision-making in this regard, I can, however, understand those stemming from the "institutional principle": the academy had been created in order to teach and not to busy itself with innovative technological ideas and "big science." But I think the real reason for the dispersal had to do with the authorities' suspicion that the members of our collective were too "individual," each with his own opinions, behaving according to his own lights. And their military training left much to be desired! Then an opportunity presented itself: many mid-level educational institutions were to be transformed into higher-level ones. So they began scattering us around the country. In the Autumn of 1947 I was given a posting in Khar′kov[6] which came with a big promotion, namely to head of the academic section of the Khar′kov Higher Institute of Aviation Technology. The appropriate rank of someone occupying such a position would normally be colonel, or, at a pinch, lieutenant-colonel, but I remained a captain.

Soon afterwards my immediate superior E. Ya. Grigoriyev was sent to Perm′[7] as deputy director of an institute, and Major Dezortsev was, like me, sent to Khar′kov. In this way those who should have been kept bunched together in the critical mass of their combined intellectual power and talent, were dispersed far and wide across the immensity of the USSR. Like nuclear fuel, good brains produce an explosion only when there is a critical mass of them!

When in the mid-1950s I was once again shown the kindness of being granted access to "secret science," I sometimes revisited my old faculty. It was unrecognizable. Although quite a few talented ones of the former "youngsters" remained, become by then dignified senior officers, the atmosphere was not the same. In the early postwar years, we had been in hot pursuit of the new: we had been absorbed by mathematical and technological innovations, competing with each other in solving all kinds of problems, and as a result progress had been rapid. But now my former colleagues fulfilled their duties as professor or dozent with mere professional aplomb.

My old faculty continued, rather like a high quality smithy, turning out highly qualified officers—which is, in fact, what the authorities expected of it. But it was no longer what it had been in the early postwar period, and as a result the country lost a great deal! The faculty had lost its soul—and the loss was exacerbated by the death of Venttsel′.

[6] Second largest city of Ukraine, located in the east of that country.

[7] A Russian city located just to the west of the Ural Mountains.

Sergey Moiseev

When I joined the Zhukovsky Academy, although its Department of Aircraft Missile Armaments had only just been created, we immediately had a great deal of work to do. In 1946 a systematic retraining and instruction in the new technology of the engineering staff of the regular Air Force contingent was undertaken. As a result I had to postpone the realization of my aspiration to test myself in science or at least recall what I had once known. Nevertheless, I began preparing for the candidature examinations, of which there were at that time as many as six or even eight. I managed to pass almost all of them over the Winter.

Just then, however, my brother's health sharply worsened, and my "return to science" had to be shelved.

Sergey had been recruited into the army in 1939, immediately following the end of his ten years of schooling. It was the time of Timoshenko's[8] reforms, and all young men aged 18 or more and fit for military service were called up; aspirations to tertiary education were to be put on hold indefinitely.

Sergey found himself at the front at the rank of senior sergeant. He was the commander of an artillery brigade equipped with a 45 mm anti-tank gun—the celebrated "forty-fiver," which had to be wheeled into an exposed position to fight one-on-one with tanks. Sergey survived the soldier's ordeal without unusually heavy losses, and in 1943, having won a raft of soldier's medals, was sent with his army unit to the Far East. I had a letter from him then, in the standard triangle shape of letters from the front. "I am already safely through the war," he wrote, "Make sure you also survive. We'll heal together after the Victory!" But life being the way it is, who knew what might overtake us in two years!

When the war with Japan started, Sergey was included in an assault force to be disembarked on the island Iturup.[9] This was, apparently, the only island of the Kuril ridge where there was any fighting. And there he was wounded. The wound was not of itself so serious, but he lost a lot of blood and lay unconscious for several hours on swampy ground. Some kind of nasty bug got into his blood, a form of streptococcus, and as a result, the condition of his heart became critical. He returned home an extreme invalid.

He nevertheless aspired to enter the university. He prepared for the entrance exams all winter and was admitted in 1946 to the Geophysics Department of the MSU Physical Sciences Faculty. Over the Winter from 1946 to 1947 he studied well and seemed to be getting better. We all began hoping for a complete recovery.

---

[8] Semyon Konstantinovich Timoshenko (1895–1970) was a Soviet military commander and Marshal of the Soviet Union. He fought in World War I, and, as a Red Army officer, in the Russian Civil War and the Polish-Soviet War. He survived the great purge of 1937–1938 to become the Red Army's senior commander. It was at his urging that army underwent mechanization and the production of tanks was increased in 1939. He also reintroduced the traditional harsh discipline of the Imperial Russian armed forces.

[9] One of the Kuril Islands, still the subject of claims on Russia by Japan.

In the Summer of 1947, however, Sergey's health suddenly deteriorated. I arranged for him to go into a clinic run by the shining light of cardiology Professor Burmin. One day he told me that Sergey's case was hopeless: he had septic endocarditis, and at that time this was practically incurable. The only hope for a cure lay with the newly rediscovered penicillin—if only one could get hold of some!

But penicillin was not yet being produced in the USSR. In certain places, they procured the drug at great cost. One cool Autumn day the professor said to me: "His days are numbered: a week or at most two." Every day I returned home from the clinic and lay awake long into the night. I thought of what my younger brother was going through, my little brother whom I had defended from the bullies who used to come to Skhodnya from Dzhunkovka to beat up the "bourgeois." For some reason I kept recalling the pages where Tolstoy describes the last night of Prince Andrey.[10]

It was precisely at that time that I unexpectedly received the order to go to my new post in Khar′kov—at once! I requested an audience with the head of personnel management General Orekhov; he alone had the power to grant a postponement. I have already related something of him, noting, in particular, his strictness. I would now experience it in full measure.

He agreed to see me. But without giving me time to put my case he began talking: "Captain, you once before refused to serve where I posted you, in the offices of the commander-in-chief. Now you don't want to go to Khar′kov and come to me with all sorts of excuses. If in three days you are not at the location where you are to serve, I'll consider you a deserter." I tried to explain that I wasn't asking for a change in my posting, and told him that my brother was dying of a wound received at the front, just nearby in a clinic on Pirogovksky St. and that he would last but a few days. I would arrange his funeral and then immediately leave for Khar′kov.

I fell silent. The general looked at me contemptuously, as at a worm: "Dismissed. You've received an order. Obey it!"

I was desperate. I couldn't possibly leave. But breaking my military oath was also out of the question. What to do? My former colleagues at Air Force headquarters tracked down the telegraph address of Lieutenant-General Khadeev, head of the Khar′kov Institute. I sent him a long telegram, the longest I have ever sent, explaining my situation in detail, including the threat of arrest for desertion. Two days later I received the laconic reply: "Waiting headquarters, Room No...., such-and-such a date, 16.00. Entry permit ordered. Khadeev."

An elderly general of moderate height. Of an unsmiling and dry mien, like his manner of address. I pass over to him the envelope containing my orders under a stamped seal. A glum middle-aged Lieutenant-Colonel is seated next to him. This turns out to be the college's personnel manager. I had guessed as much, since those working in personnel are always glum and always middle-aged! The general tears open the envelope, takes out my orders, leafs through them quickly and hands them to the personnel manager. Silence. No hopeful sign. I am worried.

---

[10] Andrey Nikolaevich Bolkonsky, one of the main characters of *War and Peace*, who dies from wounds received at the Battle of Borodino.

Then comes a question: "Where's your brother?"—"In Burmin's clinic nearby, just across the street." The general turns to the lieutenant-colonel: "Wait for me here; I'll be back soon," and, turning to me: "Let's go." We crossed Pirogovsky St., and after another 300 m or so we were in the ward. Something suddenly changed in the general. He sat down on the edge of Sergey's bed. "Hang in there, soldier."—"I'm trying, but there's nothing to hang on to." Sergey smiled guiltily.

I went out into the street so as not to start sobbing. Khadeev remained in the clinic for about an hour. When he came out he placed his hand on my shoulder: "This is how it is, Captain: from today you are considered to be in my service and entitled to all concomitant perquisites. I've spoken to the professor. The end may even come tonight. Send me a telegram when you're ready to leave."

Sergey died a few days later. There were other young men lying in the ward with him, also former frontline soldiers. On every visit I had brought old copies of the magazines *The Worldwide Pathfinder* and *Around the World*, from the time in the 1920s when they were still being published. I cannot forget my last visit with my brother. I sat alongside him and we were silent. Suddenly he spoke up: "Don't take them away"—indicating the magazines with his eyes—"The lads read them with pleasure." A lump rose in my throat.

The crematorium hall was packed with people in overcoats with shoulder-straps ripped off. These were one and all students who had begun their studies in 1946. I saw their faces. The men wept. They wept silently, and no one said anything. War had come back into our lives. A soldier was quitting life, dying when it had seemed at last to be all over and done with. The unbroken silence explained everything better than any words.

My last surviving blood relative was gone from life.[11] Next day I left for Khar′kov.

Khar′ Kov and My Candidate's Dissertation

In Khar′kov I was immediately plunged into a stressful, but in its own way interesting, life. The Khar′kov IAT[12] had been transformed into the Khar′kov HIAT; the letter H for "Higher" had been added. The original mid-level institute was still training variously specialized mechanical engineers, but was in its last months. In effect, I was one of the first appointees to a newly founded higher academic institution charged with turning out military engineers for the regular armed forces. And, in my role as head of the academic section, I had the challenging job of

[11] According to a family archive in possession of the author's elder daughter Irina, the author's father had a sister, Maria Sergeevna, who married a British army officer and went with him to England in 1919, from which time contact with her was broken off. In England Maria gave birth to two sons, the author's cousins, one of whom had two daughters, living with their children in England to this day.

[12] Acronym for Institute of Aviation Technology.

drawing up a plan of instruction in technical subjects, to be laid before some high-level council by the institute's director General Khadeev.

Thus I had to comprehend and organize what should be taught and how, and come up with a design of the academic programs.

I often had to go to Moscow—at least twice a month. There I sat in various meetings, and spent time examining the programs of study and assessing the degree of preparation of engineering students of other academic Air Force institutions. This was, in fact, work in my military specialty, requiring appropriate qualifications and resourcefulness—in a word, real work, and—why deny it?—agreeable to me. And moreover I got results: my proposals were usually accepted and, despite my risibly low military rank, I was treated with respect in the office of the administration of Air Force academic institutions.

In Khar′kov I acquired a staff consisting of a young secretary—the wife of a senior officer of the college—and a deputy, who had higher rank than me. Sometimes I had the feeling that my life, that of a career officer, was taking shape—especially when I was given a room to myself, my first personal living space. At that time many senior officers with families had to squeeze into rented closet-sized rooms, so I understood that this was Khadeev's way of showing his appreciation of my usefulness to the college—which in turn increased my pleasure in the job.

My boss was sports mad. Not for nothing was the Khar′kov IAT called "Inphyscult"[13] in jest. When he discovered that I had first-class ranking as a skier and had been a member of the second team of the CHSA,[14] he sent me to participate in a competition for best skier in Ukraine, which was to take place conveniently close to Khar′kov. I did well in my signature marathon and as a result was included in the Ukrainian national team, and even went to Sverdlovsk to compete for best in the USSR. I competed in two individual events and came out near the beginning. . . of the second half of the contestants—which wasn't so bad, all things considered! In the relay race the Ukrainian team came a solid last—but that was without my participation.

Sometimes of a Saturday I would play *preferans*[15] at Khadeev's place and usually won since no such games—neither card games nor chess—interested me, so that I didn't get worked up during play! And that's how my Khar′kov life rolled along.

Each time I travelled to Moscow, I found time to meet and talk with my former colleagues at the academy. I tried not to miss any interesting seminars, especially those of Venttsel′ and Pugachev. I also occasionally attended meetings at the Starokonyushenny Academy, at one of which, as mentioned above, I heard a talk which led to my gaining a foothold on the slopes of Mt. Olympus.

In those years ballistics specialists were preoccupied with problems of dispersion, or accuracy of fire, of unguided missiles. A simple method of calculation,

[13] A portmanteau word for Institute of Physical Culture.

[14] Acronym for the (Frunze) Central House of the Soviet Army, the main cultural and service centre of the Soviet armed forces, founded in 1927.

[15] A traditional Russian card game distantly related to Bridge; see Chap. 2.

used with success during the war—in particular, for the ground installations of the legendary katyushas—had been worked out by Khristianovich, Gantmacher and Levin. However, the application of their method to aircraft missiles yielded completely wrong results.

At the meeting in question, the future academician Vladimir Semenovich Pugachev, then head of the Department of Aircraft Armaments of the Zhukovsky Academy and recently promoted to the rank of general, gave a talk in which he explained very clearly to his audience just why that method, which had, incidentally, earned its discoverers the Stalin Prize, fails as a means of computing the dispersion of airborne missiles. The essence of the problem, as the future full member of the Academy of Sciences of the USSR quite correctly asserted, lies in the fact that the reactive force constitutes a non-stationary stochastic process. And since the part of the trajectory where the rocket engine is firing is in the case of airborne missiles rather more extended, expressing the distribution law for the points of impact of such a missile in a region of the target in terms of the parameters of this stochastic process seemed at the time in question to be impossible. That was what Vladimir Semenovich explained to us, writing the very complicated Fokker-Planck-Kolmogorov equations[16] on the blackboard.

Following the talk, Dmitry Alexandrovich Venttsel′, who was chairing the meeting, thanked the speaker and briefly summarized what he had said: "In this clear and highly scientific talk, it has been shown that we don't yet know what to measure in the stochastic process in question nor, even if we did, then how to measure it. It follows that we don't know how to compute the trajectory of a missile moving in accordance with that stochastic process. But, on the other hand, as Alexey Nikolaevich Krylov used to say: 'Formulating a problem is already halfway to its solution'."

The night following this talk I returned to Khar′kov. I could only get a ticket for a seat—no sleepers left—in a general carriage, which was absolutely crammed. It was impossible to sleep and all night I thought about the meeting I had just attended. The notion of using complicated equations for the distribution function, equations which in any case we don't know how to solve, seemed counterintuitive to me. And even if we could solve them, why do we need such detailed information? As an armourer and artillerist, I knew that in order to estimate an effective supply of armaments, it suffices to know the average quadratic deviation from the target, and for this it's enough to know this, that and the other. Thus it was that in the course of that night I arrived step by step at a chain of logical links that cast a totally new light on the problem of calculating the dispersion of a rocket engine and processing the results of laboratory tests of such a missile engine. All that remained was to submit the idea to a technical analysis. I quite quickly had the whole approach worked out, but was then overcome by doubt: the problem had turned out almost too simple for words. The simplicity of my approach prompted strong misgivings since I was not a specialist in probability theory and knew nothing whatever about

[16] Partial differential equations for stationary and transition probabilities of diffusion processes.

the theory of stochastic processes, which was then a fairly new area just beginning to be developed. So I went to Alexander Mikhailovich Obukhov, who had studied under Kolmogorov. He and I had graduated from MSU[17] in the same year, so I was not afraid of parading my ignorance before him. If I remember correctly, he had not fought in the war and was at that time already a doctor of science. All the same, like every army regular, I tended to esteem those of higher rank, and I addressed him using the formal "you."[18] For some reason he didn't immediately switch to addressing me in the same way, as behoves a true member of the intelligentsia, although he did make the switch eventually. Thus it was that as a result of that memorable conversation, we continued right up to his death conversing with each other using the formal "you."

It turned out that everything I'd done was correct. And furthermore Alexander Mikhailovich showed me a brand new 1946 paper of Doob in which that well-known American mathematician had used much the same argument, albeit to other ends.

Feeling as if I'd sprouted wings, I phoned Venttsel′ to request that a presentation by me be scheduled in his seminar. This was duly arranged, and a month later I brought the completed manuscript of my work to the seminar and gave a rather lengthy exposition to a critical audience. Professor V. S. Pugachev, whose talk at that earlier meeting of the seminar had stimulated my work, was present. Following my talk he gave a laudatory speech, adding that my manuscript would make a good dissertation and that moreover he would be prepared to act as an opponent. This was recorded in the minutes of the meeting as a submission by Venttsel′'s department. My manuscript was taken away and stamped "top secret," and the public defence of my dissertation was arranged for a month and a half thence in the Scientific Council of the Faculty of Aircraft Armaments.

In Khar′kov my scientific successes were greeted ambiguously. Some were glad for me and congratulated me, but General Khadeev was genuinely sorry: "You'll get through your defence and leave us. It won't be hard to find a new head of the academic section since it's a coveted position. I'll recommend Major Potapov"—my second-in-charge—"But who will I play *preferans* with, and who will ski the fourth leg of the relay?"

The date of my defence arrived. I had quickly taken the last of my candidate's exams, the one on philosophy.[19] I got a four and was happy with it!

I had taken the exam together with Vorovich. He was then working in the academy as adjunct, or, in civilian terminology, graduate student. As with everything he did, he had prepared for the exam long and thoroughly—beyond conscientiously. Yet he received only a three! All the more reason for me to feel that there was no stopping my dizzying career.

[17] Moscow State University.

[18] Rather than the informal, more intimate "thou."

[19] Mostly Marxism-Leninism with a nod to the history of philosophy.

So there I was at last going to Moscow for my defence, and for once not merely in connection with a routine matter to do with my job. The college had arranged for me to be granted leave for an official field trip, moreover for a whole week. But I almost failed the defence.

My second opponent was to be the dozent Colonel Boiko, of whom his students said: "Boiko gladly gives twos."[20] He was an extraordinarily pedantic and tedious fellow. He might easily have given me a "two" on this occasion. I had seen many reports on my dissertation, all of which were excellent, yet there was so far none from my second opponent. And the defence could not proceed without a second opponent.

I went to the department where Boiko worked. My legs were trembling: the defence was next day, and still no report. What was going on? His words to me were precisely as follows: "I have still not completely figured out your work. See here, Moiseev, I'll need another month or two." So there! And the defence was on the morrow!

Yet again a lucky break, once more decent people.

In the corridor I met Volodya Semenov, who had graduated from the academy at the same time as me, in May 1942. Instead of being sent to the front, he'd been kept in the academy as an adjunct. He'd already defended his candidate's dissertation and had gone ahead of me in rank: he had just been promoted to major. Subsequently Vladimir Mikhailovich Semenov would become a doctor of science, attain to the rank of general and, behind the curtain of his military career, become scientific deputy to the head of the famous Thirtieth Institute.[21]

I sniffled to him: "The defence is tomorrow, and see what's happened!"

"Let's go and ask Solovey; we'll think of something."

Major-General Solov′yev, at that time head of our faculty, was a relatively kind-hearted person, although somewhat cowardly. We go to him and tell him my tale, in response to which he shrugs his shoulders and begins talking about the unpleasantness that might ensue for the Scientific Council should the defence fail; my personal interest in the matter is, naturally, of no concern to him. I am a chunk practically cut off. But then Semenov has a brilliant idea.

"Comrade general, allow me to act as second opponent at tomorrow's defence."

"But the dissertation is secret."

"We can get around that!"

"And the report?"

"It'll be written!"

"Well, be careful you don't let the Air Force down!"

"We won't!" came the answer in one voice.

---

[20] In Russian, where the rhymes give it brio, *Boiko boiko stavit dvoiku*.

[21] The 30th Central Scientific Research Institute of the Ministry of Defence of the Russian Federation was founded in 1961 with the mandate of investigating a wide range of problems relating to the Soviet Air Force, including the development of aerospace technology and weaponry and also space technology.

The dissertation had indeed been classified top secret and was locked away under seven locks. But if a dissertation's author says he had no copies, who can believe him? Of course I had copies—and apart from that there's one's head, the securest of safes for secret documents. We went to Volodya's apartment. His wife Lena—he was already married—served us dinner and we set to work.

Volod′ka Semenov was a very capable fellow. A little disorganized and lazy—why pretend otherwise? But he was a marvellous comrade, and when necessary showed quick understanding and did what was required accurately and effectively. It took only two hours for him to understand everything, and another hour for us to compose a report together. And then, of course, we had a good few stiff drinks. I stayed overnight and next day we went together to the academy. By 3 pm I was a candidate of technological science.

The defence had gone as if to a drumbeat. V. S. Pugachev had spoken eloquently, and Volodya Semenov had not disgraced himself. He came in for praise from Solovey: he had not let the Air Force down. There were present many from civilian design bureaux and scientific research institutes. Questions were asked. I received many flattering proposals. General Zalessky, the head of SRI-2, invited me to work in his institute, and I would eventually accept his invitation.

Professor Pobedonostsev was also present at the defence. It was then that he told me what had happened to the letter I'd written to him when still a divisional engineer, and what had seemed like quirks of fate now made sense. Following the defence we spent a long time walking up and down the corridors of the academy while he insistently tried to convince me that I should—no, must—make every effort to leave the military.

Finally he said: "I am persuaded that the place for you is in academia. Get yourself demobilized and join my department in the MHIT."[22]

I was not in the least convinced, however. The security of service as a career officer continued to seem attractive. I had risen through the ranks in the normal way, and, since I was doing the job of a colonel, General Khadeev had promised to have me promoted to lieutenant-colonel sooner than usual. And I already had a room to myself for the first time in my life—although it was true that it was furnished with only a fold-out bed, a stool and a three-legged table. Still and all, it was a roof over my head.

But it's not for nothing that they say life is the best teacher!

I Return to Civilian Life

My return to Khar′kov was triumphal. Khadeev congratulated me publicly. But then of course after the holidays come the workdays.

[22] Moscow Higher Institute of Technology.

These, however, were now less pleasantly occupied than before, and in fact even less agreeable than I had imagined they would be. My conception of an educational framework for the college, now institute, had been formalized, the plan had been approved, students had begun to appear, and the routine work of an academic section began: the drawing up of timetables, the monitoring of classes, report forms to be filled out, and so on. Put briefly, there began the life full of care of a military administrator who moreover finds himself under close scrutiny by his superiors. My trips to Moscow came to an end with the ending of the recess period.

Not travelling to Moscow also, of course, meant an end for me of the opportunity of attending the scientific seminars at the Zhukovsky Academy and participating in the conversations that had lent substance to my life. In compensation, I tried to make contact with the Khar′kov Mathematical Society and University. The mathematical life of the city was then led by Naum Ilyich Akhiezer, a first-class mathematician and kind-hearted person. The Khar′kov mathematicians received me warmly enough; I even gave a couple of talks in their seminar. Their interests were, however, very different from mine. I was immersed in problems of engineering while they were all "pure" mathematicians preoccupied with classical analysis.

As had happened a few years before when I was still serving in my division, so once again, only this time more intensively, I began to feel that neither regular military service nor administrative work generally were my cup of tea! I needed to make a cardinal change in my life.

But the idea of actually quitting the military grew on me slowly. I felt some trepidation at the thought of returning to civilian life with its uncertainties and greater need for self-reliance, of having to be responsible for oneself. Nevertheless, I gradually came to understand that my leaving the military was inevitable—there was simply no alternative for me. Everything that I was presently engaged in took on the aspect of a sort of "pre-existence." Every evening I would reflect that I was already 30 years old and needed to put my life in order, needed to.... But what exactly I needed to do was not entirely clear. I simply came by degrees to the necessity of removing my shoulder straps. Even though terrifying, it had to be done! And so the day came when I composed my demobilization report.

Khadeev was grim and gruff the way he'd been the first time we met at Air Force headquarters. He stared for a considerable time at my report, reading and re-reading it. Then he took up a pen and wrote the strange yet subtle words: "Unable to be used in his specialty. Not in agreement with the reason for demobilization." But the thing was done: according to military regulations, I might now appeal to higher authorities. My report was forwarded to Moscow.

Two months later the order for my demobilization arrived at the college, what's more over the signature of that same General Orekhov whom I had twice come before. He was most probably glad to be rid forever of that pushy subordinate.

Thus I met the new, 1949, year without shoulder straps, but still in my Air Force tunic. I wouldn't be able to afford a normal jacket for some time to come.

By that Summer I had reached agreement as to possible jobs at both SRI-2 and MHIT. My candidate's defence had opened all doors to me. I began investigating the dynamics of guided missiles. This discipline was in its early infancy, so that every result obtained, no matter how insignificant, was received as a breakthrough. I often gave talks and published my findings in classified publications. I felt that next year I might submit as my doctoral dissertation the manuscript of a book I was writing. I was supported in this presumption by Professor Pobedonostsev and several of my colleagues. It seemed to me, therefore, that my life was to be linked forever with that technological world to which I had been vouchsafed entry. But fate was inclined to have it otherwise.

One fine day my stepmother was arrested, and soon thereafter everything went smash. But I have already told that part of the story, in particular, how I was forced to go to Rostov.

My Lecture at M. V. Keldysh's Seminar

As mentioned before, I had to begin from the beginning. My new family and the university in Rostov were nothing like what I had experienced in my previous life. There was one thing, however, that I knew with full certainty: I would both teach students and do science to the very best of my ability. I would not, however, do the sort of mathematics professed in the mathematics section of MechMat. I would engage instead in theoretical investigations of problems of a completely concrete physical or technological character. But exactly which problems I still didn't know. I could no longer work in rocket technology since I had been deprived of access to classified material. I had nonetheless definitely chosen the general nature of my research interests, and for his influence in that regard, I was indebted to D. A. Venttsel′.

When I turned up in the Department of Theoretical Mechanics of Rostov University, my new boss, the dozent A. K. Nikitin, at once asked me what sort of problems I would be working on; regarding me as a specialist of some sort or other, he expected a coherent answer. But how could I answer him? I really didn't know what I would work on.

In fact I've always found it difficult to answer such direct questions for the good reason that it's difficult to plan one's research activity very far ahead. I could speak only of the general direction my research would take, since specific questions would arise by themselves only in the course of the work. Sometimes I would start working on one thing but end up working on something else altogether. I believe this to be typical of every creative activity. I recall reading somewhere Simonov's words: "...and I went away to write verses on Stalin's death for the newspaper." I never read those verses but I can't imagine they were any good. I am certain that one can no more write poetry to order than one can ponder what one does not yet understand to order!

Every investigation requires thinking about something as yet unclear, and can be successful only if it becomes a part of the investigator's life, if he or she is always mulling it over. One of my university classmates, a person of very middling ability, defended his doctoral dissertation earlier than many of his evidently more gifted peers. I was surprised when told of his successes. But then another of my fellow students, the very able mathematician Alexander L′vovich Brudno, explained: "What is there to be surprised about? He never did anything else but ponder his 'flutter'."[23]

The problems that I began working on and which would some years later form the subject of my doctoral dissertation, also arose more or less by chance. At the very beginning of my time at Rostov University, I was given the task of preparing a series of problems suitable for degree theses on fluid mechanics. I knew hardly anything about this subject—more precisely, knew nothing about it, and for that reason began looking for problems related to those areas of mathematics I had studied at university. An appropriate such area was the theory of nonlinear perturbations of linear operators. And that was the route by which I came to the sort of problems pertinent to my future as a mathematician: those concerning the theory of gravitational waves in bounded regions of a liquid, or oscillations of liquids in vessels and bounded reservoirs.

I began working on this and soon enough managed to obtain a few results, which I wrote up as a moderately sized note of six or seven pages. The first person to read it was the late M. G. Khaplanov, who made so many comments on it that he effectively rewrote it. Then on one of my visits to Moscow, I met up with Professor Ya. I. Sekerzh-Zen′kovich, a specialist in wave theory and a very agreeable fellow. He approved of my note and suggested I give a talk on it in Keldysh's seminar in the Steklov Mathematics Institute.

Mstislav Vsevolodovich Keldysh was then at the height of his scientific fame. Although already a member of the Academy of Sciences of the USSR, he had not yet become its president, nor yet innominate Chief Theoretician, just as Korolev had not yet become Chief Designer.[24] However, a series of brilliant articles on the theory of non-self-adjoint operators had raised him to the status of one of the recognized "sharpest" mathematicians of the world's top team in mathematics. And that year he had attained the age of only 40. He had recently been made director of the Institute of Applied Mathematics that was in the process of being set up.

Keldysh was famous for his astonishing acumen. At seminars he understood the essence of the talk not only sooner than everyone else in the auditorium, but—so it seemed to me—sooner even than the speaker himself. And, possessing such sharpness of mind, he made no attempt to conceal his sense of superiority. It was thus not surprising that people were afraid of performing at his seminars and tended to become confused in his presence.

This general attitude towards him was exacerbated by another of his quirks. Generally speaking, people fall into two distinct types: on the one hand, those—one might label them the kind-hearted majority—who regard every person they're meeting for the first time as *a priori* intelligent and decent. Later they may be forced to conclude, albeit regrettably, that in fact not everyone is decent or intelligent. The

[23] Resonant rapid vibration of an airplane's wing.

[24] Of rockets.

second sort suspects every newcomer of being stupid and mean—and subsequently these also have then to reluctantly admit that not everyone is a fool or a villain. My long acquaintance with Keldysh has led me to the conclusion that, if he is not of this second type, then he is at least much closer to being of that type than of the first. One should take into account that M. V. Keldysh was the son and grandson of generals, and he had completely mastered the general's superciliousness air. Although in his youth he had endured all the miseries of being treated as an outcast because of his aristocratic antecedents, he made little attempt in later, still rather difficult, years to ease the fortunes of others in the same predicament.

So people were afraid of Keldysh. There was, it would seem, just one person who was not at all cowed by him. This was Kostya Babenko, later to become Konstantin Ivanovich Babenko, corresponding member of the Academy of Sciences. I'll have a word or two more to say about him later on.

Although I hadn't heard everything about Keldysh, I knew enough to be very nervous at the beginning of my talk at his seminar. When I entered the auditorium, on seeing my Air Force tunic decorated with my military orders, his lips curled in what looked like a sneer, which made me even more uneasy. By that time, 1951, I had already begun to be ashamed of my tunic. In Rostov I always wore a sweater over it. I would have removed the decorations, but in wartime fashion they were attached with screws, and removing them would have left holes visible in the tunic. And I had only the one tunic. I felt the awkwardness of my situation keenly and felt the weight of the moment on my shoulders: like a first match on a field away from home!

I was given 15 minutes to expound my result—just the bare result without any commentary. Occasionally Keldysh put questions to Ya. I. Sekerzh-Zen′kovich, whom he greatly respected. These were mainly questions of a bibliographical character involving a series of names unfamiliar to me. Then at a certain juncture, Keldysh asked me a question directly. I was on the point of answering when I heard Babenko's voice: "Once again, Mstislav Vsevolodovich, you haven't understood. This follows from. . . ," and so on and so forth. Keldysh withdrew into his shell. And Kostya Babenko was then, like me, a mere candidate of technological science.

The seminar lasted for around an hour. At its close, Mstislav Vsevolodovich was quite laconic: "The theorem is a simple one, but useful. I am prepared to present your note to the *Reports*.[25] Prepare a version for publication." I said that I had a suitable text with me. He then asked Yakov Ivanovich if he'd seen my text, and, on receiving an affirmative answer, wrote "I hereby present this" on it.

It this way I took one more step up Mt. Olympus. I now had my first publication in a journal of the Academy of Sciences of the USSR.

[25] *Doklady Akademii Nauk SSSR*, the prestigious mathematics journal of the Academy of Sciences of the USSR, devoted to announcements of significant new results.

Sobolev, Vinogradov and my doctorate at the Steklovka[26]

In scientific research there is always a chain of successive influences linking the investigations the researcher has done earlier to those he's presently engaged in, affecting his choice of problems and their formulation, the methods he uses in trying to solve them, and so on. And along this "chain of time intervals" are strung secondary circumstances. This is how scientific research acquires a certain logical structure. Although we ourselves often fail to acknowledge the sequence of influences, it nevertheless exists and percolates up into our consciousness as it were independently of us.

My duties at Rostov University included the supervision of student seminars on fluid mechanics. The students were required to read original works in the area and produce summaries. Thus I was confronted with the problem of choosing works appropriate to such essays without myself having much knowledge of the discipline beyond its basics. Given these circumstances, I made a decision that ultimately turned out to be perhaps the most reasonable. I proposed that we start with the study of classical works in the area.

At that time the collected works of Nikolai Egorovich Zhukovsky were in the process of being published, and there one could find problems stated precisely and in simple, clear language. Added to which there was Zhukovsky's mastery of analysis. In a word: back to the classics! The students were bound to learn a great deal from such good examples. And maybe even I would learn something! My proposal was approved by the department.

Among the works I chose for summarizing at my student seminar in the Autumn of 1950, was the famous article of N. E. Zhukovsky devoted to an investigation of the motion of a rigid body with cavities completely filled by an ideal liquid. In this article he shows that such a system is dynamically equivalent to that of a certain different rigid body, whence it follows immediately that the presence of the liquid adds no further degrees of freedom.

By then I already had my final-year students looking at problems concerning the oscillation of a liquid in a vessel. Now a vessel is also a rigid body containing a liquid, except that now the liquid doesn't completely fill its cavity, but has a free surface on which waves may appear. The oscillation of the liquid is due in some way to the movement of the vessel, and influences its motion in turn. Zhukovsky's theorem was not, of course, applicable to such a system since the number of degrees of freedom is now infinite. But what if such a system—a rigid body with an oscillating liquid contained within it—has some other special properties? I began to think about this question.

[26] Popular name for the V. A. Steklov Mathematics Institute of the Academy of Sciences of the USSR, a prestigious Soviet scientific research institute founded in 1934. Its director from its founding to 1983, except for the years from 1942 to 1944, was the number theorist Ivan Matveevich Vinogradov. From 1942 to 1944 it was headed by Sergey L′vovich Sobolev. See also e.g. pp. 9, 39.

I approached the problem as a mathematician: I produced a mathematical formulation of the system via a rather general operator and began a detailed investigation of the properties of the class of linear operators to which this one belonged. I discovered unexpectedly that such an operator decomposes into an infinite-dimensional one that is always positive definite, and a finite-dimensional one whose properties may be quite arbitrary. I saw at once that this very simple, purely mathematical fact might well have a variety of physical and technological consequences. Not for nothing was I an engineer who had it drummed into him at the Zhukovsky Academy that one should look for such consequences.

The first consequence I drew was that for such a system to be stable it is necessary—and, as it turned out subsequently, also sufficient!—that a certain different rigid body be stable. This represented a generalization of Zhukovsky's theorem. It was a fact, if you like, one ready for the textbooks. But there was also the purely practical aspect: the active segment of a rocket, the engine, is in effect a vessel containing a liquid, and therefore constitutes a system with infinitely many degrees of freedom. A natural question is then: how can it be guided?

The fact of my formulating this question shows the aforementioned chain of influences at work. Independently of me, my consciousness was oriented towards those very problems of the dynamics of rocket engines constituting my scientific "original sin." Circumstances over which I had no control had forced me to think of other things, yet when the first analogy with my former interests presented itself, they automatically came to mind.

And all the more *à propos* in that it followed from my theorem that one can guide such a system containing a large amount of liquid in the same way as a standard system with only finitely many degrees of freedom. To this end, it suffices simply to introduce suitable new variables.

Years later I would be awarded the Stalin Prize for this work—or rather the State Prize as it had been by then renamed.

That evening I at once saw the perspectives revealed by what I had discovered, its implications for wider science and its technological applications. I formulated on the spot a plan for future research—research that would keep me occupied, perhaps, for several years. Pleased with myself, I returned home and at the door greeted my wife with the words: "Would you believe that I have in my hands a doctoral dissertation!?" My late wife was wont to be sceptical about such announcements, and with good reason. To tell the truth, she tended to be sceptical about my scientific preoccupations generally! I have several times been prey to unfounded optimism. Problems would turn up, it would seem to me that their solution would change the world, and then they would quietly disappear. Most often they would turn out to be far more complicated than I had initially thought. No wonder then that in the present situation she simply waved me inside to sit down to dinner. While my mood was not dampened, I nonetheless resolved to verify what I had done thoroughly.

At Voronezh[27] University there was a mathematics professor by the name of Selim Grigoriyevich Krein. Although he was about the same age as me, he had managed to do much more than I in science, and, naturally, knew functional analysis much better than I. I wrote to him asking if I might consult with him and he arranged for me to give a seminar talk. My audience was of medium size. Of well-known mathematicians apart from Krein, my talk was attended by M. A. Krasnosel′sky, chairman of one of the mathematics departments of that university. The seminar went along smoothly and without incident. On the whole they approved of my theorem. No gaps were discovered in my proof. Concerning the perspectives I had entertained, I was advised to consult with Academician S. L. Sobolev.

It would be at least another half-year, however, before I got to speak at Sobolev's seminar. I felt much more sure of myself on this occasion than I had at Keldysh's seminar. This was attributable not so much to my growing experience as to the quite different atmosphere surrounding Sergey L′vovich. He was a person of an altogether different make-up from that of Keldysh. For one thing, he positively glowed with benevolence. During my talk I had no sense of any cautious, suspicious hush. Sobolev himself prompted me with the formulations, commented on the calculations, and closely followed what I wrote on the blackboard; in a word, he acted not like a menacing judge but like a participant in the talk. This was my first meeting with him, i.e., as a researcher presenting his own work; I had several times before been among the audience at his seminar.

Sobolev, elected to the Academy of Sciences at the early age of 31, made a powerful impression on me. He was tall, well built, and seemed much younger than his years—almost boyish, although at the time in question he was well over 40. I first set eyes on him at the famous Petrovsky-Sobolev-Tikhonov seminar. Each meeting of this seminar was reckoned a big event in Soviet mathematical life.

At one of that seminar's meetings at which I happened to have been present, there occurred an incident typical, so I was told, for the seminar. The speaker was proving something deep and I had the feeling that no one understood anything of what he was saying. When he came to the end of the proof of his theorem, there was an awkward silence, finally broken by Academician Petrovsky:[28] "I don't understand why. . . "—and he formulated a question. Academician Sobolev responded, evidently wanting to be supportive of the speaker: "Look, Ivan Georgievich, this is how it is. . . ," and he went up to the blackboard and gave an outline of the proof. This seemed to convince Tikhonov.[29] In any case, he launched into a positive commentary on the theorem in question. Then something comical occurred. It appeared that everyone present except Petrovsky (and me, of course) had understood everything, and all together,

[27] An historical city about 460 km south of Moscow.

[28] Ivan Georgievich Petrovsky (1901–1973) was a Soviet mathematician working mainly in partial differential equations. He was rector of Moscow State University from 1951 to 1973.

[29] Andrey Nikolaevich Tikhonov (1906–1993) was a Soviet and Russian mathematician and geophysicist. He made important contributions to topology, functional analysis and mathematical physics.

in a chorus, they began explaining to Petrovsky the crux of the matter and how trivial it all was! Could there be anything at all difficult to grasp here? Petrovsky, however, continued failing to understand. Finally, mumbling inaudibly, hunching his shoulders and seemingly humiliated by his failure to understand, he went to the blackboard and. . . constructed a counterexample showing that the theorem was, for elementary reasons, not true.

No one seemed embarrassed by this except Tikhonov (and me).

However, at the meeting of Sobolev's seminar at which I spoke there was no such incident, everything went swimmingly. Afterwards Sergey L′vovich came up to me and told me that my results would form a sound basis for a doctoral dissertation and what's more that he was prepared to recommend my acceptance in the doctoral program at the Steklov Institute. Next day he took me to meet the director of that institute, Academician Ivan Matveevich Vinogradov. This meeting was so funny, I shall never forget it. I had long known Vinogradov. As a student at MSU I had attended a course in number theory taught by him. He lectured not just badly but very badly. We could hardly hear him; he lectured exclusively for certain very attentive "fivers."[30] We were made to feel ashamed. It was said that Vinogradov was a great mathematician, in particular because he had settled Goldbach's conjecture.[31] We, however, were convinced neither by decanal exhortations nor by someone by the name of Goldbach who lived no one knew when or why, and we voted with our feet. But here I was years later standing before the great scholar while he examined me with curiosity. I was glad that I had thought to wear a sweater over my Air Force tunic.

Sergey L′vovich began telling Vinogradov about my work, which, it soon became clear, was of absolutely no interest to him. He quickly turned the conversation from science to a completely different topic. He evinced great interest in the question as to whether or not I was a Jew. I told him I wasn't. Then he began to test me, enquiring after the surnames of my grandfathers and grandmothers, especially the maiden names of my grandmothers. His ears pricked up when he heard that one of my great-grandfathers went by the name of von Sperling. The conversation halted while he looked me over minutely. I took this in my own way, and said with a grin: "But he's a 'von'," to which Vinogradov replied quite seriously: "There are Yids[32] even among the 'vons'." I understood that he was playing out some sort of standard comedy, or, to be more blunt, playing the fool. Sobolev followed the conversation with a knowing smile on his face.

But the finale was completely unexpected.

---

[30] Students apt to get a "five," the top grade, in the course, Vinogradov's devotees.

[31] This states that every even integer greater than 2 can be expressed as the sum of two primes. The conjecture emerged in the course of an exchange of letters between Christian Goldbach and Leonhard Euler in 1742. By radically refining a method of G. H. Hardy and J. E. Littlewood, Vinogradov was able to prove in the 1930s that every sufficiently large odd integer is expressible as the sum of three primes.

[32] In the original the word *Zhidy* is used, a pejorative name for Jews in Russian.

Clearly Vinogradov had enough of the game he was playing. He gave a *résumé* of the proceedings: "Nevertheless, there must have been a small-scale Moses among your ancestors. You all came from him." Then after remaining silent a few seconds: "OK, let's stretch our muscles a little." He removed his jacket and placed an elbow on the table.

Vinogradov would then have been around 70. His arm was strong and sinewy—that of a peasant rather than a mathematician. However, at that time I was very active in sport and moreover half his age. I obligingly met his challenge and was able to slowly and smoothly force his arm down, adding at the end: "Ivan Matveevich, you are very strong."

I understood that the game was over. Vinogradov turned to Sobolev: "Well, do we accept Nikita?" And from that time on he would always address me using my first name and the intimate "thou." Sergey L′vovich and I left together. He parted from me very warmly, and I understood that I had played my part with discretion in the performance in the director's office just ended—not overplayed it on either side, but with appropriate moderation.

A participant in the doctoral program had the right to spend up to two years at the Steklovka, but... at the expense of the university where he or she had previously been working, so in my case at the expense of Rostov University. By law I was to continue to receive my dozent's salary while being completely free of my teaching duties. However, the situation at Rostov University was difficult: there was absolutely no one available to teach fluid mechanics nor was there anyone to organize topics for final-year students' degree theses. So a compromise of sorts was reached: The university, in the person of its rector Professor Belozerov, released me to the doctoral program at the Steklovka but with the proviso that I spend one week of every month in Rostov. Over that week I was to give four or five lectures and work with the graduating students before returning to Moscow.

Academician Leonid Ivanovich Sedov agreed to act as my scientific consultant in the doctoral program. This suited me perfectly. In the past I had attended his seminars. I had been impressed by the precision of his thinking and the down-to-earth way in which he formulated problems and approached their solution, for which he was well known. In his attitude towards theoretical research he reminded me in some ways of D. A. Venttsel′. With respect to theoretical work—although, it's true, not to his own—he would give vent to a certain irony. Once, following one of my talks, he summed up his reaction to it as follows: "For higher science there are here rather too many assumptions, and for real applications too much complexity." In that particular instance the assessment was justified, and the work remained unpublished. Throughout my life I have striven to adhere to this principle, but not always successfully.

I was put on my guard, however, by Leonid Ivanovich's well known snobbery.

One day I met him on a trolleybus on the way to the CSRIAED.[33] where he had a laboratory. For some reason he seemed embarrassed and to my surprise began

[33] Central Scientific Research Institute of Aircraft-Engine Design.

justifying himself, explaining in so many words more or less that they had not sent a car in time so he had been obliged to use public transport. He was then still a vigorous young man, much less than 50, and an explanation of this sort, moreover to someone he hardly knew, seemed odd to me. I too began to feel awkward.

So I resolved to tread carefully around Sedov, and we didn't become friends despite the fact that we might well have, given that in a great many ways, especially in our evaluations of scientific work, we had practically the same attitudes.

I didn't remain in the doctoral program very long since I already had the basic results. All that remained was for me to prepare a few papers for publication and write up my dissertation.

One amusing remark in conclusion. That year, thanks to my being awarded a prize by the university equivalent to a dozent's salary for one month, I was able to buy a German-made *Erika* typewriter and my first civilian suit. How much has changed since that distant time! What I now receive monthly as a full member of the Russian Academy of Sciences is scarcely enough for half a suit. And the idea of buying a typewriter or computer belongs to a looking-glass world. So how do present-day dozents fare? They don't even have old army tunics to wear!

I Become a Doctor of Physico-Mathematical Science

The defence of my doctoral dissertation before the Scientific Council of the Steklov Institute took place a little over a year following my memorable conversation with Ivan Matveevich Vinogradov.

In the early 1950s this was a quite rare event, so the whole of the Steklovka areopagus—all of its luminaries—gathered to the fray. Academician Lavrentiev sat in the first row, and, strange to say, listened attentively—which fact would play a not insignificant part in determining my future. Keldysh was also there in his capacity as member of the Council. He and Sedov sat up near the back. They were engaged in a lively discussion and hardly paid attention. Judging by their jolly faces they might have been talking about women—a not unreasonable suspicion, especially in the Scientific Council, and especially amongst vigorous men in their prime already become academicians. Keldysh was then only 43 or 44 and Sedov only two years older. Just the age for talk of women! Reminiscences may be less joyful later on.

I was moderately brief in my exposition. I spoke for at most twenty minutes and I think the Council was appreciative of this. My opponents were very distinguished: the academicians Sobolev and Vekuya and the future academician Ishlinsky.[34] There was a entertaining moment when Sobolev read his short positive report. At the very end he was suddenly overcome by doubt as to the veracity of my main theorem, the result that had originally impressed him so much that he had taken me by the

[34] Alexander Yulevich Ishlinsky (1913–2003), highly influential Soviet and Russian applied mathematician, scientific organizer and pedagogue.

hand to meet Vinogradov himself. An argument ensued in which I took no part since Vekuya fiercely took my side. In his unmistakeable Georgian accent he began: "Now, wait a minute, Seryozha. . . ." Neither Keldysh nor Sedov reacted to the scene; perhaps they were oblivious to it. They were apparently wholly absorbed either by pleasant memories or pleasant perspectives. I would happily have exchanged places with either of them.

In his report Ishlinsky spoke about an analogy with the situation of an oscillating pendulum—a somewhat attractive idea, I felt, though not very relevant. However, an opponent is allowed to say whatever his heart desires; after all, he is not the one defending the dissertation! Between Ishlinsky and myself there reigned special, very warm relations. In the first place, Alexander Yulevich, then an assistant, had been in charge of conducting tutorials in theoretical mechanics in the academic group I belonged to in my third year in MechMat. He led those sessions exceptionally well—and, strange to relate, we really did learn mechanics from him, something I didn't really come to appreciate till I myself began to teach theoretical mechanics. Not even my years of military service could erase from my mind those approaches to problem-solving shown us by Ishlinsky.

There was, however, another arena of mutual activity: volleyball. Back then I was playing in the faculty's first team and Ishlinsky, I think, in the third. And—why pretend otherwise?—I took this as a pretext for feeling a sense of superiority to our favourite teacher. Artem Grigoryants, the main offensive player of the first team, seemed to me back then of much greater consequence than a talented science candidate playing on . . . the third team.

To be brief, my performance ended safely, and I was voted doctor of science unanimously.

There followed a banquet in the Petrovsky Passages. Of the Great Ones only Sedov attended. There he told me that he hadn't actually read my dissertation—which explained why he had earlier asked in surprise: "Why were all those mathematicians praising you so?" He then also set to praising and congratulating me, with apparent sincerity and fellow feeling. Although our scientific paths would from that time on tend to diverge, we maintained friendly relations throughout our lives. He followed my scientific career and more than once showed signs of attentiveness. And in return I tried to lend him support whenever I could.

I got joyously and thoroughly drunk, though still managing to notice that two or three bottles of champagne remained unopened. Next morning I discovered these in my briefcase, and instead of taking morning coffee, I continued celebrating my defence with my father-in-law. The substitution of champagne for morning coffee, especially if the champagne is already payed for, would seem to be unlikely to excite disapproval, so my respected mother-in-law's objections, mild though they were, seemed to me inappropriate. In those days I had many friends, so the celebrations ended only when there was just enough money left in my pockets to save me from dying of hunger on the way back to Rostov.

The official confirmation of my doctoral degree came through very swiftly even for those times: two months later I received the certificate attesting that I was a doctor of physico-mathematical science.

Once back in Rostov, I began fulfilling the duties of head of department in place of the dozent Nikitin. A few months later, however, I left Rostov, a city dear to my heart, forever, and returned to Moscow. There were many reasons for this, the most compelling of which was the insult I received from the new rector O. A. Olekin, a corresponding member of the Academy of Sciences. I was at first quite hurt, but then understood that the rector's actions issued from his ignorance. The essence of what happened is as follows.

In the Spring there was to be a competition to fill the vacant position of head of department, the associated duties of which I was presently carrying out. In the advertisement for the post, it was stipulated that the hiring would be at the rank of dozent. This meant that, should I be hired, I would have no claim to the rank of professor. I went to the rector and asked him to change the specified rank to that of professor, giving as reason that I was already a doctor of science so naturally wanted to become a professor. By way of answer, however, the rector said that mechanics is neither mathematics nor physics and the university is not a polytechnic institute, so that for the Department of Mechanics a dozent is plenty good enough!

It so happened that just at that time I had received a flattering invitation from Professor Lavrentiev to take part in a competition for a professorship in his Department of "Explosion Theory" in the Moscow Physico-Technological Institute.

It was already 1955. Stalin had been consigned to oblivion, my step-mother had returned home from the camp at Taishet, and I had once more been given access to classified work—although, it is true, not at the same high level as when I was at SRI-2 and MHIT. Thus Mikhail Alexeevich's offer of a job in his department had come at the right time. Most appealing in connection with the offer was the stipulation that I teach two courses concurrently, one on fluid mechanics—which by now I had considerable experience of, teaching it in my own fashion and not by the textbook—and the other on the theory of functions of a complex variable. The latter course was run officially by the Department of Mathematics. This pair of courses had been taught before by many famous professors, in particular Lavrentiev and Sedov.

Such an offer was impossible to refuse.

Then in Autumn I received the offer of a deanship in the Aeromechanical Faculty. I have already described the dramatic circumstances surrounding my confirmation in that position.

And that is how, in my 38th year, I reached the summit of Mt. Olympus. From afar it had seemed that gods with doctoral degrees and professorial ranks dwelt there, and now here I was amongst them. Amongst doctors of science certainly, but gods? This had yet to be determined!

Chapter 6
On the Intelligentsia, Its Fate and Responsibilities

Will I Become a Member of the Intelligentsia?

Rather early on in my life I had become aware of a sense of belonging to the intelligentsia. The awareness was there long before I even began thinking about what the word meant. Later I'd often ask myself to what extent I had the right to count myself a member of that subset of the citizenry. Here the word "citizenry" is important, since a member of the intelligentsia must be a citizen in that lofty sense of the word that has come down to us from antiquity. A member of the intelligentsia is, as I understand it, not merely an educated citizen, but a person of certain definite moral principles.

My first apprehension of myself was as a member of my family: a Russian son of my father, a grandson of my grandfather, with their perceptions of Russia—most of all of Russian culture. And that first intimation of the intelligentsia, of the kind of intelligence peculiar to it apprehended in childhood, has remained unchanged throughout my life, even though much else has changed radically in my views and conception of reality since then.

The sense of belonging to the Russian intelligentsia was the first manifestation of myself as a social being. I considered much of the world around me from the point of view of that portion of Russian society and of the tragedy that overcame it before my very eyes. For that reason I was never able to accept Bolshevism or Stalinism, even though I was a member of the Party for almost half a century. But I did, on the other hand, internalize the ideas of socialism; I have always felt deep sympathy for its precepts, and Christian ones likewise, and for a long time I could envisage no alternative other than socialism to the form of life described by Dickens, Balzac, and other great authors who wrote about the capitalist society of the 19th century. The memory of an excursion to Ivanovo-Voznesensk (now simply Ivanovo[1]) has

[1] A city about 250 kilometres northeast of Moscow.

N. N. Moiseev, *How Far It Is to Tomorrow...*,
https://doi.org/10.1007/978-3-030-96651-5_6

remained with me my whole life: I saw workers' barracks there with their blind windows, stale air and double bunk cots in narrow rooms like pencil cases where two families had to live.[2] And just nearby there were well-lit villas with fancy gateways.

I could never disavow a feeling of the most profound injustice at the core of capitalist society, never forget that it was built on the woes of millions of ruined peasants evicted from their lands, on the misfortunes of people with nowhere to go but the factories. It has always seemed to me that there should be a different way of achieving industrial development.

Even though I travelled to the West and saw how very different life is there from the standard capitalist sort described in nineteenth century novels and twentieth century propaganda, I remained unable to change my sympathies. I felt that socialism could not but be attractive to a person of real intelligence. I still feel this way, although I understand that socialism is no more than a utopia. But there have always been utopias. People must have stories! May socialism continue to be one of them.

The fate of my family was extremely grim, and I myself endured sufficient hardship to be able to see the horror of the Soviet reality enveloping us. I tried hard, however, not to attribute that horror to socialism, and thought a great deal about what should be changed and how. Most dreadful to me was the helplessness of the individual before those in power, the monopolism which brings to the top poor, spiritually flawed types whose intellectual and spiritual deficiencies prevent the nation from blossoming, from developing its true potential. But even more terrifying, I now understand, is the *lumpen* individual wearing the toga of a democrat.

While my views were very close to Marxist ones, I never connected what was happening in our country directly to Marxism, considering our reality as an extremely dangerous perturbation of that doctrine. It was only when in the 1970s I turned to working on problems of the evolution of the biosphere, the theory of the self-organization of matter and universal evolutionism in the widest sense of the phrase, that I began to understand how jejune were many of our ideas, especially those relating to the Marxist philosophy of history with its concept of an ordered historical progression of socio-economic configurations.

My views were significantly influenced by one particular episode, not so significant in itself perhaps, but, as so often happens, diverting the course of my thinking into another channel altogether.

At the beginning of the 1939–1940 winter, when the Finnish War had been launched, I was recruited into the army as a skier. I didn't participate in the fighting, thank God, but spent three months in northern Karelia[3] training groups of skiers. In the evenings there were long, and I would now say rather comical, discussions held in the commander's barracks. I should mention that those in command of

---

[2] But presumably this was when the Soviet socialist system had prevailed.

[3] A region overlapping with Russia, Finland and Sweden, originally peopled by a Baltic-Finnic people called Karelians.

the groups of skiers were for the most part enlisted students, so people of some education. Once the following question was discussed very gravely: Why don't the Finns surrender? After all, we had come to free them from the capitalist yoke! Then someone remembered Babel,[4] in one of whose stories a Red Army soldier asks a similar question during the war with Poland. But the literary types felt the need to shush talk about Babel, since he had just been shot. Then when the battalion commissar showed up, someone asked him to explain the Finns' lack of class solidarity, ending with: "Doesn't the catchphrase 'Proletariats of all nations, unite!' still apply?"

I don't recall what the commissar said in answer. Doubtless something unsatisfying, since, lying on my bunk, I kept turning this theme over in my head and was for a long time unable to fall asleep. Does it follow, then, that the proletariats don't particularly want to unite, and class solidarity is not after all a magnet sufficiently powerful to draw together people of a single class but different nationalities? From this it was a short distance to the seditious question: Is it possible, then, that nothing is as they've been teaching us?

In March 1940 I was demobilized, but that discussion stayed with me.

I gradually began to understand how much more complex life is than any schemes dreamed up seated at a desk, no matter how logical they might seem, and my own system of views took form little by little. Nevertheless, much later, when I had fully grasped the necessity of dispelling the illusions of socialism, I experienced a feeling of sadness—like that of a child when a pleasant and engaging story comes to an end.

However, my ideas of the intelligentsia, absorbed as they were from earliest childhood, remained unchanged; they were fleshed out with greater substance and gradually led to an appreciation of the role played by the intelligentsia in the development of society.

The theme of the intelligentsia and the evolution of my thinking on the structure of society were for me indissolubly linked. The beginning of a revision of my views or, to be more precise, the beginning of their formation, I date back to my reading of a book by Karl Kautsky,[5] that very Kautsky whom Lenin called a renegade. A Russian translation of his work *The Ancient World, Judaism and Christianity* was published in St. Petersburg in 1909. In it he describes in detail the metamorphosis of the communism of the first Christian communities into the rigid hierarchical system of the Catholic church with its dogmatic canon and inquisitorial burnings at the stake. He ends the book with a question: "Will contemporary communism, like Christianity, develop a dialectic transforming it at some future time into a new organism of exploitation and domination?" How clear-sighted was that "renegade"!

[4] Isaac Emmanuilovich Babel (1894–1940), was a Russian-language journalist, playwright, translator, and writer of short stories. Called "the greatest prose writer of Russian Jewry," he fell victim to a Stalinist purge in 1940.

[5] Karl Johann Kautsky (1854–1938) was a Czech-Austrian journalist and Marxist theoretician, recognized as among the most authoritative promulgators of orthodox Marxism from Engels' death in 1895 till the outbreak of World War I in 1914.

Perhaps there exist laws analogous to biosocial ones governing the evolution of human organizations. Cicero wrote, after all, that monarchy inevitably degenerates into despotism, aristocracy into plutocracy, and democracy into chaos. But, on the other hand, to the biosocial laws humankind has been able to oppose morality, civil rights, and governmental regulations, in this way curbing their severity. Might the intelligence and talent of those capable of seeing what lies ahead perhaps suffice to prevent the apparently inevitable degeneration? My thinking along these lines brought me progressively to the idea of the responsibility of those whom one might call "the far-sighted ones"—those who have the requisite knowledge, who are not content to stay shut up in their shells, and for whom the phrase "ethical principles" is not an empty one—in a word, the intelligentsia.

When in France some years later, I happened to read Hayek's[6] book *The Road to Serfdom*, and this likewise made a great impression on me and gave me much food for thought. At that time I was much more independent in my thinking and there was much of what he wrote that I couldn't accept, and some things I was prepared to dispute. It seemed to me that in some ways Marx and Hayek complement one other. I'll have more to say on this later on.

Thus the concept of the intelligentsia with its characteristic kind of intelligence that gradually took shape in my mind is different in many respects from the one generally held. It was precisely the combination of responsible citizenship with strong moral principles and a humane system of judgments as to the best structure of society that for me characterized a member of the intelligentsia. To my mind this definition captures the concept to a much greater degree than that of a mere "intellectual."

Gryzlov and Lunarcharsky

I most probably first heard the word *intelligent*[7] pronounced by my father, moreover in a negative context, as an accusation brought against someone for *not* having the characteristics of a member of the intelligentsia. In the 1920s there was in Skhodnya a family by the name of Simenkov or Simenenkov who had a fine dacha. In my memory of it, I see spacious well-lit rooms, elegant furniture, a piano—someone in the family liked to make music—and a great many books with expensive bindings. The Simenkov family was evidently well educated. There was a boy in the family

[6] Friedrich August von Hayek (1899–1992) was an Austrian-British economist and philosopher best known for his defence of classical liberalism.

[7] The Russian word for a member of the Russian intelligentsia. The g is pronounced hard and the stress is on the last syllable.

of about my age. I now think that it was most probably the family of a successful Nepman.[8] At the end of the 1920s the whole family quit Russia.

Notwithstanding the external respectability of the Simenkovs, my father spoke several times of how unlike *intelligents* they behaved: how they lied, even in trivial matters, how they were at heart boorish, and how disrespectful they were towards the kinds of work done by others. I would be hard put to say what reasons lay behind such judgments, but I do remember that my father was reluctant to speak badly of anyone. In any case, I rather early in life understood that one cannot equate merely being educated with being an *intelligent*, who was a being of a certain higher category. An *intelligent* moves in a special spiritual world and satisfies particular spiritual requirements, he has the ability to esteem and respect the spiritual world of others, even though this be very different from his own. Members of the intelligentsia may come from a wide variety of social classes and professions. That was my introduction to what it meant to be an *intelligent*, hedged about with a whole range of concrete circumstances.

In our house in Skhodnya there were three stoves, which in those distant 1920s were wood-burning. All three stoves had been installed by the stove-maker Ivan Mikhailovich Gryzlov. He was a master of his profession and charged a lot for his work, which he insisted on being evaluated at its true worth. Once he rebuilt one of our stoves for us. It appeared that the work was done, and Grandfather was getting ready to pay him, when something about the stove began to bother Ivan Mikhailovich. He stopped my grandfather with the angry words: "Wait, Sergey Vasilievich. You don't understand a thing about stoves. But what if someone who does understand should come and ask who set it up? What will you reply? Gryzlov." And next day he redid the whole job.

My grandfather liked Ivan Mikhailovich, however, not just for his decency and the way he honoured his profession. He would say of him that he was both highly intelligent and had the best qualities of an *intelligent*. Grandfather took pleasure in talking to him about what was going on in the world. They, the elderly railway engineer with the rank of general and the far from young stove-maker, would converse in leisurely fashion in long sessions, each paying very close attention to the other's words. My grandmother also loved Ivan Mikhailovich, and, whenever he dropped in, would serve him tea and join the conversation with evident pleasure. But once his conviviality and liking for tea cost Ivan Mikhailovich dear.

On that occasion, Grandmother offered him tea as usual, and he eagerly accepted. He liked to drink his tea with lumps of sugar, very strong, and, especially when the conversation was interesting, in large quantities. But now I have to interrupt the flow of my narrative briefly to explain that Grandmother, like a great many elderly people, had a pretty strong stomach. Before going to bed she used to drink a decoction of Alexandrian senna, which they now call simply "senna." There was always a teapot

---

[8] A *nouveau riche* member of Russian society from among those who prospered during the period 1922–1928 of Lenin's New Economic Policy (NEP); see also Chap. 2.

of that brew standing in the kitchen. It looked just like strong tea, but the taste? Well, there's no accounting for taste.

The conversation on that particular day must have been very interesting to Ivan Mikhailovich, for he drank cup after cup, praising the taste—until he'd drunk the whole teapot. When he'd gone, Grandmother suddenly realized that she'd been pouring him Alexandrian senna instead of ordinary tea. How the story ended, how the elderly Ivan Mikhailovich coped with such a large dose of laxative, I unfortunately never found out. But our family's relations with him remained good.

I have more than once been convinced that among simple Russian folk one often comes across people who belong to the intelligentsia in the deepest sense, with a worldview developed by long reflection and natural subtlety. Such a one was my sergeant-major Eliseev, with whom I lived side by side over several difficult years at the front.

And here's yet one more episode from among those that have influenced my attitude to the question of the intelligentsia.

My father had been excluded from his position at Moscow University immediately following the revolution, but continued hoping that someday he would be able to return to his teaching and scientific duties. One day, acting on the advice of some of his former university teachers, he wrote a letter to Lunacharsky[9] asking that he be reinstated in the university under any conditions. Those were different times from now, and, despite their severity, members of the government sometimes answered letters. Lunacharsky answered, inviting my father to come to his dacha. Thus one Sunday my father went to meet him, full of hope and feeling as if he'd grown wings. But when he returned late that evening, he was very upset, trembling from the insult he'd endured.

No serious conversation such as Father had hoped for had taken place. He had not been given the chance to say anything at all to Anatoly Vasilievich, although the latter was reputed to be a great *intelligent*, practically the conscience of the Party.

Lunacharsky had met with my father in the hallway of his big two-storey dacha, after having kept him waiting there a long time. When at last he descended to meet him, he was wearing a barely fastened bathrobe. Holding the letter in one hand and the lapels of the bathrobe in the other, and not even inviting him to sit down, Anatoly Vasilievich began speaking at once in a high tone: "How dare you write such a letter to me! Can it be possible that you and your intermediaries don't understand that you're not needed by anyone here, that no one will ever trust you to educate students again?! Be thankful that the Party still tolerates you."

Father's hopes were dashed—forever! And he was not yet forty. He was still a strong and energetic man, and his greatest hope had been to serve Russia. Imagine the state he must have been in! I see the scene of his return home before me: he's seated at the table hunched over, while Grandfather is standing in the middle of the room. Father is saying something like: "He's a minister, after all. Is it possible that

[9] Anatoly Vasilievich Lunacharsky (1875–1933) was a Russian Marxist revolutionary and the first Soviet People's Commissar of Education; see also Chap. 2.

a minister doesn't understand that without literate people Russia will perish? And how could he talk to me like a master to his servant? He's said to belong to the intelligentsia!" Grandfather approaches and places a hand on his shoulder: "Take it easy, my boy"—Grandfather always addressed him thus—"What sort of *intelligent* could he be!? He's a knave, and that's all there is to him. Back in 1903 I showed you something he wrote about Chekhov,[10] where he says that all Anton Pavlovich can do is drool over the fates of three sisters.[11] Is it possible that a member of the intelligentsia, moreover a Russian, could so fail to appreciate Chekhov? And who was it who starved Blok[12] to death?[13] He's the same narcissistic scoundrel he's always been. But you ... minister, Russia's interests. ..." etc. etc. Among other things he mentioned Gumilev.[14]

I remember how the conversation ended. My father asked: "Will Russia ever have a government run by the intelligentsia?" Grandfather reacted to this question very positively: "Don't you worry. That time will come! That time will come! Only your precious intelligentsia will again make such a mess of things that even you won't be pleased with them!" I have already related how Grandfather regarded the February Revolution and the ending of tsarist rule by members of the intelligentsia as Russia's chief misfortune.

Thus as a ten-year-old boy I already knew that the stove-maker Ivan Mikhailovich Gryzlov was an *intelligent*, but Anatoly Vasilievich Lunacharsky was a scoundrel and a crook without any connection to the Russian intelligentsia. Thus did I grow up viewing the intelligentsia from various perspectives. And overall, I would now say, these views were accurate or almost so. I at least clearly understood back then that it wasn't so much education as adherence to a particular ethical principle that was essential for a person to have the right to be counted among the intelligentsia. But an *intelligent* should also have certain definite interests close to his or her heart yet lying outside his or her profession, family, and everyday life.

Smashed to Pieces

In 1959 it fell to me to spend two months in Fontainebleau, at a "Centre" dedicated to problems of control or guidance of technological systems. At that time France was

---

[10] Anton Pavlovich Chekhov (1860–1904) was a Russian playwright and short story writer, considered to be among the greatest exponents of short fiction in history.

[11] One of Chekhov's best-known plays is *The Three Sisters*.

[12] Alexander Alexandrovich Blok (1880–1921) was a major poet of the Russian symbolist school. See also Chap. 4.

[13] In 1921 Lenin and Lunacharsky initially forbade Blok, who was, in his own words, "suffocating" under Bolshevism, from going to Finland for medical treatment, but then relented—too late.

[14] Nikolai Stepanovich Gumilev (1886–1921) was an influential Russian poet of the Silver Age. The speaker may be hinting that Lunacharsky played a part in the decision to execute him in 1921. See also Chap. 2.

still a member of NATO,[15] which was in charge of the Centre. No one, including me, understood why I was not just permitted to go but actually *invited* to go, or why NATO was paying my expenses—moreover so handsomely! I surmise that it was all a chance outcome of over-bureaucratization, of which there was as much in France—and perhaps also in NATO—as in the USSR. I didn't uncover any secrets there, incidentally; in fact their knowledge of the science of control turned out to be significantly inferior to ours, to a degree I had not anticipated. And at that time we were also better at using computers than the French. In fact everything they were doing there was of little real interest, although it was presented as the very last word.... But of course the French are good at self-presentation generally. We might well learn from them how to transform rachitic and flat-chested female citizens into princesses, and deep-frozen cod into *lapardan*,[16] extolled even in the time of Peter the Great. But it would have been rude to complain about anything since I was surrounded by attentiveness and concern.

From a scientific point of view, no especially interesting business was conducted there, although I met many quite well-known people. The American Kálmán[17] was present, a man of about my age; he was super-famous for his invention of the "Kalman filter." He seemed to me not very well educated, at least by Moscow mathematical standards, puffed-up and with a huge opinion of himself—a very common American characteristic. In conversation he wasn't very cordial, and I tried my best to avoid him.

To be frank, for me personally the most significant portion of my time in France was that I spent outside the Centre. Even though I was at the Centre five or sometimes six days of every week, I nonetheless seemed to have plenty of time for long-forgotten idleness.

I was well set up. I was accommodated in the Latin Quarter, fed *gratis* in Fontainebleau, and on top of that given a daily allowance of 60 francs, which at that time was very liberal. To get an idea of just how liberal, I note that a French full professor would then have had a monthly income of around 3000 francs with which to run a household, including family and food. But the icing on the cake was the car they provided me with! It was, apparently, too expensive a proposition to transport me every day between Paris and Fontainebleau. "Would Monsieur le Professeur be agreeable to driving a rented car himself? Alternatively, we could arrange accommodation in a hotel in Fontainebleau, in which case a car wouldn't be necessary. As Monsieur le Professeur wishes." Need I mention the choice I made?

---

[15] Starting in 1959, de Gaulle gradually withdrew France's armed forces from NATO's command. France was returned to full membership by President Sarkozy in April 2009.

[16] In English, haberdine.

[17] Rudolf Emil Kálmán (1930–2016), was a Hungarian-born American electrical engineer, mathematician and inventor, most noted for the co-invention and development of the "Kalman filter," a mathematical algorithm widely used in signal processing and control systems and in guidance, navigation and control.

The car, which I drove without a license, was a Renault-5 or 6. Although of middling quality by European standards, by comparison with my own Moskvich-406 rattletrap it was pure luxury. Each day, on arriving at the Centre I gave the car key to some lady, and she—the car, not the lady—was washed and refuelled, so I had none of the care associated with running a car.

Under such conditions, busying myself with the "Kalman filter" or optimal control methods seemed, to put it mildly, unreasonable—and all the more so in that Fontainebleau lies on the route from Paris to the Chateaux of the Loire and other famous landmarks, known to every Russian from the novels of Dumas.

Looking back from my present perspective, I now see that I conducted myself absolutely correctly all those 30 plus years ago when all our movements were so cramped by the hard regulatory chains of the "communist codex." That was to be the only time I would be abroad under conditions of both freedom and material sufficiency. I lived through the few months that fortune had vouchsafed me as if in someone else's unfamiliar skin, so to speak. I absorbed myself in this fantastic unfamiliarity, in an effort to understand things. How the understanding I acquired turned out useful to me in the future! It helped me to become my own self, my "I."

The activities at the Centre were so organized that it was necessary for me to make contact not only with mathematicians but also electronic engineers, and among these I came across many with Russian surnames. They were mostly people of the same age as me or a little older, who had left Russia as children and been educated in France, but still spoke Russian well. There were also people somewhat older who had been ejected by the Soviets in the middle 1920s.

My acquaintance with these people was not haphazard, although for a considerable time they were cautious in their approach to me. My own uninhibitedness, my fluent though imperfect French, and the fact of my turning up at the NATO Centre, all made them suspicious. But Russians are Russians no matter how much they may look like Frenchmen! They gradually opened their hearts to me, and received me into their Russian "technological brotherhood": they invited me to their homes, went with me on excursions and to the theatre.... I conversed with Russian experts responsible for many of the successes of French electrotechnology and electronics.

And I managed to come into contact with a different world, that of Russian humanist thought.

Once during a lunch break I was strolling in the grounds of Fontainebleau Palace when, on overtaking two middle-aged people immersed in conversation, I heard Russian being spoken. Excusing myself for interrupting, I asked them something in Russian. They answered in Russian, and we gradually began a conversation. One of the two turned out to be the current curator of Fontainebleau Museum; this was Professor Rozanov, a relative of the famous Vasily Vasilievich Rozanov.[18]

---

[18] Vasily Vasilievich Rozanov (1856–1919) was a controversial Russian writer of the pre-Revolutionary epoch who tried to reconcile Christian teachings with ideas of healthy sex and family life.

We became friends. Our friendship began by his giving me books to read, mostly by *émigré* Russian authors. It was then that I first read works by Berdyaev[19] and Ilyin.[20] I also read Hayek's *The Road to Serfdom* in French, and read Russian newspapers published in France; it was all devilishly interesting! I began taking tea with the Rozanovs on Saturdays. They lived in a government-owned apartment in one of the palace wings. We gathered on an open veranda, where Mme Rozanov treated us to a spread consisting of real Russian tea with hot water from a samovar and her own home-made *varenie*;[21] there were also *pirogi*.[22] These get-togethers were especially pleasant; they reminded me very much of the Saturday evenings at home with my family in the 1920s.

There would always be other guests, making for a modest gathering of Russian society. I think that these gatherings were not so much traditional as occasioned by my presence: the guests were being treated not just to home-made *varenie* but also to a real Moscow professor.

Thus was I given an unique opportunity to get to know some of the "shards" of what had been "smashed to pieces." The people I met! I especially remember my meetings with the daughter of the great Russian microbiologist S. N. Vinogradsky,[23] the founder of the Institute of Experimental Medicine in St. Petersburg. She knew many Russian naturalists, among them Vernadsky[24] from the time of his visit to Paris in the 1920s. She had attended Bergson's[25] seminars with him, and even remembered Le Roy's[26] proposing the term "noosphere"[27] at one of those seminars. I hadn't heard the term before and hardly knew anything of Vernadsky.

I occasionally drove that lady, who, incidentally, was getting on in years, to her home in Paris. At our next, and last, meeting, she related many interesting details

[19] Nikolai Alexandrovich Berdyaev (1874–1948) was a Russian political and Christian religious philosopher. He was expelled from Russia by the Bolsheviks in 1922.

[20] Ivan Alexandrovich Ilyin (1883–1954) was a Russian religious and political philosopher. Together with Berdyaev and others, he was expelled from Russia on the so-called "philosophers' ship" in 1922. See also Chap. 3.

[21] A kind of diluted jam used as a spread or taken by the spoonful with tea.

[22] Russian pies.

[23] Sergey Nikolaevich Vinogradsky (1856–1953) was a Russian, and for a brief period, Soviet, microbiologist. In 1891 he became director of the newly founded Institute of Experimental Medicine in St. Petersburg. In 1922 he moved to France as director of the Division of Agricultural Microbiology of the Pasteur Institute.

[24] Vladimir Ivanovich Vernadsky (1863–1945) was a Russian, Ukrainian and Soviet mineralogist and geochemist, considered one of the founders of the disciplines geochemistry, biogeochemistry and radiogeology. He is most noted for his book *The Biosphere* in which he elaborated on Eduard Suess' idea of the "biosphere" as one of the geological forces shaping the Earth.

[25] Henri-Louis Bergson (1859–1941) was a French philosopher, best known for arguing for the significance of processes of immediate experience and intuition over abstract reason and science.

[26] Éduard Louis Emmanuel Julien Le Roy (1870–1954) was a French philosopher and mathematician. He was a friend of Teilhard de Chardin and a close disciple of Bergson.

[27] The sphere of human thought, in lexical analogy with "atmosphere" and "biosphere." The term was introduced by Pierre Teilhard de Chardin in 1922 in his book *Cosmogenesis*.

about life in occupied France, in particular, how Sergey Nikolaevich Vinogradsky and a young American who'd become stranded in wartime Europe had carried out experiments in the former's home laboratory somewhere on the outskirts of Paris. I couldn't help thinking how different the German occupation of France had been from what we'd experienced in Russia, where the idea of doing experiments, moreover in a home laboratory, would have been inconceivable. She also told me that Sergey Nikolaevich had written a textbook on microbiology which after the war he'd sent to the then president of the Soviet Academy of Sciences in the hope that it would be published in Russian as a textbook for Russian students. She had many other things to relate about Sergey Nikolaevich, and others of my famous compatriots whom she had known. Thus did I learn, for instance, of V. A. Kostitsyn's[28] participation, despite his advanced years, in the *Résistance*,[29] and of his last sad years during which the Soviet government continually denied his request to be repatriated.

To one of the Saturday gatherings on the veranda they brought along Alexander Benoit.[30] He then had just one more year to live. He dolefully recounted how he'd been transformed into an emigrant. If I understood correctly, he had simply been refused a visa to return from an official trip abroad. And in those days he had been first curator of the Hermitage. Thus the Soviet government had simply decided not to allow the director of the Hermitage to return home to his Motherland from an official visit abroad.

I had spent the evening prior to this meeting at the *Opéra*, where I had looked at paintings by Chagall.[31] Frankly speaking, I am not an admirer of the later Chagall; in particular, I don't very much like his little flying Vitebsk[32] people. So all the more out of place did they seem to me in France's premier theatre. At the gathering next day, I said that I was astonished at how the French were gradually losing their characteristic taste and artistry. My judgment was well received by the members of the little tea-drinking society. This micro-incident served as yet one more thread joining me to these expatriots: we were from the same clan-tribe. I was filled with an agreeable feeling: the Bolsheviks have come and will go, but Russia will remain!

---

[28] Vladimir Alexandrovich Kostitsyn (1883–1963) was a mathematician, astrophysicist, ecologist and political activist. He attended a Gymnasium in Smolensk, and in 1909 enrolled at the Sorbonne. In 1916 he returned to Russia to take part in the war against Germany. From 1919 to 1928 he taught in the Physico-Mathematical Faculty of Moscow University. From 1928 he lived and worked in France.

[29] In June 1941 Kostitsyn was arrested by the German occupiers of France and released in March 1942.

[30] Alexander Nikolaevich Benua (or Alexandre Benoit) (1870–1960) was a Russian painter, art historian and art critic. In 1926 he left the USSR for Paris where he worked in Diagilev's "Ballets Russes" as artist and author-director.

[31] Marck Zakharovich (or Marc) Chagall (1887–1985) was a Russian-French artist of Belorussian-Jewish origin. Art critic Robert Hughes called him "the quintessential Jewish artist of the twentieth century."

[32] Chagall was born in Belarus not far from the city of Vitebsk.

To summarize: I had been granted a tremendous opportunity to read a great many pages of the astounding history of the Russian intelligentsia—although somewhat superficially, as might be expected of a young barbarian. But in fact I was not so young: I was in my fifth decade. Nevertheless, at the time I took everything I heard and saw for a kind of exoticism supplementing the touristic impressions afforded me by that two-month official visit to the land of D'Artagnan, of which all that remained were confused memories.

On resurrecting those conversations and impressions in memory, however, I now understand that I had by chance been brought into contact with people who had succumbed to a deeply tragic fate. Although all my new acquaintances were comfortably off in the material sense—and by Soviet standards positively wealthy—their lives were far from easy. Possession of a French passport and a few francs in one's pocket doesn't make one a Frenchman or Frenchwoman. In France they were all aliens. They were always thinking of Russia and living through Russia, just as my friends and I do now, as we have always done. It is precisely in this respect that they so differ from the current batch of emigrants, who flee from the high cost of living, from the "dearth of baloney," hurrying to get naturalized without a backward thought for Russia. My acquaintances of that earlier time, on the other hand, had no intention of becoming real Frenchmen and -women.

Many of them contemplated returning to Russia, and some even discussed the idea with me and sought my advice. What could I tell them? If it was a question of being employed in their field of specialization—as engineers, say—then it was perfectly clear to me that since our Soviet engineering complement was then immeasurably superior to its French counterpart, there would be no particular demand for their services, although most of them would nevertheless eventually find a decent place to work in the appropriate section of the Soviet military-industrial complex. But would our all-powerful government organs allow them access to such work? I told them of the difficulties involved and, remembering my own troubles in that respect, advised them not to be in too much of a hurry.

But humanist members of the intelligentsia were an altogether different matter: such people were in catastrophically short supply in the USSR. In those years Soviet humanities—I'm not referring here to the nascent movement of 1960s' activists nor to the activities of various individual silent intellectuals—were in a pitiful state. Opportunists were in charge in both science and the arts and would be loth to welcome any competition. So much broader was the outlook and so much higher the culture of the humanists I met in France, that competition from them would spell the end for those entrenched Soviet mediocrities.

And there was also socialist realism to contend with: would the Central Committee allow, even during the current "Khrushchov thaw," a rebirth of liberalism and dissent? There was only one possible answer, so I tried to avoid discussions touching on the problems involved in repatriation; I didn't want to disappoint my amiable hosts.

But one thing I understood perfectly back then: when the end of Bolshevism at long last arrives, as is inevitable, we will find ourselves lacking most of all in humanist culture.

Government and the People, Foundation and Superstructure

Over a period of many years a false idea of the intelligentsia and its place in society has emerged. Of course, the concept of the "intelligentsia" is inseparable from that of intellectual activity—not from intellectual *labour* as such, but from the spiritual life of individuals and society. There are many intellectuals who are not *intelligents*, and conversely. But on the other hand the connection between the intelligentsia and the "superstructure" of society is indubitable: the intelligentsia is its medium.

The mutual relations between society's foundation and its superstructure have in our country been interpreted in very primitive terms. The superstructure, i.e., the intellectual and spiritual life of society, was long considered not only of secondary importance relative to the foundation, but essentially as a by-product of it. The role of the spiritual principle and of folk history and traditions was, in consequence, considered solely from a utilitarian perspective: the intelligentsia was just a social "stratum," whose task consisted in fulfilling assigned work for which it was qualified. Here the word "assigned" is important. It was reckoned that the determination of the content of that work, including the creative sort, is the prerogative not of the intelligentsia, nor of the creative individual, but rather of the "worker and peasant" government, which automatically knows what the people need. The interests of the people were thus identified with those of the government, and the intelligentsia had to fulfill the "social order" given it. Such was the generally accepted doctrine.

But in actuality it was all infinitely more complicated. In the first place, the interests of the government and the people are, in fact, completely different. In a liberal society the government is just one of the institutions of civil society, and clearly cannot reflect the whole palette of societal interests. This is all the more the situation in a totalitarian society, since there the government acts according to a specific doctrine and, whatever that doctrine may be, it can be just only with regard to a particular and rather narrow set of people. Thus no matter how a society is organized, the interests of its government cannot be identical with the "interests of the people"—which, incidentally, is a concept needing to be made precise. Even in a best-case scenario, a government's interests will coincide only with those of one or another circumscribed group of people.

In the second place, under appropriate conditions, the spiritual attitudes of a society, that is, the system of consolidated ethical norms or scale of values, the formation of which is often inscrutable, may turn out to be not just an effect but also a cause of profound readjustments of the social order. And these changes may over many years determine the development of the foundation on which the superstructure is supposed to be based.

The influence of the superstructure on the structure of the foundation, and so on the life of the people, is especially marked during certain critical periods. That is why the history of the next few decades, including the development of our economy and our living conditions, will to a very large extent depend on those ideational and moral principles now in the process of being formed.[33]

It follows that now the onus of grasping the present situation and finding an alternative developmental route falls not on our contemporary rulers but squarely on the intelligentsia. It is precisely the intelligentsia who must now sort out what is happening and why, and what should be preserved of the past: global nihilism is extremely dangerous since it withers the people's spirit, excites rancour, and deprives people of one of the most remarkable of human characteristics, namely, the ability to forgive. This conviction provides the context for a critical rethinking of many of the moral strictures that have been taken for granted for three-quarters of a century.[34] Note in this connection that one should not consider morality the business merely of the superstructure and social consciousness: it is to be sought deep in our subconscious, in the fundamentals of human society as such, notwithstanding the fact that each nation, class and even social grouping has its own peculiar behavioural norms. For, in the last analysis, the concepts of good and evil and of eternal truths represent in concentrated form the experience of humanity from the first stages of anthropogenesis to the present day.

I believe with many others in a material basis of the universe. This is just my general philosophical position, however. In our everyday perception of reality, matter and spirit are merged into a whole by means of several mediating associations. And these associations have an ambiguous, and sometimes even paradoxical sense.

Although civilization and morality are not at all synonymous, they are nonetheless inseparable. Morality is the core of civilization. Whether or not everyone agrees with me on this point, I take it as axiomatic, as a basic postulate, since I am completely convinced that any civilization that has lost its morality or its spirituality, or even just had its moral foundations weakened, is doomed to degradation, to gradual degeneration, and can be expected to disappear from history's stage. There are many historical instances in support of this assertion! It suffices to recall the history of ancient Rome. Reading Kautsky will convince the doubter.

Nor is civilization identical with "culture." Culture is but one of several components of a civilization, that determining the behaviour of its people. It is interwoven with the civilization's morality, serving as it does as one means, perhaps the most important, of curbing the savagery and aggression bequeathed to us by our distant forebears and, alas, encoded in our genes like biosocial rules of which they are the material counterparts.

It must never be forgotten that the universal ancestor of all people presently alive, Cro-Magnon man, was formed many tens of thousands of years ago in an environment populated by formidable beasts, so that in those distant times his psychological constitution was adapted to a precarious existence. Genetic evolution

[33] The author presumably has in mind the "critical period" leading up to the collapse of the USSR in 1991.

[34] That is, during the years of the existence of the Soviet Union.

of humankind ceased at that very time, i.e., at the end of the Old Stone Age.[35] Hence those of the modern human being's psychological and physiological characteristics encoded in the genes we have inherited, are the same as those evolved to adapt their bearers to the conditions of life in the preglacial epoch. They are not at all suited, however, to modern conditions of technological superhumanism and the headlong growth of knowledge in a great variety of fields. The modern human being has to mitigate in a timely manner the effect of the biosocial laws written into his genes, laws which regulated the life of the primordial tribe but which today are at odds with the circumstances of life, changed as it is by a new morality and the decline of communal life. If not, then misery and catastrophe must ensue! In this consists the significance of the social phase of the evolution of human society, which must now contrive to bring hunters of the Mammoth into the nuclear age.

And that's why modern societies cannot survive without "spiritual civilization," without culture, in particular, the fine arts. And that's why laws of civilized behaviour are essential to humankind, laws incorporating the totality of prohibitions, or taboos as they were called at the dawn of civilization, or legal and moral norms, as we have become used to calling them. The role of all this non-economic "fine tuning" in the fortunes of humanity must grow no less quickly than the complexity of life, the complexity and quantity of what we now call the basic requirements of life.

Through its constituents—morality, culture, and laws (or legal norms)—civilization secures generational continuity, continuity of people's modes of behaviour and thought. This may be viewed as a kind of folk memory, the memory of the positive experience that humankind has accumulated since ancient times, but which always remains open to the future. The riches of civilization do not emerge readily when the choice of individuals' actions is regimented. In this regard the latter are completely different from conditioned reflexes, serving rather as our points of departure in seeking to overcome the difficulties by which we are continually beset. They represent landmarks indicating the direction of the ford across the stream of events comprising history.

Civilization is at the same time a filter separating out lies of all kinds from the truth, which we do not always recognize but which is as necessary to people as air. The first sign of decline in a civilization and the degradation of its people is the proliferation of lies. Once become a society's norm, lying, like a product of metastasis, begins to pervade people's behaviour and consciousness, depriving them of strength, hope, and optimism.... It robs them of belief in people, in their capacity for purposeful collective action.

Civilization is a fine thread connecting the past and the future. If that thread is broken, then the people are forced to begin from the beginning—or almost from the beginning. If this is not feasible—which has evidently often been the case—

[35] Relatively recent research based on an examination of the human genome has provided strong evidence for the continued evolution of human beings right up to the present day. See e.g. Nicholas Wade, *A Troublesome Inheritance*, Penguin Books, New York 2015.

then the society will plunge back into ignorance and savagery. This explains why a weakening or even an overly precipitate change in the web of peculiar societal relations and social behaviour, woven over many generations, leads in most cases to catastrophe and a gradual fading from history's stage.

After spending a great deal of time analyzing the consequences of nuclear war, and thinking hard about what might happen as a result of ecological cataclysms, I finally arrived at the conception of an ecological imperative and its essential attendant idea: a *moral* imperative. I am convinced that the breaking of the thread joining the past to the present represents, both for particular peoples and for humanity at large, a mortal danger, no less dire than the destruction of the connection between humankind and nature. Such reflections force me to the understanding—or rather to the feeling—of how delicate and fragile are the constructions of civilization, culture and morality. It must never be forgotten that these are like a thin membrane impeding the roiling current of human passions, and it would seem that an accidental and apparently insignificant eddy might suffice for this current to wash away the tenuous barrier protecting civilization, allowing full rein to man's primordial nature. Isn't that what happened in Iran, when over a period of a few months that country was hurled from the twentieth century back to the 8th? And didn't something similar happen in our own history? Didn't Mendeleev[36] warn us of precisely this in his *Treasured Thoughts*?[37] That is why today, in our troubled times, I feel so anxious about those teetering scaffolds linking the Russia of, say, the Silver Age[38] with the present, post-Bolshevik, Russia. We didn't simply lose three-quarters of a century, however, but also gained something. It would be exceedingly risky for us not to concede this, and discard what we acquired over that period along with all the terrible things we endured.

Civilization is never faceless. It always has deep national and historical roots in its language, culture, religion, and ethical principles, and the people will reject any new ideas or new dogmas in conflict with these. This is illustrated by our own example. Someone I'd considered far from stupid once said to me: "October threw out the culture that was alien to the Russian people, namely, that of the Russian intelligentsia." This statement contains at least two errors, closely related, lending themselves to an entirely incorrect interpretation of the problem of "October and the fate of culture." Above all else, the revolution rejected Russian culture, and in the process rejected its bearers, the Russian intelligentsia. It was the revolution that rejected Russian culture, not the people! And in any case there was no such thing as a culture peculiar to the intelligentsia and opposed to that of the people.

---

[36] Dmitry Ivanovich Mendeleev (1834–1907) was a Russian chemist and inventor, most famous for his far-sighted version of the periodic table of the elements.

[37] In this book, published in 1905, Mendeleev gives his views of the best path for Russia to follow in the geopolitical, economic and scientific spheres.

[38] The period embracing the last decade of the nineteenth century and the first two or three decades of the 20th, during which Russian poetry flourished. It was dominated by the artistic movements known as Russian symbolism, Acmeism and Russian Futurism. The term "Silver Age" was first used by the philosopher Nikolai Berdyaev.

To repeat: there was a single Russian culture, and its exponents were the members of the intelligentsia. There were two sources of this culture: the societies of the countryside and the town, both uprooted by October.

This rejection was inevitable. Recall the history of the French revolution: much the same rejection occurred there. They even invented new names for the months and a new chronology. Every revolution is an attempt to erect something utterly new, to realize on the same territory a new way of life, a new culture, a new philosophy, and a new worldview, to affirm new values opposed to the old ones! As a result, it becomes essential to reject everything deriving from the spiritual world of the people subsisting in the town, the countryside, and amongst the intelligentsia. And an attempt to replace these discarded values by new ones born out of the euphoria of victory, is probably typical of every revolution.

The links between the culture, the people, and their leading representatives, namely the intelligentsia, are multifarious and paradoxical. It's true that the intelligentsia has always been somewhat removed from the people: the main body of the people has always had difficulty entering into the ideas, hopes and aspirations of the intelligentsia. Recall, for instance, the "going to the people."[39] It was no accident that, prior to the revolution, the "lords" and the "smerds"[40] were so different in their dress. And it is worth remembering that in the years 1917–1918, along with landowners' homesteads, peasants burned marvellous libraries and art collections without the viewers and organizers of the conflagrations evincing the slightest regret. It's enough to recall what happened to Blok's homestead.[41]

But at the same time culture and civilization as a whole has deep roots in the people, so that the intelligentsia, together with its culture, cannot be divorced from the people. If the latter are likened to a tree that has grown for centuries on the same land, then the intelligentsia is represented by the leaves of this tree, and a creative sap flows from the roots to the leaves; if not, the leaves must fade. But we should remember too that there is always a flow in the other, downward, direction, a flow strengthening the trunk and roots, without which the tree likewise cannot survive.

Although the intelligentsia is the bearer of culture, one should not equate intellectuals or servants of culture with the intelligentsia. I know a great many people, both here and abroad, employed in radio-electronics or programming, exceptional masters of their profession, worthy of the fullest respect, indisputably intellectuals, who at the same time I would never regard as belonging to the

[39] A movement begun in 1861 of students and *narodniks*, seeking to bridge the gap between the intelligentsia and the *narod* (people) and spread enlightenment and revolutionary fervour among the peasant masses. The movement is believed to have had little effect on the mass of peasants it was aimed at.

[40] A *smerd* was originally a worker of a noble's land of the period from the 11th to the 14th century, lacking the freedom to leave that land. They were differentiated from serfs in being able to own small holdings which could be bequeathed to their sons.

[41] Called "Shakhmatovo," and situated in the Moscow District, the homestead was built by Alexander Blok's grandfather in 1874, and rebuilt by Blok himself in 1910. It was burned down by peasants in 1921.

intelligentsia. Such has been the progress in technological applications of science that the borderline between white-collar and blue-collar workers has largely been erased. People are no longer so distinguishable by their way of life, the way they dress, or even by their material circumstances. This fact does not at all mean, however, that the intelligentsia has grown significantly, although its importance as a determinant of the fate of humankind *has* grown.

Everyone knows that the concept of "the intelligentsia" is of purely Russian origin. But its meaning has tended to evolve even with us. No European language had a precise word for the concept. The widely used word "intellectual" captures its meaning not at all. Amongst intellectuals capable of conversing on sophisticated topics, I have met not only *intelligents*, but also candid miscreants, entirely absorbed in the arrangement of their own petty affairs, whose behaviour in no way meets the criteria of admission to the intelligentsia. In eighteenth century France the word *philosophe*[42] was bandied about. I believe that the meaning given to that word at that time is more or less close to that of our contemporary notion of an *intelligent*.

An *intelligent* is someone always engaged in a quest beyond the framework of his or her narrow professional or purely class interests. It is characteristic of members of the intelligentsia that they ponder the fate of their people relative to universal values. They are capable of looking beyond the narrow horizons of their personal and professional lives. One of the first Russian *intelligents* whose name has come down to us was, I would say, the famous archpriest Avvakum.[43] It goes without saying that every nation has its intelligentsia.

In the Russia of the second half of the nineteenth century, a rather substantial stratum of *intelligents* crystallized out, marked by its spiritual leanings, its traditions, its weaknesses and its strengths! Linked with its emergence are the soaring flights of Russian culture, bestowing on the world so many remarkable writers, artists, musicians and composers, and, of course, scientists. The intelligentsia of that time was comprised of people from various social classes: aristocrats, high-ranking public servants, and merchants like Tret′yakov[44] and Mamontov.[45] The Russian scientific and technological intelligentsia was largely made up of commoners.[46]

But although the intelligentsia has its roots deep within the people, it is nonetheless to a large extent split apart from the people; there is always a certain barrier separating it from them. It cannot be otherwise, since people come in all stripes and it's not given to everyone by fortune or nature to be capable of abstracting

[42] Philosopher.

[43] Avvakum Petrov (1620–1682) was a prominent Russian orthodox churchman, social activist and author of a great number of polemical works. For his opposition to reforms of the orthodox church begun by the Moscow patriarch Nikon and Tsar Alexey Mikhailovich Romanov, he was first exiled, and then imprisoned and executed.

[44] Pavel Mikhailovich Tret′yakov (1832–1898) was a Russian businessman, patron of the arts, collector, and philanthropist. The famous Tret′yakov Gallery in Moscow was founded by him.

[45] Savva Ivanovich Mamontov (1841–1918) was a Russian entrepreneur and patron of the arts.

[46] People outside the nobility, emerging from the middle or lower classes and possibly the peasantry.

themselves from everyday worries in order to think about things "not of immediate concern to them." And then there's education, a by no means redundant attribute! In this way a rift between the intelligentsia and the people gradually makes itself felt.

Yet the intelligentsia nonetheless derives, flesh of their flesh, blood of their blood, from the people. Furthermore, any general rise in the level of culture of the people—in its way of life, its level of thought, in the character of its interests, and, most importantly, in its creative, spiritual and moral potential—depends primarily on the intelligentsia. The intelligentsia is represented, in fact, by more than just the foliage of the tree by which we are representing the people, but by the upper portion of its energy distribution system, the apparatus that takes in energy from the sun's rays and, with the aid of chlorophyll, nourishes the rest of the plant. And all that is new and useful, in particular, a renewed sense of beauty and harmony, even if it should emerge from the midst of the people, is filtered through the intelligentsia for the sole purpose of becoming, at the appropriate time and in completely different guise, an achievement of the people overall. The role played by the intelligentsia as so described is clearly visible in the evolution of the forms of life, character and customs of the people.

The disappearance of the intelligentsia or its exclusion from the spiritual life of a society constitutes a national tragedy, one that may end in complete moral demise, or, at the very least, in the gradual withdrawal of the nation from the forefront of history's stage into its depths. Once an intelligentsia has been lost, its restoration takes many years. It makes no sense to talk of artificially "cultivating" an intelligentsia. The development of an intelligentsia—its establishment—is a process qualitatively different from the training up of a qualified worker, engineer, or nuclear physicist. It is rather a natural process taking place within the self-development of the nation.

At all times, and in all countries, the intelligentsia has aroused and continues to arouse suspicion among denizens of the higher strata of society, especially among those wielding power. And not without reason! It is part of its nature to constitute a sort of *fronde*[47] opposed to the current regime, to always feel unsatisfied with whatever's occurring, and to be incapable of fully adapting to its environment or unreservedly accepting the way things stand at the moment, however that may be. This characteristic, naturally, tends to foster wariness in certain quarters.

But it is precisely in this that its strength lies! For the feeling of dissatisfaction and the aspiration to find alternatives to the established way of life, to comprehend the means to its improvement, are precisely what ensure progress.

The intelligentsia sometimes produces rebels and revolutionaries, but never tyrants. From its midst emerge thinkers and artists, but never Stalins or Hitlers.

One sometimes hears the intelligentsia referred to as a stratum of society not especially related to any social class, and this is indeed the case. Although one

[47] The name for a series of civil uprisings in France between 1648 and 1653 by a combined opposition by the nobility, the *parlements*, and most of the people, to policies pursued by the king Louis XIV, who ultimately prevailed.

hears the phrases "the bourgeois intelligentsia," "the proletarian intelligentsia," etc., such expressions make little sense since the interests and aims of the intelligentsia are not in the least related to those peculiar to one or another social class. I might cite many instances where representatives of the so-called noble or bourgeois intelligentsia acted as spokespersons for the proletariat, and, conversely, where certain workers advocated for the bourgeoisie. The emergence of an intelligentsia depends on the whole people, on its social and biological nature—although, of course, each *intelligent* may hold different views and traditions, either class or national, according to their upbringing and subsequent path through life.

There is yet another propensity of the intelligentsia which has on occasion irritated heads of state. This is its inclination towards internationalism, or cosmopolitanism, if you like. An *intelligent* is by nature endowed with the ability to think in terms of universal questions, independent of whatever nation he should belong to, faith he should profess, party affiliation, or skin colour. While it's not possible to present oneself as an *intelligent* without a nationality, even if one should call oneself "a citizen of the world," *intelligents* of different countries nonetheless have no trouble finding a common language and interests. I myself have discussed Russian music with a Japanese, and Islamic fundamentalism with someone from Latin America, and in each case my interlocutor and I each found the other interesting and our way of thinking rather similar. At least I found the Japanese *intellligent* much easier to understand than a Party apparatchik or a modern young businessman!

On a flight from Brussels to New York towards the end of the 1970s, I met a Belgian *intelligent* working as a doctor. When the conversation turned to his party affiliations—he was reading *Le Drapeau Rouge*, the magazine of the Belgian Communist Party—he said that although he belonged to no party, he would nevertheless say of himself that he was pinkish-green. Adopting his terminology, I said I was reddish-green, adding that I was working very actively, on both the scientific and societal fronts, on problems of the environment. At once perceiving that our grounds for reflection and action were largely the same, he began confiding in me ideas that were, in fact, close to my own.

"Do you know," he said, "what separates us, the more or less leftist European intelligentsia, who are sympathetic towards countries with socialist leanings, from all of you living behind the Iron Curtain? You probably don't! It's the words of your communist anthem:

> Du passé faisons une table rase;
> Foule esclave, debout! debout![48]

[48] The first two lines of the *International*, the revolutionary anthem written by Eugène Pottier in 1871 following the suppression of the Paris Commune, and set to music by Pierre Degeyter in 1888. An English translation is as follows:

> We'll make a tabula rasa of the past;
> Enslaved mob, arise! arise!

(The standard Russian translation of these words conveys a somewhat different meaning:

> We'll destroy the world of violence
> Utterly, and then
> We'll build a new world, ours;
> Whoever had been nothing will become everything.

Compare this with the original French version, which recommends destroying not just the world of violence but the whole world of the past.)

After quoting these words to me, my interlocutor continued: "In order to have proper dealings with us you need to understand us. We feel great affinity for the ideas of socialism, but do not want the *Chateaux de la Loire* to share the fate that befell many of your architectural masterpieces immediately after the revolution; we wouldn't want the Rheims Cathedral to be replaced by a pond of chlorinated water.[49] It is our wish that the treasures of our thousand-year civilization be bequeathed to our descendants. So it's either us or the clean slate. And in any case we no longer have slaves: where nowadays could you find Lyonnais weavers?"[50]

How much these words and others in the same vein were in harmony with my worldview, with what I had imbibed as a child at those Saturday gatherings at home!

And nowadays such thoughts don't need any commentary or validation; over the last hundred years much has changed in the world, and the end of this century[51] is utterly different from its beginning. The role of intellectual principles in the fortunes of humankind has grown by leaps and bounds, and along with it that of the intelligentsia, of culture, and of spiritual and humanist principles.

But it is of prime importance—as, I hope, the intelligentsia of the end of this century appreciates—that nothing ever be destroyed irretrievably; only a gradual and very cautious evolution is capable of preserving the most splendid and most fragile creations of humankind's genius: its civilization and culture, and with these its hope for the future of "human existence."

In that model of socialism they attempted to impose on half a planet, the role of our cultural heritage, national traditions, and national mentality were not just undervalued, but not even seriously taken into account. Was it really possible to mistake the semi-literate chatter of the proletarian culture-mongers and their equally illiterate opponents for an attempt to understand what a people's culture might mean? This has yet to be fully comprehended in terms of the people's spiritual life, traditions, science and religion. It still has to be grasped that the various manifestations of culture are all interconnected by means of unbreakable threads, that it has not all come about by chance but is a natural condition of the natural life

[49] This is a reference to the razing of the Cathedral of Christ the Saviour in Moscow, ordered by Stalin in 1931, and the erection of all-weather public baths on the site. It was rebuilt between 1990 and 2000.

[50] Major revolts by Lyonnais silk workers occurred in 1831, 1834, and 1848, provoked by effective reductions in their wages.

[51] The 20th.

of the people. In those post-revolutionary years, however, only the half-educated were allowed to speak, and intellectual Russia had to keep silent. This is one of the reasons for the nihilism with regard to spiritual values, to the intelligentsia, and to intellectual and spiritual life generally, which was ultimately, and in most tragic fashion, to play a part in the fate of the nation.

How otherwise can one explain the destruction of the Cathedral of Christ the Saviour, built in memory of the expulsion of Napoleon from our land and financed in part by contributions from the people? How may one explain the liquidation in the late 1920s of the "movement of local historians,"[52] and the liquidation or expulsion from the Soviet Union of a great swathe of the intelligentsia at the end of the civil war?

The expulsion of the intelligentsia is often explained as a consequence of their refusal to accept Soviet rule.

I believe, however, that the reasons behind the destruction of the intelligentsia were more profound than that. They arose from the prevailing doctrine, and the logic of the struggle to achieve "final ends," which as a rule were, however, never formulated in detail but given a certain utopian slant.

In fact, the attitudes of the intelligentsia to Soviet rule were far from simple. Of course, no genuine *intelligent* could ever arrive at an inner acceptance of Bolshevism, which thus remained unacceptable to the intelligentsia throughout its existence. However, the vast majority of members of the intelligentsia, especially those of the scientific-technological and engineering-technological spheres, were prepared to collaborate, and did indeed collaborate, with the Soviet powers-that-be, just as my own family did. Those working in the scientific-technological sphere were largely immune to ideology, and in any case they might take the view that they were serving their country rather than the regime in power. For the old intelligentsia was at heart highly patriotic—far more so than the contemporary "democratically inclined intelligentsia"[53] which without a murmur, in fact often gleefully, took to destroying our great thousand-year-old state.

But here I'm not justifying the banishment of the intelligentsia. Can it possibly be considered just to deprive the people of their intelligentsia, of their "forward-looking," most talented and thoughtful individuals, merely because they criticize the government? In reality it was a struggle for control of the hearts and minds of the people, and for the Bolsheviks someone like Rozanov or Berdyaev was in this respect a million times more dangerous than dozens of counter-revolutionary atamans.

The reason for the exile of our intelligentsia was the same as that behind the exodus of the German intelligentsia from Germany during the rise and rule of Nazism. There the process was, by the way, far less tragic than in our case.[54] The

[52] Aimed at preserving local history and culture in various parts of Russia.

[53] Presumably the author is here referring to those who engineered Yel′tsin's rise to power in 1991.

[54] Here the author is presumably comparing only the effects of emigration on the two intelligentsias.

German intelligentsia was also given to "interfering." It interfered with the creation of a government concerned only with itself, obstructing that government's attempt to realize certain principles, and this redounded to its detriment.

Once during an interview, the host—it was Vladimir Molchanov—asked me the following question: "In the first years of the revolution, there were many members of the intelligentsia in the government. How did this fact affect the fortunes of our country, above all its culture?" I took it upon myself to argue against the very phrasing of the question.

It's true that in those early years there were many impressively educated people in the government—for instance, that very A. V. Lunacharsky who played so negative a role in the fortunes of my family. I have already mentioned that episode, and also the fact that the extent of one's education must not be equated with the extent to which one is an *intelligent*. One must rather take into consideration what people do in practice, and, in particular, their attitude towards the intelligentsia. The initial acts of the Soviet government already said a great deal about them.

Repudiation of Russian culture and its bearers, that is, of the intelligentsia, was one of the main principles of Bolshevik doctrine. The implementation of this principle in the pre-Stalin period was one of the chief causes of our moral and scientific decline. I again recall Lunacharsky's actions, revealing so clearly his attitude towards Russia's culture and traditions, an attitude so disastrous for the fortunes of our country back then, and which, regrettably, continues to affect them even now.

In Lunacharsky was reflected, as in a "mirror of the Russian revolution,"[55] the attitude to Russia deriving from Marxism, which continues to characterize certain sections of society engaged in trumpeting the "demise of Russia," as one democrat recently put it.

May historians of literature correct me if I'm wrong, but it seems to me that Lunacharsky's credo was fully determined already at the very beginning of the twentieth century. It's enough to read his articles on Russian literature for it to become obvious that one of the most remarkable traditions of the Russian classics was alien to him, namely that displayed in, for example, Gogol''s *The Overcoat*, namely the tradition holding that every person, no matter who, is still a person, and as such always worthy of concern, empathy, and at the very least an audience for his grievances.

At the beginning of the twentieth century, Bolshevism declared war not only on the Russian intelligentsia and its associated traditions, but also on the deep, genuinely folk, aspects of our culture. Unlike protestantism, according to which "grace" is affirmed as the mark of worldly success while sufferers are reckoned to be "beasts in men's clothing," and unlike Judaism, which takes sickness and adversity to be God's punishment, early Christianity, and in its wake Orthodox Christianity,

[55] Taken from the title of an article by V. I. Lenin: "Leo Tolstoy as the mirror of the Russian revolution," referring to the reflection in certain of Tolstoy's works of the situation in Russia leading up to the revolution of 1905.

asks for mercy. This principle became part of the flesh and blood of our people, and without eradicating it there was no possibility of inculcating Bolshevism in them, i.e., deep within them and not merely via a husk of words foreign to them. And in that far from cold war, Lunacharsky played his dreary part.

It's my belief that without Lunacharsky, proletarian culture, so-called, would never have arisen, nor its offspring, socialist realism, which, thank God, has almost been forgotten. There were other actions more or less associated with Lunacharsky, for instance, the exile of leading humanists from Russia. Admirers of the People's Commissar[56] for Enlightenment[57] have been heard to say that he thereby saved many scholars and scientists from a worse fate, namely that which would have befallen them in 1937.[58] But could Lunacharsky possibly have foreseen what would happen to the nation 15 years later? No! The exile of the cream of Russian intelligentsia was inevitable. It was dictated by the logic of warfare. Berdyaev, for instance, was dangerous not at all because he could not stomach Bolshevik doctrine, but because he represented a lightning rod of Russian culture and because his powerful intellect was capable of developing and expanding the cultural tradition.

In the 1920s and the early 1930s, Russian culture and science lost a whole pleiades of exponents: Chichibabin, Ipat′yev, Gamov, Kandinsky,[59] Chagall, Benoit, Shalyapin, Rachmaninov, and many, many others. And Lunacharsky, Bukharin and the other Party intellectuals of those pre-Stalin times were unable to understand or intuit that the slogan "whoever doesn't sing with us today is against us," which in those days hung on the façade of the Lenin Museum, and its enactment, would come to represent one of the most fundamental causes of our present lag behind the developed nations! As is well known, the actions of the "Party intelligentsia" were aimed at "razing to the ground, and then...."

One may embrace or not the various *avant garde* and quirky works of art produced by the artists and poets of the 1920s, even the "Nichevoki."[60] Like them or not, one should in any case keep in mind that, while in their ranks there were indubitable geniuses bringing glory to their Motherland, there were also many mere talentless followers of fashion. I'm even prepared to believe that the latter were in the great majority, and am no less certain that such cultural incrustations are very quickly swept away by the wind. But taken altogether, the mass of *avant garde*

[56] The Council of People's Commissars was formed by the Bolshevik government shortly after the revolution, evolving to become the highest government source of executive authority in the USSR.

[57] That is, for education.

[58] That is, at the height of the Great Purge.

[59] Boris Alexeevich Chichibabin (1923–1994) was a Russian poet living in Ukraine. He was imprisoned in the Gulag for five years. Vladimir Nikolaevich Ipat′yev (1867–1952) was a Russian and, from 1930, American, chemist. Georgy Antonovich Gamov (or George Gamow) (1904–1968) was a Soviet, and from sometime in the 1930s, American, theoretical physicist. Vasily Vasilievich Kandinsky (1866–1944) was a Russian painter and theorist of art.

[60] A Russian literary group affiliated with the Rostov Poets' Union in 1920. In 1922 they published a manifesto entitled "All hail the last international to the world of Dada." Following the arrest of its leader, the group disbanded in 1923.

output represents the necessary context from which the path of rational development of art will emerge. Without such casting about, society and its culture would be bound to find itself in a dead end. And to say in advance which of the proposed paths of development is precisely the one we need, is much more difficult that predicting the weather a month ahead. I make this comparison in the capacity of a professional who really does know how difficult it is to make such a prediction!

But in our former society of "evolved socialism" it was taken as axiomatic, beyond debate, that every kind of creative activity of the intelligentsia should be guided by one or another political program, formulated independently of the nature of said creative activity. The question of the extent to which any particular creative activity slots into the political program was to be decided by functionaries, by people, that is, without direct experience of creative activity of any sort.

Initially, this promulgation of the "monocultural principle" and the resultant narrowing of creative potential, was carried out using "kid gloves," if one may characterize in such terms the expulsion from the country or preemptive arrests of a creative *élite* naturally incapable of conforming to the imposed pattern of thought. Well, and then . . . then the kid gloves came off and the "scything" began.

The establishment of a monoculture by means of "scything" had been invented by English gardeners over 300 years earlier: on their lawns they systematically cut out everything that, even ever so little, failed to grow as the lawn's owner wished. The result was a lawn that might be stamped on as much as you like without spoiling it.

However, is it worthwhile writing in such detail about all that? So much has already been said on that theme, that it's very unlikely I'll add anything new! The crux of what happened is that the tradition of the unfettered thought—the thought "free in itself," as Vernadsky liked to express it—was gradually replaced by judgments based on definite rules and taboos. And this took place not only with respect to humanist thought and the fine arts. For in no sphere of human activity is creativity compatible with a system of rigid directives or a monopoly imposing the belief in just one "correct" path to knowledge and social development.

This primitive, so-called "pragmatic," attitude towards the meaning and place of science and culture in the new society, led naturally to a depreciation of the role of the intelligentsia. The intelligentsia became diluted, its role was reduced catastrophically, and its renewal would proceed extremely slowly.

To prepare cadres of highly qualified technicians and engineers was, as it turned out, not all that difficult. In this respect we were rather successful, as was demonstrated by the Great Patriotic War,[61] the postwar period of recovery, and the history of the development of our rocket and nuclear potential. However, the formation of a body of thinkers, that is, of an intelligentsia—or philosophers, as the French called them in the eighteenth century—takes generations. And such a body arises not at the behest of the Party or government, but via a natural process of self-development of a society. And, as the post-perestroika period showed, when people

[61] That is, World War II as waged by the USSR from June 22, 1941 to May 9, 1945.

come to power claiming to be *intelligents*, in fact there usually turn out to be none such amongst them!

The presence of an intelligentsia represents one of the most important guarantees of the viability of a society. Only through it can a society cope with the continually impinging new and unexplored, deal not only with new material conditions but also new morals and ways of thinking. The growth of our technological lag is a direct consequence of the lack of the qualities inhering in the intelligentsia, at the very moment when these were most urgently needed. And they are needed more than ever now, when the technological basis of social development is changing so swiftly, when we are beset with difficulties arising from new modes of thought, and an understanding of the ecological imperative and the necessity to formulate a new morality have become urgent.

One indicator of this is our present journalism. Although they say that *perestroika* represents a new kind of thinking, it's really the result of a search for an intelligentsia, and although our new publicists are given credit for being *intelligents*, the situation is in fact somewhat different: the intelligentsia has not yet really shown up in the pages of our wider press. A couple of swallows do not make a Spring. Furthermore, the stream of articles revelling in our misfortunes, miscalculations, and adversity generally, don't augur much for their authors' belonging to the intelligentsia. For the time being it is all merely derivative of the political struggle. Or is it possible that in fact many journalists have lost the ability to rise to the level of the intelligentsia?

However that may be, I would still like to end this "essay" on an optimistic note. Although our history has been punishing and we have had to endure great losses, the connection over time has not been broken. A *tabula rasa* was not achieved. We didn't after all become rootless Ivans, all kinship obliterated, but remain the heirs of a great culture. Our folk memory, the memory of the intelligentsia, was not entirely snapped off. It was preserved thanks to such titans as Vernadsky and Timofeev-Resovsky,[62] and activists like Sakharov,[63] Losev[64] and the many others who managed to "keep the candle burning" and pass on the baton. Now all we need

[62] Nikolai Vladimirovich Timofeev-Resovsky (1900–1981) was a Soviet biologist and geneticist. After a period of study at Moscow State University, and teaching and research at various institutions in Moscow, he was invited to work at the Institute for Brain Research near Berlin, where he remained from 1925 to 1945, first as an assistant and then as head of the Department of Genetics and Biophysics. In 1945 he was returned to Moscow and imprisoned, and in 1946 sentenced to 10 years in the Gulag. In 1947, on the verge of death by starvation, he was removed to a site where research into radiation genetics was being carried out. In 1950 the Swedish Nobel committee wished to award him a Nobel prize for his work on mutations, but the Soviet government refused to say whether he was alive or not. He was cleared of all charges in 1955.

[63] Andrey Dmitrievich Sakharov (1921–1989) was a Soviet nuclear physicist, dissident, and activist for disarmament, peace and human rights.

[64] Alexey Fedorovich Losev (1893–1988)was a Russian philosopher, classicist, translator, and proponent of Russian culture. Sentenced to hard labour on the White Sea-Baltic canal in 1930 for belonging to a Christian-church monarchical organization, he was released at the urging of Maksim Gorky and others in 1933.

do is create the appropriate atmosphere for that guttering flame to burn brightly again, as has happened before following various troubled times in our history. And it's the members of the intelligentsia who have to take on this task. One would like to hope that at some time, instead of engaging in the struggle for "chairs," which they are bound in any case to lose, the members of the intelligentsia will occupy themselves with the business for which they are most fitted!

At the moment, the main thing is not to let the flame go out: a step backwards, and once again we may find ourselves wandering for many decades far from the main road.

Lessons of the Past

I am convinced that there occurred no Thermidor,[65] or Stalinist revolution (or counter-revolution—either word will do) at the turn of the 1920s into the 1930s. The basic principles remained as before. The process of realization of the original Bolshevik program continued, one of organizing a government not concerned with the flourishing of the country or its people, but with reaching certain goals unrelated to any such flourishing. An unstated principle at work, most probably from the very beginning, was the following one: different aims for different people. Lenin, Trotsky, Bukharin and the rest founded a government "for themselves," that is, according to the ideals of their group, and Stalin likewise "for himself," that is, for himself personally. And being an Oriental, he built an oriental despotism, which was in fact no more cynical or harsh than the government by "educated" Bolsheviks had been, despite their pretensions to membership in the intelligentsia.

Unlike Lenin, Stalin had occasion to take account of the Russian people and their traditions and courage—not out of love for them or their culture, but in a time of greatest need, when without their tenacity and blood his rule could not have endured. History proceeded against the background of an uncompromising struggle for power. Stalin's personal qualities played a part, although he continued relying to a great extent on the state apparatus created by Lenin. That is why it was crucial to further elaborate the bureaucratic command system set up in the first post-revolutionary years. The ending of the NEP was predetermined by this requirement, although it occurred rather sooner than had been planned. Civilization was sacrificed to the original basic principle.

Yes, I really did mean to say "civilization" here!

The English historian Toynbee considered Russia to be a civilization in its own right, not Western and not Eastern, but one of a kind! It is bordered on one side by a region of the world where the ideals of Western Christianity hold sway and on the

[65] On the 9th of Thermidor of Year II according to the French Republican Calendar (July 27, 1794), Robespierre and other radical revolutionaries came under concerted attack in the National Convention. They were arrested that night and guillotined next day.

other by Islam, a civilization qualitatively different from the Western one in its ideals and worldview. And this our dualism has lasted since the ninth century. The spiritual heritage of Byzantium has left its mark on the whole of Russian history, while in the east, on the other hand, there was, by virtue of Russia's expansion eastwards, exposure to the influence of the Islamic world. This had begun already by the time of the Volga Bulgars,[66] who adopted Islam long before Russia became Christian. And the "forest culture" of the Ugro-Finnic peoples, ultimately assimilated by Russia, did not simply vanish, but also made its contribution.

As a result of this highly complex mixing of cultures, there arose a unique world, made up for the most part by "peasant Rus′" with its peculiar way of life and scale of values. And it was precisely out of this world that an intelligentsia emerged, with its particular culture, its understanding of the world, and which, in spite of all barriers, be they national, religious or linguistic, propagated its spiritual values into Western Europe. It's impossible to fully understand Pushkin, Gogol′ and Dostoevsky without knowing the characteristics and spirit of orthodox Russia, Ukraine, Belorussia—of all that emerged from Kievan Rus′.

The level of industrial development in our country was much lower than that of the USA, England, Germany.... But an industrial revolution had begun, and was proceeding at its own pace, one determined by the general logic of the development of our society and its productive capacity, in sync with our country's particular characteristics. As had been noted as early as Klyuchevsky,[67] unnaturally sharp deviations from the natural path of development, taking no account of that logic, led ultimately to negative outcomes. Peter the Great spurred the nation mercilessly, and it would appear that eighteenth century Russia made great strides as a result. By the end of that century, she had the best army in Europe and a navy that even Britain might envy, and produced more iron and steel than the rest of Europe. But for all this she had to pay a terrible price.

In the eighteenth century in Russia serfdom was made more rigid even though it had practically disappeared in the rest of Europe. In order for the imperial powers to realize "Peter's *perestroika*" it was necessary to create a very powerful bureaucratic apparatus, which became an autarkic force binding the nation with hoops of steel. This was the reason for Russia's being unable to respond to an imperative to participate in the industrial revolution, which required the freeing up of initiative and the implementation of the *laissez faire* principle. Allowing anyone the independence to do anything they liked was precisely what was beyond the Russian bureaucracy, so that Russia began precipitously falling behind a rapidly developing Western Europe.

[66] Volga Bulgaria was an historic Bulgar state existing between the 7th and 13th centuries around the confluence of the Kama and Volga Rivers, in what is now European Russia. The Volga Bulgars adopted Islam in 922, some 66 years before Russia adopted Christianity.

[67] Vasily Osipovich Klyuchevsky (1841–1911) was a leading Russian historian of the late imperial period.

The military might of the empire wasted away. Industry was not even able to provide the army with guns with rifled barrels. Transportation was unable to satisfy the demands placed on it. Our defeat in the Crimean War wasn't a matter of simply losing. We were fired on from distances beyond the range of our cannon. No greater heroism on the part of Russian soldiers or competence of our officers could possibly have changed the course of that war. It wasn't Russia that lost the war, but the system created by Peter the Great! And despite the "revolution from above" of 1861[68] and the subsequent development of capitalism, the system bequeathed by Peter was never finally broken with. Let me cite Max Weber, who closely studied the Russian revolution of 1905. He maintained that the chief impediment to Russia's entry into the path of all-European development was not the throne, nor the nobility, nor a weak bourgeoisie, but rather an all-powerful bureaucratic system for which any change was mortally dangerous.

Something similar to Peter's revolution took place in October 1917: certain persons, having wrested power to themselves, went ahead with their program heedless of the logic of development of our civilization or the specifics of the spiritual life of the people. And since Western variants of socialism preached by utopian socialists such as Dühring[69] and the like were alien to them, we were dragged along the Bolshevist rut.

Russia—or, more precisely, Russian civilization—was, perhaps more than any other civilization of the pre-revolutionary period, primed to be receptive to socialist notions of social security, equality and justice, especially in the villages, with their communal outlook. However, the system of socialist ideals and values created by the social-democratic *émigrés*[70] took no account of Russian peasant culture, which they largely despised. The totality of ideas comprising the model of socialism which the Bolsheviks imported onto Russian soil, could be accepted only by a lumpenized part of the population, not by the people as a whole. In order to entrench that model in our country it was necessary to exterminate the most active part of its people: the intelligentsia, the entrepreneurs, the independent peasants.... Such was the logic of the situation!—stemming, in fact, from a Marxist analysis.

Russia had been pregnant with the idea of revolution; Peter's system was completely worn out. In one or another form revolution was inevitable. What happened thereafter should have been determined by the peculiarly Russian way of life and traditions, but for this an entirely different social structure was needed from that which had given rise to Peter's bureaucracy.

---

[68] In 1861 Tsar Alexander II had serfdom abolished in Russia, and instituted other liberal reforms. He was assassinated in St. Petersburg by means of a terrorist's homemade bomb in 1881. His son and successor Alexander III governed, by contrast, very conservatively, undoing many of his father's reforms in order to maintain political stability.

[69] Eugen Dühring (1833–1921) was a German positivist philosopher, economist and social activist, highly critical of Marxian socialism. Friedrich Engels' book *Anti-Dühring*, published in 1878, was the latter's major contribution to the exposition and development of Marxist theory.

[70] Such as Lenin.

The rejection of our traditional mode of thought and scale of values by the Russian social-democrats and Bolsheviks was not restricted to the post-revolutionary period. Recall, for instance, the attitude of the social democrats to Russian philosophical thought, in particular, their practically total ignorance and rejection of the ideas of Russian cosmism, and the oneness of the people and society, etc. Now, on the edge of an ecological crisis, we see the loss to us, and perhaps to all Europe, occasioned by this lack of understanding of the deepest currents of our culture, by the attempt to extend everywhere and over everything inapt revolutionary ideas and standards emerging from the bosom of Western industrial civilization.

I would like to mention also that the notion of socialism, that is, of a "real socialism for the individual" conceptually close to the original socialism of the first Christians, found much more fertile ground in Russia than in the West. In order to redirect the country along that path of development, however, a much deeper understanding of the essence of the interrelationships between natural developmental processes and the worldview and way of life formed in Russia over millennia would have been necessary. And it would have been especially important to understand that there is more than one model of socialism, in much the same way that every form of human cohabitation comes in many variants, that there is a multitude of human worlds and there is no single standard of development of peoples.

Since the search for a new way of life is going on at present, all that I have said above about the past has actuality in the here and now. The lessons of the past must be mastered: not in the form of the standard schemes so dear to the hearts of the dogmatic and the bureaucratic, but as showing the manifold modes of human being. The concepts of social fitness, of social orientation, appropriate in Lutheran Estonia and Islamic Tadzhikistan will certainly be very different. What they should have in common is a human form of cohabitation capable of ensuring the social security of the individual.

But back then, at the beginning of our post-October history, nothing of this was understood. Standards for life, worked out by mere speculation, were simply imposed on people everywhere: in Russia, in the Caucasus, and in Central Asia—standards which even Russian village civilization found impossible to accept, so that it collapsed once and for all. There are no longer any peasants like those who, in the period 1922–1923, given the good will of a kind wizard, might have resuscitated the agricultural economy and fed a devastated Russia. What will replace that peasant civilization? Who will take the place of the Russian peasant? Very difficult to say.

I might call this my little meditation a discourse on the collapse or doom of civilization—our Russian civilization. It was not only Toynbee who thought about our civilization. Before the revolution Russian civilization was a riddle to many, and remains so now. Should we perhaps not bother discussing it? No, we must, it's essential! We have to solve the riddle since it is one of the keys to our future, the past having not yet fully receded.

Russian civilization is an extraordinary amalgamation and stratification of different cultures and traditions, whence derives its multifariousness and the way it manifests a variety of mutually contradictory values. On the one hand there was its splendid European culture, the extremely high level of education of its intelligentsia, cheek by jowl with the vast masses of the peasantry living in communes and observing their time-honoured laws, much as, indeed, in all northern countries.

Is it possible to imagine European culture without Mendeleev, Tchaikovsky, Tolstoy, or Kandinsky, to name but a few? Or, on the other hand, Russia without Leskov's characters,[71] so difficult to understand for Western Europeans? Christianity came to us from Byzantium at a time when the separation of the Eastern and Western Churches was already more than merely formal. But a hundred years before that, Islam was being consolidated on the country's eastern frontier, as a result of which Kievan Rus′ fell under the influence not so much of Arab culture as Persian. We are also, therefore, the heirs, via Byzantium and Persia, of the ancient Hellenistic worldview, so thoroughly assimilated, evidently, by the inhabitants of Kievan Rus′.

Note that the development from that Hellenistic beginning was quite different in the Catholic and Orthodox worlds. This difference was the source of that mental boundary dividing the Slav peoples, which is presently making itself felt in the blood-letting in Yugoslavia,[72] and by virtue of which Russian Moscow is far closer to Kiev than Catholic L′vov, which, although now it's in Ukraine,[73] was ruled for many centuries by Catholic kings.

Furthermore, Russians, and Slavs more generally, have very little in common culturally speaking with the Teutonic knights or the Catholic and later Lutheran missionaries. As they migrated gradually towards the northwest, the Slavs assimilated great multitudes of Ugro-Finnic and Turkic peoples, adopting many features of their pagan cultures. In us, Great Russians,[74] there flows as much Ugro-Finnic blood as in present-day Finns and Estonians. Yet how dissimilar we are to them both! Several centuries of subjection to Lutheran rulers has almost entirely erased from the Finns' and Estonians' memories what has been preserved in the memories of the people living in Russia's northwest.

We lived for centuries among "foreigners": Tartars, Chuvashes,[75] Mordovians,....[76] The worship of a variety of gods by people living close to one another led to an inevitable mixing of cultures. Even now in certain regions of the Volga and its tributaries one hears songs not traceable to a single folk. There is so much consonance in them—but not just incidentally!

---

[71] Nikolai Semenovich Leskov (1831–1895) was a Russian novelist, short-story writer, playwright and journalist. Many of his stories deal with the Russian peasantry.

[72] In the period 1990–1993, when the author was writing this book, Yugoslavia was undergoing violent disintegration into its ethnically different regions.

[73] L′vov was in Poland till 1945, when at Stalin's behest that country's eastern and western borders were moved several hundred kilometres westwards, leaving L′vov in Ukraine.

[74] *Velikorossiya* (Great Russia) and *Malorossiya* (Lesser Russia) were terms used formerly to refer to Russia and Ukraine respectively, no longer in favour with Ukrainians.

[75] A people of Turkic tribal origin. The present Chuvash Republic is in the centre of European Russia, in the Volga upland.

[76] A Ugro-Finnic people originally living in the region between the rivers Oka and Volga.

This extremely complex intermixing of cultures and traditions, from ancient Hellenistic to orthodox Christian to Islam, and ending with the old pagan traditions of the Ugro-Finnic peoples and the Slavs, gave rise to our particular rustic civilization, a melding of cultures giving rise in turn to the Great Russian Culture.

In recent years the intelligentsia has begun to revive a little. A warm breeze has begun to blow, the fetters have been loosened, and people are starting to think again. But only starting. Real thinking is still some way off. Now, however, the grounds from which thought takes flight are different. I believe that rustic or village culture, and in its train that special civilization that Toynbee was talking about, has been completely destroyed. Will something new arise in its place? If so, what of the past will enter into it? These are very difficult questions, but answering them is essential since otherwise we are in for new experiments and yet more destruction.

To conclude: Society is at present undergoing a very rapid evolution. Polemics are growing in stridency. It seems to me, however, that they still have a merely speculative, abstract character. The discussion as to the right way to live should not be conducted in *general* terms. The organization of a society, its stability, the well-being of its people, the perspectives open to them—these are all closely dependent on that society's history. Each people, each country, and even each region comprising a country, has its own peculiar path forward. There are, however, certain universal patterns of development, especially in our day and age, when we find ourselves on the brink of an ecological crisis, when the means of production have grown to such an extent, and when people are becoming more and more interconnected.

The time has come for the Russian intelligentsia to stop being "Westerners"[77] or "slavophiles" (or "back-to-the-earthers"), time to confront the actual contours of our contemporary reality, give up the principles of the "great utopias," personal political ambitions, and the "thirst for revenge," and understand that in the world, as in a mountain stream, there is a main current, and woe to the swimmer who doesn't swim with it.

[77] A traditional dichotomy in Russia between westward-looking and eastward-looking intellectuals.

Chapter 7
Work, Searchings, and a Change of Decor

Computer Technology and Symptoms of Insecurity

In recollecting the first decade and a half of my professional life in Moscow, I find myself unable to single out any especially vivid impressions; it was just work, work, and yet more work! The Computing Centre (CC) of the Academy of Sciences, where I was offered the position of department chair in parallel with my job at the Moscow Institute of Physics and Technology, was one of those rare academic scientific institutions prepared to actively collaborate with research and design organizations involved in the production of new aviation and rocket technology. There was no need to go looking for problems to work on: they fell on our heads out of the sky, moreover in significantly greater numbers than we could cope with. They were precisely the sort of problems to my taste, combining as they did physics and engineering interpretations with appealing but difficult mathematics.

Our main partner institute was the Design Bureau with General Designer my old acquaintance from MHIT[1] V. N. Chelomey, although now I would be working also with Korolev[2] and Yangel′.[3] When they encountered difficult problems requiring the intervention of the Academy of Sciences, I always preferred to work on these at my "home base," that is, at the CC, where I had computers at my disposal as well as the expertise of colleagues. There were always employees of so-called "client" institutions assisting us in this work; there were times when as many as 30 engineers from various design bureaux and scientific research institutes might be seen at work in the three rooms comprising my laboratory. From the middle 1950s we found ourselves—though not for long, alas—at the centre of a whirlpool of problems, the solutions of which were always needed "as of yesterday." The problems given us

[1] Moscow Higher Institute of Technology; see also e.g. Chap. 1.

[2] "The father of Soviet space technology." See Chap. 1.

[3] Mikhail Kuz′mich Yangel′ (1911–1971) was a leading Soviet designer of missiles.

N. N. Moiseev, *How Far It Is to Tomorrow…*,
https://doi.org/10.1007/978-3-030-96651-5_7

were completely new, requiring both new approaches, new mathematics, and always a large dose of inventiveness. It was a sort of scientific banquet.

The 1950s and the first half of the 1960s were, generally speaking, a marvellous time for our applied science intelligentsia. Its energy, talent and ability were all in great demand by the people, the country and the government. The reasons for this are well known and they were known to us back then, but this in no way lessened our enthusiasm—on the contrary, we felt very keenly that we were participating in the establishment of a Great Power. Is there anything to compare with being needed, being in demand? Are there other stimuli of optimism and a desire to work of comparable power? And added to this was the fact that with Stalin's death the feeling of dread had begun to fade and people were beginning to relax.

Rereading the memoirs of the dissidents, I see how different the post-Stalin world was from what had gone before. "Kitchens" and "kitchen discussions" had ceased being a standard feature of our world. We were now able to talk relatively freely about what interested us, not only in kitchens but at seminars and conferences. And we no longer needed to be so careful about how we expressed ourselves, especially after the 20th Congress.[4] New "rules of the game" were gradually worked out, accepted, and observed by the majority. They included, of course, certain taboos: Render unto Caesar the things that are Caesar's, and unto God the things that are God's.[5] In fact, the "Caesars" bothered us very little since we hardly thought about politics. We lived in a world of science and technology, and there we had complete freedom and "even more." We were absorbed in the technological race with the West and had not the least intention of losing. Today, in the reigning epoch of "hopelessness," it would do no harm to recall that attitude and the realities of those years. The attitude was shared by the vast majority of the technocrats of the era, in particular by our great citizen and future dissident A. D. Sakharov.[6] At one time during those years I happened to spend a few days in Arzamas,[7] and dined with Andrey Dmitrievich twice. We were then absorbed in our work in equal measure, I would say. But when I met him again in Moscow ten or twelve years later, I barely recognized him. He was a different person. I think that to some extent we were at that time all suffering from having lived through first the consolidation and then the destruction of our interior postwar world. And—why pretend otherwise?—that had been a world of youth, a world of belief in one's country and of ambition and hope for the future. We hardly gave a thought to the Bolsheviks, the Party, or the communistic tomorrow. All of that and its like must come to an end, but Russia will

[4] At the 20th Congress of the Communist Party of the Soviet Union, which took place in February 1956, Khrushchov delivered a "secret speech," so-called, denouncing the personality cult and dictatorship of Stalin.

[5] From Mark 12, Verse 17, with similar versions in other gospels.

[6] See Chap. 6.

[7] An historic town some 410 km east of Moscow and 112 km south of Nizhny Novgorod. Sakharov was at that time working on the hydrogen bomb at a research institute located there. He was exiled to Nizhny Novgorod in 1980, following his public protests against the Soviet intervention in Afghanistan in 1979.

remain. We used to talk in such terms very openly, no longer especially intimidated by anyone.

During those years I often travelled abroad to give courses of lectures or talks on my work, speaking in Russian in every country I visited except France, since I could speak French. My audiences were always large and very attentive. I saw that in the area in which I was working, our progress was at least on the same level as America's. And sometimes it even seemed to me that there would come a time when the Russian language would be confirmed as the second international language for communication in science.

Illusions are, when all is said and done, a good thing: they foster belief in the future, and provide energy and distraction, and thence new stimuli and new ideas.

All the same, symptoms of insecurity appeared very early on, more than 30 years before the initiation of *perestroika*. Although we couldn't help noticing them, we continued hoping they were not as yet the harbinger of a fatal illness, believing that there was still hope, that they could be eliminated by the will of those on whom the fortunes of the country depended. That these fortunes should depend on a smallish set of concrete individuals was then considered axiomatic. Trust in a kind and intelligent tsar has always been part of the Russian mental makeup—another bitter utopian state of mind, and one inculcated in us not only by the Bolsheviks. But how this trust simplified life: it's enough to educate that all-knowing one appropriately and everything will fall into place!

Among the symptoms we perceived, perhaps of most significance was the situation with our computer technology. The history of its establishment and difficulties relating to its development reproduced as in a microcosm the complete bankruptcy of our social organization and the inability of our society to stop the downward slide towards inevitable catastrophe.

Jumping ahead somewhat, I would like to observe that the reason for the subsequent decline of our computer science school was the result not of our somehow missing out on the upsurge in scientific and technological progress, but rather of our inability to take part in it *in principle*. Academician M. A. Lavrentiev, among many others—including the author of these reflections—had been maintaining as far back as the middle 1950s that the establishment and development of industry must be carried out using the latest technological means. All our leadership was capable of, however, was pushing harder, and that we failed to understand. The outcome for the implementation of computer technology in the USSR vividly illustrates the particular defects of its system of ministerial branch monopolies.

In the form of a marvellous apparatus of vacuum tubes (or radio valves), the computer was born in the Soviet Union at about the same time as, and independently of, its appearance in the USA. We knew nothing whatsoever of the work of the group of American engineers and mathematicians led by John von Neumann, since it was kept securely under wraps. While leaving the details surrounding this epochal invention to the historians, I can affirm with confidence that the impetus for it lay in the coincidence of the development of technology on the one hand and of the need for computational capability on the other. The need for computing technology arose from the 1940s into the 1950s in connection with the arms industry. And as long as the need persisted, as long as we didn't have military parity with the US, our progress in computing technology went neck and neck with that of the West.

At the end of the 1950s, I was a member of the first or one of the first groups of Soviet specialists to be shown around the computing centres of Western Europe. The impression I had from that trip was: Nothing new here! We saw the same terribly unreliable vacuum tube monsters, the same magician-engineers in white lab-coats fixing errors in the work, and essentially the same processing speeds and machine memories. And the problems they were using them for? It seemed to me that in the Soviet Union we were doing more sophisticated things. And our algorithms were clearly more robust.

That curious fact is now well known. Practical work, especially on problems thrown up by the military-industrial complex, had greater prestige in the Soviet Union, and there was a large complement of talented (or, as they say, promising) mathematicians working enthusiastically in various secret organizations there. The situation in the West was completely otherwise. There on the whole the gifted young preferred university careers and occupations good for their souls, removed from practical applications. In other words, in the sphere of computational mathematics we were competing with the West's "second best team" of mathematicians, and clearly beating it. The how and why of all that was taking place was painted for me in bright colours by Richard Bellman,[8] whom I befriended at the beginning of the 1960s, and with whom I remained friends up to his death in the early 1980s.

Thus I returned from my first trip to the "far beyond" full of optimism and confidence in our prospects: our country had muscle, and on the world's scientific and technological stage our chances were not so bad—a case of drawing conclusions based on incomplete information! Our situation was, in fact, completely different from my estimate of it, and we began sensing this rather quickly!

The fault lay not in us, the mathematicians and computer scientists. Already by the early 1960s, when I happened to be in France again, the situation was completely reversed from what it had been three years previous. Yet even then I, together with most of my colleagues, still failed to grasp that everything that was going on with us reflected the basic inability of our by then sclerotic political and economic system to achieve any real improvement. The fact of our lagging behind in the development and implementation of computer technology was in fact only one symptom, one indicator of many, of a fatal disease. But at the time hardly anyone acknowledged this. My own diagnosis, in particular, was off track.

So what exactly was happening in those fateful years of the early 1960s? That was precisely the time of the transition from vacuum tube computers to those using transistors. But why was it that a single technological invention—the replacement of electronic valves by semiconductors—changed the world situation in so qualitative a fashion? How was it that this invention led to our ejection from the club of

---

[8] Richard Ernest Bellman (1920–1984), American applied mathematician making important contributions to various areas of mathematics. One such was the introduction of dynamic programming in 1953.

technologically advanced nations, prefiguring the collapse of our great state to a far greater degree than all the actions of all possible dissidents? It seems to me that even now many remain at a loss to explain exactly what went wrong.

Vacuum tube computers, with their continual malfunctioning and computational errors, were extremely unreliable. Their operation required the employment of a highly qualified team of engineers and mathematicians, and they were useful only for certain special kinds of computation—which explains why they were used only for computational tasks that could not be carried out in any other way even in principle! The computations needed in missile and nuclear technology were predominantly of this sort. No one wanted to risk sending Gagarin[9] into space in the absence of a means of guiding his trajectory.

But then the almost completely reliable semiconductor technology appeared on the scene, with the result that digital means of processing data—doing calculations, in particular—became accessible to the masses. It soon became clear that the new instrument was more useful in trade, business, and mass production than in matters of pure defence. In the latter sphere the computer is needed to enhance the nation's security and prestige, but in business it actually makes money! With the new technology, furthermore, computers became a determining factor of success under conditions of free market competition, in the economics of commercial and industrial enterprise, and this was more important than solving military problems! Free entrepreneurs quickly grasped the money-making potential of computers, thereby determining the future of computer science.

As soon as this situation was grasped by Western business interests, a boom began in the West. Much has been written on this topic, so there's probably no point in rehashing it here. I'll make just one remark: the computer revolution heralded the beginning of a new cycle of scientific and technological progress. The energy crisis, brought on by a series of sudden steep increases in the price of oil and other energy resources played an important role in this respect. Capitalist countries were compelled to undergo a fundamental restructuring of industry; energy-saving technology appeared, along with personal computers and so-called higher technology, that is, precision equipment reliant on in-built electronic instrumentation. The face of Western industry was radically altered in the space of two decades.

Our Soviet economy, on the other hand, bureaucratized and stuck in its branch-monopolistic rut, was unprepared and in fact unable to heed the call of the scientific-technological revolution, a revolution not only unanticipated but fatal to us. There began a rapid technological and economic falling behind, and not only relative to America and Japan. How many even now recognize that call as having prompted Gorbachov's *perestroika*? It was this lag, exacerbated by the ambition and impotence characteristic of politicians, which brought the country to its present

[9] Yury Alexeevich Gagarin (1934–1968), Soviet pilot and cosmonaut, was the first person to fly into outer space. On April 12, 1961 his Vostok spacecraft completed an orbit of the Earth in 108 minutes. He was killed in 1968 when the plane he was piloting on a routine training flight crashed.

state.[10] I think that even M. S. Gorbachov himself had no idea, even as far back as the beginning of the 1980s, of the real reason for the loss of muscle of our Great State. Had he understood this, his *perestroika* would have proceeded entirely differently.

But then we, the representatives of science and technology, also failed to grasp immediately what was afoot. We alerted the government to the prospects for electronic technology, and talked of the need for extreme measures like those taken earlier in connection with the creation of our rocket and nuclear potential. V. M. Glushkov[11] and G. S. Pospelov,[12] along with the present author and many others, addressed notes to the government—to the Central Committee—gave addresses at various conferences and at meetings with representatives of the military-industrial complex, wrote articles in newspapers, and gave voice to many impartial concerns. It was all in vain, however. And, in doing all this, we failed to realize that things simply couldn't be otherwise, that our attempts to be heard were bound to fail!

Our state, taken together with its political and economic system, was uniquely the product of Nature. Just Nature! It had not been created according to any thought-out plan, but was the evolving resultant of internal forces of organizational development emerging from the initial stimuli that had been set in train in the course of the revolution. Even if Stalin had not existed the system could not have developed differently, since he was a creature of the system as much as one of its founders. The system would have collapsed with or without Gorbachov. Sooner or later, in one way or another, it would inevitably have failed since it was simply unequal to the struggle for a place in the sun on our sad little planet. But the form of its collapse might have been different. There used to be a popular illusion that there was no competition in Soviet society. It's true that as far as the production of goods is concerned there was scarcely any, since monopolism reigned there by law; everything needing to be done, who was responsible for what—all such matters were prescribed in detail for the ministerial branches of production. But although the System strove to impose the "principle of the screw" in order to convert every phalanstery into a human version of a termite mound, people remained people, with their passions and desires. The biosocial laws continued in place, ensuring that competition in the goods market was replaced by a different sort of competition. A different kind of market arose whereby people were selected not for success in business or production or trade, as happens in a free enterprise system, but according to a principle of service to the System, or, more precisely, service rendered to those one rung up the ladder. And the chief service to be rendered was that of maintaining peace and quiet. Stability. "Do not disturb!"—that was the command providing the basis for the ideal, the chief characteristic, according to which people were evaluated. The restless and agitated

[10] The rather chaotic state of the first several years following the collapse of the USSR.

[11] Viktor Mikhailovich Glushkov (1923–1982), outstanding Soviet mathematician and cyberneticist.

[12] Germogen Sergeevich Pospelov (1914–1998), Soviet scientist working in the area of guidance systems and founder of the Soviet school of artificial intelligence.

were culled by the System and exiled to the fringes of society in much the same way as a capitalist system culls—throws overboard—unsuccessful businessmen. In this way every nook and cranny of economic and political power was gradually filled with people capable of ensuring conditions of comfortable existence for those above them, and moreover only in the immediate future, since no assessment of longer term perspectives was involved. This explains why workers in the ruling apparatuses—Party, ministerial, or regional—had the psychology of the temporary worker. Pressure from heads of the military-industrial complex gradually eased once we'd achieved arms parity, and thenceforth it was a matter of simply maintaining the status quo and not rocking the boat. And for this no need to think up anything supernatural!

Thus we began to copy Western "tried and tested" variants; that route was easier and more dependable. This was a simpler course incurring less risk of immediate error. The system was such that more and more often ideas produced from within were not pursued, and not only in the field of computer technology. And there was hardly any thought for the future, no seeing that this approach represented in effect a program of planned obsolescence, or that it must lead to a significant reduction in the nation's intellectual potential.

Even worse, however, was the fact that the monopolism prevailing in industry meant that outdated technology was preserved everywhere, in particular, in the military sphere, with the result that the world's second most powerful nation was transformed into a sort of archaic bureaucracy, powerful only in memory. We, specialists working in the military-industrial complex, understood this only too well, and agonized over finding a way out.

I believe that *perestroika* was launched precisely when at last the fact of our having fallen behind in the military sphere penetrated to the consciousness of the topmost echelons of power. I would not especially wish to cast aspersions on those who instigated that restructuring: who at that time would have been able to predict the sequel? It is only in hindsight that we have grasped the predetermined nature of the collapse of our system, and of the part played by that collapse in the general world crisis.

One way or another, the process of gradual degradation of our industry, including the production of military armaments, began way back in the 1960s. Equipment began to age, the production of replacements slowed. The decline was first felt among those engaged in experimental development: interest in technological innovations and new ideas began noticeably diminishing.

By the beginning of the 1960s "military parity" had been achieved. It was not at all easy to explain, however, what that convoluted phrase actually meant. We're still unable to say exactly what it means. I think that even the military people had no clear conception of what precise sense to give it. Perhaps the best interpretation is the usual one: arms parity meant that the two great powers could, in the space of an hour, annihilate each other—together with every living thing on the planet. I myself understood it as signifying that level of accumulation of arms rendering a war between the two great powers synonymous with their mutually assured self-destruction.

Following the achievement of arms parity, our military and political leadership developed a sense of self-sufficiency, a sense that in that regard also there's no need for alarm. This in turn led to the defence industry's being less interested in fundamental research, and so to less intensive collaboration with academic collectives. It was no longer a case of industry approaching us with requests to undertake this or that research, but of us, academic theoreticians, trying to interest industry, since our very existence depended on their influence and the funds at their disposal. The time when industry could not do without us had passed. Apparently forever!

In that unfolding of events there was a rather important underlying cause at work, likewise having to do with the system of branch monopolism. The various design and technological institutions began trying to establish their own theoretical sections and in the end succeeded: by the beginning of the 1960s, theoretical research groups in the scientific research institutes and design bureaux were already of considerable strength.

This was yet one more turn of events having a very negative effect on the situation with our computer technology. The defence industry opted for building special-purpose electronic computers for use in the various institutions, as a result of which the general-purpose computers needed primarily by researchers were no longer at the centre of attention of manufacturers. Work on domestically developed processors, which at the dawn of the computer age had allowed us to carry out the computations needed for the building of nuclear weapons and sending a man into space, had little by little gone to zero! They were replaced by so-called Unique Series models, which were unreliable copies of obsolescent IBM computers. Our own capable computer designers began turning to drink. What alternative is there for a gifted artist if all he's given to do is copy others' paintings?

Even worse was the project to apply electronic technology in the country's administrative, trade and agricultural spheres, with its promise of greater productive efficiency. Of course, something along those lines was carried out, but more from social pressure than productive necessity.

If there's no competitor, if you're the unique producer of a good, then there's no point in improving anything, in making any further effort; there being no choice, they'll consume it anyway! And now add to that the principle "Don't rock the boat!"—in particular, by introducing management technology based on information processed by a computer, which is bound to make waves. Computerization always brings with it, after all, the necessity to learn something new, to undergo retraining, moreover in one's already advanced years. And, what is most frightening for any office worker, such technological change incurs an inevitable reorganization of the administrative structure, involving the replacement of some employees by others. This is especially painful for any organization, and every clerk would be happy to pay handsomely to avoid such restructuring.

That is how things gradually worsened. We at the Computing Centre and the Institute of Physics and Technology very soon sensed the change. We began having to look for new projects to work on; our approach to both research and teaching altered. The theme of space and rocket technology began to attract less interest in the Academy of Sciences. This was perhaps inevitable, since the nature of our work began to change from investigative to routine engineering practice. In his role as President of the Academy of Sciences, M. V. Keldysh was absolutely correct when he spoke of the necessity of finding applications in the civilian sphere for the mathematical apparatus we had devised in the course of our work in the Military-Industrial Complex. He called upon us to do new research, although he also sensed, perhaps better than anyone else, the "beginning of the end."

The postwar boom began to run out of steam. The System went over to a stationary state, later characterized as "stagnation." This was, however, its natural, and irremediable, state. At the time we didn't fully understand this, and tried to buck the trend by appealing to common sense and to science, with well known results.

These changes all affected me with great immediacy. I was awarded the USSR State Prize for my theory of the motion of a rigid body containing a liquid, or, in other words, for determining a theoretical basis for the dynamics of a liquid fuel rocket. For devising asymptotic methods for computing the trajectory of space vehicles, yielding stability in output with minimal error, I was made a member of the International Academy of Astronautics. Thus while all my research activity and associated successes were in rocket science, the perspectives for large-scale academic research in that area were becoming more and more limited.

I had two choices: return to "pure" engineering, or look for a use for my particular skills in the academy.

The first option was the simplest to realize since I had a good reputation in industry. And, furthermore, I had received flattering offers from both Chelomey and Yangel′. Once I even accepted such an offer, although it's true I was rather under the weather at the time.

On the outskirts of Dnepropetrovsk, Yangel′ had a property, a house surrounded by a tall forest. It was a marvellous place, right next door to the famous Southern Design Bureau. I was there as his guest one early Autumn, which is very beautiful in Novorossiya.[13] The purpose of his invitation was to persuade me to transfer to Dnepropetrovsk. And, under the influence of copious libations—Yangel′ liked everything, in particular rockets and strong drink, to be on a heroic scale—I agreed. Next morning I had a terrible hangover, and after sitting down with a group of his leading engineers and trying to penetrate to the essence of the problems they were describing to me, I understood that I could not take leave of the free-ranging kind of thinking I had enjoyed in the academy. So, even though I understood that I was in some way making a big sacrifice, I rejected the offer, plumping for the second option.

Keldysh was highly displeased with my decision. It turned out that the invitation to Dnepropetrovsk had been at his initiative.

[13] A large historical and cultural region on the north shore of the Black Sea, acquired by the Russian Empire following the Russo-Turkish wars of the second half of the eighteenth century. Long considered part of Ukraine, it has since 2014 become the scene of a continuing conflict between that country and Russian-backed separatists.

Although I had never had especially close relations with M.V. Keldysh, he had nevertheless on several occasions tried to help me up the ladder to high-level administrative positions. But I had rejected every such offer.

Operations Research: Germeier, Bellman, Zadeh

In the early 1960s, intensive research into methods of optimization was begun in the Soviet Union—in Moscow, Leningrad and Kiev. This represented a new stage in the working life of a rather large collective of scientists—mathematicians, engineers and economists—entered into with many high hopes but ultimately subject to many disappointments. The search for optimal solutions of problems has always been a central mathematical concern,[14] not least because many an engineering problem reduces to that of optimizing one quantity or another. The advent of electronic computers opened up new perspectives in this direction. To many of us, including the author of these reflections, it had seemed that work in the area of optimization, more specifically in the theory of optimal control, rather than remaining just another purely mathematical exercise, promised more than anything else to open a new chapter in the history of our country. I don't believe this development had anything to do with Marxism, since everywhere in the West optimization was already the rage.

Traditionally, since the time of the great Euler, physics, mechanics and then the practice of machine construction have been the main source of variational problems. Then at the end of the 1950s, our rocket technology opened up a new field of activity divorced from the traditional interests of pure mathematics and routine engineering practice. Sending a load into orbit demands an enormous expense of energy. The problem becomes crucial of choosing that trajectory of the initial stage of the rocket allowing an additional kilogram of useful load to be sent into orbit with the same amount of fuel. The first to grasp the essence of the problem was D. E. Okhotsimsky, who published an article on this theme in 1946.

It turned out that the problem of locating an optimal trajectory was beyond the scope of the classical Calculus of Variations of Euler and Lagrange, requiring new mathematical approaches like that contained in Okhotsimsky's remarkable paper. However, the decisive step was, alas, not taken by him, and only a few specialists remember his work.

This came about through the publication by L. S. Pontryagin[15] five years later of his "maximum principle." Pontryagin proposed a very simple and elegant con-

[14] Partly because of the "Principle of Least Action," according to which every mechanical system behaves so as to render some quantity—the "action"—stationary. This principle dates back to Leibniz and a little later Euler and Maupertuis, but has ancient Greek antecedents. It remains basic to modern physics, although there are now explanations of it at a more fundamental level.

[15] Lev Semenovich Pontryagin (1908–1988), Soviet mathematician. Blinded in an accident at the age of 14, with the help of his mother he nevertheless became one of the foremost of twentieth

struction allowing the reduction of non-standard problems of analysis to solutions of boundary condition problems of ordinary differential equations—still difficult problems, but soluble using classical methods of numerical analysis. It is my deep conviction, however, that the decisive step had, all the same, already been taken by Okhotsimsky: it was he who first showed, albeit by an example, how one might solve such problems. To this end he used "needle variations," so-called, eliciting certain special properties of the optimal trajectories. Incidentally, "needle variations" were invented by Legendre at the beginning of the nineteenth century, but who remembers that?

One way or another, Pontryagin had the last word. And that's the "absolute truth!" I have always regretted that the "Pontryaginites" never referred to the foundational work of the young MechMat master's student Dmitry Evgenievich Okhotsimsky. But that's the way of our mathematicians: not acknowledging work done by others.[16]

It has always seemed to me that the main thing in pursuing science is to grasp the essence, the root idea, and give it an interpretation making it stand out in relief. A rigorous proof and the possibility of later far-reaching generalization are also essential, by way of fixing what is known. But real scientific development is determined by interpretations. These are more essential than rigorous proofs since they furnish the understanding necessary for the generation of new ideas.

I remember, for instance, how, at the beginning of the 1950s, Andrey Vasilievich Bitsadze[17] gave several remarkable examples illustrating the property of strong ellipticity.[18] Later, however, these results became associated in mathematicians' consciousness with a certain Professor Vishik,[19] who apparently constructed a general theory of such systems in his doctoral dissertation. No matter how important Vishik's work might be, it remains true that the discovery of the property of strong ellipticity and the interpretation of its peculiarities were in the first place Bitsadze's achievement, his contribution to mathematics. No purely "sportsmanlike" result, no mere technical surmounting of obstacles, such as are especially esteemed by mathematicians, but an understanding of the "soul" of the problem at hand—that is what has always attracted me most of all. And that is why I so esteem Okhotsimsky's work, and why I quit pure mathematics.

It was for this reason that, when at the beginning of the 1960s I began teaching a course on methods of optimization at the Institute of Physics and Technology, I decided to go to the original sources of the maximum principle and try following the path taken by Okhotsimsky to its end. I didn't try to create a rigorous theory.

---

century mathematicians, working mainly in topology. Later in life he became a controversial figure for alleged anti-Semitism.

[16] A barbaric custom observed not just in Russia, but widely in the mathematical world.

[17] Andrey Vasilievich Bitsadze (1916–1994), Soviet mathematician of Georgian origin.

[18] Of differential operators.

[19] Mark Vishik (1921–2012), mathematician from L′vov specializing in partial differential equations, professor at MSU from the 1960s, and later member of the Russian Academy of Sciences.

In any case, by that time everything had long been thoroughly understood from a purely mathematical point of view, and all the basic results obtained. What I wished to do was communicate to my students an interpretation of the theory allowing them to see how simple the principle is in essence, how naturally it is related to classical mathematical analysis, and that the maximum principle can be derived essentially in the traditional way, using only the ideas of Lagrange and Legendre.

In expounding optimization problems, I have always tried to formulate them in the widest terms possible, considering them as natural forerunners of the more general problem of constructing a theory and finding methods of deriving rational solutions. In other words, I considered it essential that, at an institution such as ours, oriented as it was towards practical solutions of applied problems using computer technology, we should study optimization problems in the context of the discipline that in the postwar years they were already calling operations research. In the Computing Centre an operations research laboratory was organized, and I invited my old friend Yury Borisovich Germeier to run it. As schoolboys we had attended Gelfand's Circle together, then as MechMat students shared a room in the student residence in Stromynka, and finally, at the end of the 1940s, been colleagues at SRI-2.[20] working under Chief Designer Dillon.

Around that time, Academician A. N. Tikhonov was setting up a Faculty of Applied Mathematics and Cybernetics at MSU.[21] I introduced Germeier to Andrey Nikolaevich. They seemed to get on; in any case, Germeier was shortly thereafter appointed head of the Department of Operations Research in the new faculty.

There soon arose an effective cooperation on optimization problems between the several actively engaged laboratories in the CC and the two departments: my Department of Applied Mathematics at the MIPT[22] and Germeier's department at MSU. For me personally, however, this was more than mere cooperation: my old friend was once again in my orbit, a friend to whom I could disclose all my thoughts and communicate any idea, no matter how silly.

Work on the theory of optimal control led naturally to a whole series of related mathematical problems, resulting in a significant widening of my circle of contacts. I began organizing regular all-union[23] mathematical seminars devoted to optimal control. These were held at a wide variety of places in the Soviet Union: Moldova, Estonia, along the Volga, Siberia,.... We also invited foreigners, though principally from the socialist bloc. An idiosyncratic collective began taking shape, under the auspices of which an interesting group of applied mathematicians began emerging, dissertations were defended, and monographs published. Such well known specialists as Mikhalevich, Pshenichny, Dem′yanov, Evtushenko, among many others, passed through these seminars one way or another. The seminars' topics weren't restricted to any particular area of application; people of very different

[20] Scientific Research Institute No. 2; see e.g. Chap. 1.

[21] Moscow State University.

[22] Moscow Institute of Physics and Technology; see e.g. Chap. 1

[23] That is, open to mathematicians from all republics of the Soviet Union.

interests attended. And not all of them worked in applied mathematics; there were participants interested in pure mathematical problems. All this intercommunication led to the formation of a distinctive kind of national school bringing together first-class mathematicians with physicists, economists, mechanical engineers, etc.—a very rare combination in world practice. In time many of those taking active part in our joint activities became known abroad and achieved rather high international repute, as did the activities themselves.

In contrast to the work done under the umbrella of the Military-Industrial Complex, which was mostly of the classified sort, research in the theory of optimal control and, more generally, into methods of optimization, opened up a range of possibilities for establishing international contacts and collaboration. We participated in a great many conferences on a variety of themes and in connection with various international projects, and travelled abroad to give courses of lectures. We began to attract high-level foreign graduate students, which in turn brought further friendly contacts, some of which have played a rather significant role in my life. At the very least these demonstrated clearly that a shared spirituality, scientific interests held in common, and coincidence of views on the meaning of what one is doing, bring people much closer together than merely having the same nationality or political views.

In that connection, I would now like to recall two of my friends: Richard Bellman and Lotfi Zadeh.[24]

I first met Zadeh in Moscow at a conference held at the Institute of Control Sciences of the Academy of Sciences. I think it was towards the end of the 1950s, when he was already a professor at Berkeley and in the US had become a leader in the field of control of technological systems. He was born in Baku, but apparently his first language was Russian. His father had been an Azerbaijani of Persian citizenship working as a journalist, and in 1927 the whole Zadeh family had been obliged to leave the Soviet Union, first going to Iran, and then America. Lotfi had nonetheless preserved a superb Russian. Since I don't speak English, when I gave lectures in Berkeley he acted as interpreter. On the occasion of our first meeting in Moscow, I invited him home and we spent a very pleasant evening. His wife Fanny also made a great impression on me as a charming and cheerful lady. One could not help liking her, and not only because she was elegant and interesting, but also for her kindness and benevolence.

Sometime in the 1980s, during a stopover in San Francisco I phoned Zadeh's apartment in Berkeley. He was out of town, but when Fanny heard that I would be at the airport for only a few hours, she insisted on coming to see me. This involved a long drive, and there was her age.... She was then, alas, over 60. I tried to dissuade her, but Fanny would be Fanny!

---

[24] Lotfi Aliasker Zadeh (1921–2017), was a mathematician, computer scientist, and researcher into artificial intelligence. He invented so-called "fuzzy" mathematics. Born in Baku to an Azerbaijani father of Persian citizenship, he studied at the University of Tehran and Columbia University. Worked principally at the University of California, Berkeley.

She parked her car on a steep incline, and we went into a cafe and immersed ourselves in reminiscence. She remembered the names of all her Moscow acquaintance, recalled their various troubles, and asked me to say something on her behalf to each of them. On taking my leave I kissed her hand and a tall American standing by expressed his astonishment at the act. Fanny smiled and responded: "This gentleman is not an American; he's a European," prompting the further splendid response: "If I were escorting such a beautiful lady, I also would be a European!"

My closer acquaintance with the Zadeh family began two or three years following that first Moscow visit, in Dubrovnik, at an international seminar devoted to optimal control problems, whither I had been invited as professor. I was accommodated in a small house together with Zadeh and Richard Bellman and their wives. Bellman, Zadeh's wife and I were all of about the same age, but Bellman's wife Nina, an attractive blonde girl, seemed very young to us, which somehow made us all very attentive to her. The house stood right above the seashore. To reach the water it was a matter of going down about 200 steps. At Nina's initiative, every morning we all went swimming. Not bad exercise for people no longer in the first bloom of youth! Incidentally, none of us was then suffering from age-related ailments, so that the 200 steps upwards represented merely a good constitutional before breakfast.

We were always accompanied by a big Alsatian dog belonging to the house's owner, called Yashin[25] after the famous goalkeeper of the Soviet Union's national soccer team. She was agreeable and very gentle. Dogs often take on the personality of their masters, and our host, a retired sailor, was much like his formidable watchdog. The story of how he had been given the name Yashin was as follows. Every morning, whenever anyone appeared on the veranda, Yashin would bring a pebble up to them, drop it, retire 2 m or so, and, assuming her goal-keeper pose, wait till someone kicked the pebble with their foot. As a rule, he managed to catch the pebble, and then, satisfied, allowed his victim to pass. If he failed to catch the pebble, he would whine pitifully and then go and fetch another pebble.

And that's the company we formed, us five and Yashin.

In the evenings we sat on the veranda, directly above the sea, sipped light wine and conversed about ... mathematics. That was when I really began to esteem my new friends. We were not only of about the same age but thought alike on just about every subject.

My views of mathematics and its place in the system of sciences and the life of humanity, and of science quite generally, were formed under the influence of my teachers, among whom I would single out D. A. Venttsel′, I. E. Tamm, and those other titans life had brought me into contact with: M. A. Lavrentiev, N. N. Bogolyubov, and S. L. Sobolev. Engineering practice in solving concrete problems arising in the aerospace industry has also had a large influence in that respect. I had developed an ambiguous attitude towards mathematics. On the one hand, I greatly esteemed the subject itself and those whom I considered great mathematicians,

[25] Lev Ivanovich Yashin (1929–1990), Russian football goalkeeper, considered by many the greatest of all time.

the first among whom was for me Poincaré. On the other hand, I detested the mathematical snobbery that I had been exposed to at university, and that was the main reason for my rejecting the representation of mathematics as sufficient unto itself so characteristic of the Moscow school of mathematics.

I had once been certain that any theory lacking a mathematical formulation is somehow incomplete and had striven to translate every merely verbal description into a mathematical one. But at the same time I understood the tenuousness of my position, since the fundamental postulates underlying mathematical formalism are highly conditional. And I eventually grasped that it is far from always possible to get a description in mathematical terms. It is in fact impossible to penetrate to the absolute in any field, in particular, that of strict mathematical deduction, which, in view of Gödel's results, has become relative even in pure mathematics.

To put it another way, in every field of endeavour a sense of humour is essential: in mathematics, in one's own preoccupations, and above all with respect to oneself! I was taught this principle by D. A. Venttsel′, who responded with irony to the sententious assertions stemming from what I had imbibed from another of my teachers, D. E. Men′shov, under whom I had written my master's thesis in the Department of Functional Analysis in 1940. The critical attitude whereby one acknowledges the merely relative value of all that one is capable of inventing and understanding, does not at all lessen one's enthusiasm for doing research. It simply puts everything in its proper place. That attitude causes the absolute of any situation to be replaced by an interpretation of it, thereby changing one's scale of values. This frame of mind, which I now use as a guide in all I do, I associate with the name of Niels Bohr.

And there's another principle to go along with that one: "Various kinds of mothers are necessary and they're all important." For a person to do something really well he or she needs to be interested in it. And only things that have been done well accumulate in humanity's treasure chest. But it is not easy to understand why one person finds one thing interesting, and another something else. Just human nature.

In the evenings, my new friends and I discussed from this point of view our own lectures and those of other participants that we had paid close attention to. The conversation was carried on in a strange mixture of Russian, French and English: Zadeh and Bellman spoke to one another in English, and I spoke to Bellman in French and to Zadeh in Russian. Fanny was always there with us and since she spoke every language, both real and imaginary, she often extricated us from problematic situations.

These my new acquaintances were all highly gifted people, but of quite different fortunes. Zadeh immediately identified himself as belonging to the engineering domain. He never claimed to be a mathematician, even though he was perfectly at home in probability theory and algebraic methodology. He had very early on achieved recognition for his work on the control of technological systems, and only when he had made a name for himself did he turn to more abstract interests. In that very year he had begun creating a theory that was later to be called "fuzzy set theory." I valued his creation, even agreeing to be a member of the editorial board of the associated international journal. It seemed to me that his methods would find their most interesting development, perhaps, in the theory of filtrations of a random process by a nonlinear operator. I even started to work out an appropriate elaboration of this idea, but without success.

Bellman's fortunes were completely different. He regarded himself first and foremost as a mathematician and wanted recognition from mathematicians. But American mathematicians, alas, refused him this: he didn't come within the conventional norms. At first he had worked in the Rand Corporation, and only much later started teaching at the University of Southern California. He conjured up methods and began applying them without much thought for checking the validity of his ideas. To even his purely mathematical results he gave proofs that were somewhat inaccurate by the standards of higher mathematics. He wrote his books rapidly, sometimes leaving them unpolished. They were, however, snapped up, and translated into many languages. They were read by engineers, physicists and economists rather than mathematicians. He was much better known in the USSR than in the USA. In our country he was especially famous for his invention of dynamic programming.

The history of the concept of dynamic programming is convoluted. I myself was involved in it to a certain extent.

At the end of the 1950s, I found a method for determining, among all trajectories of a guided missile avoiding a certain forbidden region, that trajectory allowing largest load for a given quantity of fuel. The idea behind my computer algorithm was very appealing, and I was proud of it. Following my talk on the discovery at our departmental seminar, however, V. G. Sragovich informed me that the Kiev mathematician V. S. Mikhalevich had solved a similar problem, and that his solution had, moreover, already been published. I went to Kiev, and verified that this was indeed the case. Although, it's true, he had solved only the problem of profiling the desired trajectory using an analogous procedure for roads, and had not derived any differential equations for it, his method of computing it numerically was the same as mine. It would seem that we had conceived this method at the same time, but Mikhalevich had published his work earlier. Furthermore, my paper was published as a classified report, so that for a long time no one knew of it. And that is why, when I included this method of solving optimization problems in the textbook I wrote on the theme, I called it the "Kiev broom" and mentioned Mikhalevich as its first inventor.

But there's more to the story. It turned out that about two years before the above events, Richard Bellman had published the very same method in America, calling it "dynamic programming." We got hold of Bellman's book and translated it into Russian. We discovered that the method of the "Kiev broom" was a version of dynamic programming; it was not as universal as Bellman's method, but had certain advantages for numerical computation in the context of the specific problems that Mikhalevich and I had solved.

Which explains why it was so interesting for me to meet Bellman and spend almost a month with him in Dubrovnik. We gave our respective cycles of lectures in parallel, each day comparing notes on what we had talked about. We became firm friends, and our friendship lasted till his death.

Towards the end of the 1970s, a tumour was discovered in Bellman's brain and he left the Rand Corporation, while retaining his position at the University of Southern California. Despite a series of operations, the tumour proved resistant. Although he was no longer able to work, the university kept him on at full salary. That was insufficient, however, to cover his medical expenses, and his once wealthy family found itself in financial difficulties. They had to sell their house and live very modestly. Mutual acquaintances informed me that his last year had been especially hard, with Nina devoting every possible effort to easing her husband's suffering, hoping for a reprieve to the last.

Planned Development, Programmed Design, and the K-K Economics Model

Every diversion has its term. In particular, research in the field of optimal control began to lose its lustre little by little. There was no shortage of problems, of course: every theory can be expanded indefinitely. Interest in continuing with it, however, gradually decreased to zero. For this there were definite reasons.

Most of all it was due to the fact of our having got to the point of knowing how to solve rather effectively the problems arising in engineering practice, especially subsequent to the development of optimization systems using person-machine dialogue. The implementation of these allowed many problems, such as, for instance, the minimization of the weight of construction materials for a given structural robustness, to be solved in routine fashion. However, in dialogue systems there is little overlap with the traditional work of a mathematician; at their core they depend on the skills of research engineers or physicists well versed in practical details. Having at his disposal a package of programs realizing the latest suite of mathematical algorithms used for solving optimization problems, the research engineer sits down before the monitor of his computer, which displays information not only in numerical but also graphical form. Before his or her eyes the whole process of searching for the required form of the desired construction and its characteristics can be seen unfolding. Via an iterative process involving the application first of one algorithm and then of another, the engineer sees the result of each step of the procedure, and makes appropriate adjustments as he goes along. In this way problems of design or planning which not so long ago furnished material for candidates' dissertations, can now be solved in a matter of minutes.

The second reason for my declining interest had to do with the loss of many of my illusions about the contributions that mathematical methods might be capable of making to economics and public administration.

In the mid-1960s, among mathematicians and others involved in computer science, principally those occupied with methods of optimization, there was a sudden surge of interest in economics. The work of L. V. Kantorovich,[26] one of the originators of linear programming, played a significant part in this. Of greater impact, however, were the successes we had in developing efficient algorithms for solving various optimization problems, e.g. in the design of a program for sending a rocket into space or the choice of an airplane's optimal flight-path or the optimal control of one technological process or other. It had seemed back then that the transfer of these ideas and methods into the realm of societal processes must surely yield the perfect system of public administration, above all of a country's economy. Thus arose the idea of programmed management methods.

The idea of programmed management of a process subject to input from an unpredictable external world, was in and of itself completely reasonable. Imagine you have some external situation modelled by a plausible abstract scenario, and you want to distribute a particular resource in such a way that over a given time interval you approach your desired goal in the best possible way. To achieve that end, all you need is a model of the process in question, i.e., one couched in mathematical terms, since, at the time in question, methods for calculating optimal programs (or trajectories) had already been thoroughly worked out. And then, in order to compensate for unforeseen hitches, that is, for departures from the given scenario, you would need to implement a feedback mechanism to keep the process moving along the programmed trajectory. This would appear to be a completely rational method, and in fact engineers use it to this day to compute the trajectories of their rockets.

The first time I heard the expression "programmed management method" in connection with the national economy was from my former student Yu. P. Ivanilov, when he had risen to a full professorship at the Moscow Physico-Technological Institute. Academician G. S. Pospelov devoted a large monograph to an exposition of the method, and Academician V. M. Glushkov also came under its spell. To be honest, I must confess that I also took an interest in, and actively spread propaganda on behalf of, the programmed method.

In fact, there's nothing wrong with the method itself. It can be successfully used to solve a variety of problems arising in both socialist and capitalist economies. The mistake lay in the representation of the method as a sort of panacea, in entertaining the possibility of a fully centralized economic system developing according to a given program, in the idea that a developmental trajectory is a sort of economic law that will, or should, be unswervingly observed. Mistaken, and crucially so,

---

[26] Leonid Vitalievich Kantorovich (1912–1986) was a Soviet mathematician and economist, known especially for the theory and practice of the optimal allocation of resources. He and the American G. B. Dantzig are considered the founders of linear programming.

was the concept of systematic planning, formulated as a law of development of the socialist state. What seemed to us then as being most obvious, namely the goal of development, is precisely the stumbling block to a realization of the programmed method, since that goal is always a compromise among the millions and millions of aspirations and desires of the people. We ignored the fact that each separate person is capable of acting, and acts, according to his or her own individual interests, to his or her personal idea of what really needs to be done in each concrete situation. The programmed method is not a method of managing *real* development, therefore, but only provides estimates for the possibilities of development.

By the early 1970s, I had gone to the other extreme: I had gone over to complete scepticism as to the value of mathematical economics, and began expressing my attitude towards that subject in highly ironic terms. At that time, I'd begun to work on problems of evolutionism and self-organization, and I was engaged in a profound internal process of re-evaluation of the values I had subscribed to hitherto. I was leaving Marxist dogmatism and rationalism further and further behind in the quest for a new paradigm.

In the discussion following a seminar talk in Moscow by the American economist Professor Tjalling Koopmans[27] of Yale University—a "Kantorovich analogue," so-called—I mentioned the term "K-K economics," having in mind not only an abbreviation of "Kantorovich-Koopmans economics." Pronounced in Russian that term expresses rather more than just dubiety,[28] but of course Koopmans did not understand the play on words and was proud at being grouped together with Leonid Vitalievich. Those two were indeed incommensurable quantities. Koopmans understood the hierarchy very well—unlike the Nobel Committee, which had ranked Koopmans at the same level as Kantorovich.

My little joke had unexpected consequences.

I subsequently received an invitation to spend a month at Yale University as visiting professor. And then, a year later, when Koopmans had already been awarded his Nobel Prize, I was invited for another three months. In the absence of any standing on my part in the field of economics, I suspect that I owe these two invitations to my joke, which had flattered Koopmans and foreshadowed his Nobel prize.

Scientifically speaking, my sojourn at Yale University was not very productive. I was not at all encumbered with work. I gave a few lectures, wrote reports on a couple of dissertations, and participated in seminars. As a scientist, Koopmans himself was of no great interest to me. His economic ideas seemed rather trivial, and as far as analytic methodology went, he was far from having that depth we had been used to from contact with people of Kantorovich's class. Talking to him, however, was

[27] Tjallings Charles Koopmans (1910–1985), was a Dutch and American mathematician and economist. Joint winner with Kantorovich of the 1975 Nobel Memorial Prize in the Economic Sciences for their work on the allocation of resources.

[28] *K-K ekonomika* would be pronounced *káka ekonómika*, and since *kaka* is Russian children's talk for something unpleasant such as "pooh," the term might be taken as meaning "crappy economics."

pleasant. He was a real European, born in Holland, and spoke French well, so there was no linguistic barrier to communication. He was very cordial and attentive to my needs—although, it's true, when he discovered one day that I wasn't Jewish, his attitude towards me cooled.

Once I was asked to give a public lecture, i.e., one not intended just for students. It was advertised as dealing with the problem of controllability of an economy, and it was the first time I was to formulate the basic ideas of the view of that topic that I continue to hold even now. As it is usually understood, an economy is not subject to guidance, and the very term "planned economy" is a sort of linguistic nonsense. In some form or other, a market is essential. Furthermore, even in situations where the economy is declared to be fully planned, market relations of one form or another still arise. The only economic goal that can be objectively realized is that of societal homeostasis, requiring only the imposition of appropriate restrictions. Society inevitably throws up interdictions, taboos, and for policing these the government is responsible. And the more the forces of production are developed, providing greater power to a civilization, the harsher will those interdictions be, necessitating a greater degree of guidance of economic development and thence greater participation by the government in that development. And not only the government, but the whole of the citizenry. Not total planning, therefore, not economic control, but directed development capable of avoiding economic crises.

My lecture was well received, but I was afraid that it might resonate back in my own country, since the position I had expounded was very far from the accepted doctrine and style of thinking of our economists. However, my lecture went unnoticed at home, even though for me personally it represented a new stage in my life: I had abandoned all illusions as to the controllability of an economy, all the dogma surrounding Marxism in the USSR, and had gone on to live with a new paradigm. Incidentally, neither was there much of a reaction to my lecture "with them,"[29] and I didn't get a Nobel Prize. Koopmans didn't like my lecture at all since it was at odds with his pseudo-Marxism.

I did, nevertheless, have my reward, for which I was obliged not to Tjalling Koopmans, and not to society, but to a young couple who invited me to travel across America to California with them.

In John Steinbeck's wonderful book *Travels with Charley in Search of America*, he recounts how, in the company of his dog, he crossed the continent in a car, driving along the border with Canada and then south to Los Angeles. I travelled along the same route with an empathetic young couple—a marvellous reward for my joke and the perfect conclusion to my work on K-K economics.

[29] That is, in the West.

Pavel Osipovich Sukhoy and the Automation of Aircraft Design

Late one evening, the famous aircraft designer Pavel Osipovich Sukhoy[30] phoned me at home, requesting that I come to see him next day at his Design Bureau.

Sukhoy was one of our most interesting and gifted designers of warplanes. By the end of the 1960s he would have been around 80, but when I came into his office, I saw, not the wise but decrepit old man I had been expecting, but someone more like a man of middle age in good physical shape, holding himself straight in the manner of an officer of the old Russian army, and of clarity of mind and impressive erudition. He was also energetic, full of projects and plans.

He explained to me at great length why the traditional methods not only of design, but of the whole process of construction of airplanes was no longer appropriate. Fundamental restructuring was in order. As particular instruments of warfare, airplanes had become so complex that their production process—design, the construction of test versions, flight tests, and final construction—took up so much time, was becoming so drawn out, that the end product was already obsolete when it was rolled off the production line. Over the previous 12 or 15 years, science and technology had advanced so far so fast that the latest model was always practically an archaism. A qualitative perfecting of the whole production cycle of an airplane was therefore essential—not merely a speeding up of the design process, but the implementation of a means of continually adjusting the on-going production process. And it would also be necessary to learn how to simultaneously plan several alternative variants.

To achieve all that, however, was impossible without using contemporary methods of computer science to create automated design systems. And of prime importance was the immediate necessity of switching to multiple-variant designing. But how is an engineer to be trained to simultaneously plan several variants of one and the same airplane, how should such a process be organized, what tools should he have at his disposal? All these questions remained unanswered. Only one thing was clear: the problem was impossible to solve without a powerful computer and a radical reorganization of the process of building an airplane so as to incorporate features of greater computing complexity.

Sukhoy proposed not only that we take part in this work, but that we both organize and lead it. In connection with the project we would have *carte blanche*, as he expressed it, in his Design Bureau. I thanked him for his trust, noting that we would need to train a new generation of design-engineers, preferably chosen from among recent graduates. Their training would have to be a collaborative effort, and this would tend to cause difficulties since some instructors would be unacquainted

[30] Soviet-Belorussian aerospace engineer. Designer of the SU series of military aircraft. Founder of the Sukhoy Design Bureau. Lived from 1895 to 1975.

with airplanes, while others knew no computer science. And quite apart from that neither sort would know what or how to instruct!

I very much liked the clarity of thought and precision of expression of our General Designer. The project seemed complicated yet feasible. I began gathering together a moderately sized group from among those under me at the Computing Centre of the Academy of Sciences. To my surprise, however, I met with an uncooperative attitude to the initiative from the director of the institute Academician A. A. Dorodnitsyn. It had seemed to me that collaborative work with Sukhoy's Design Bureau would be of interest to him, since he had spent his working life in aviation. In the present situation, however, his stance was purely territorial: that's not your business; for such projects we have the CAHI.[31] The fact that he was then a deputy director of CAHI doubtless had something to do with his attitude.

I should say that this turn of events was not exactly unexpected. In the past, Anatoly Alexeevich had reacted to the majority of my initiatives with scepticism. But then when any of them worked out he was quick to embrace it: "In my institute...." Likewise in this case: when ten years later we were symbolically crowned with laurels[32] in the Kremlin for having created a system of automated design and construction of airplanes—implemented, by the way, in the construction of the SU-25[33] and all succeeding planes from Sukhoy's Design Bureau—Dorodnitsyn raised no objections to being included in the list of authors of the system. Following the official ceremony in the Kremlin, we bought champagne and went to my apartment, where my wife made dinner. I didn't invite Dorodnitsyn, however, so as not to put him in an embarrassing position.

That dinner was still a decade away, however.

Following my discussion with P. O. Sukhoy, I used my standing as Dean of the Applied Mathematics Faculty of the MIPT[34] to choose a group of capable graduates, and united them with a group of young engineer-designers freshly graduated from the Moscow Aviation Institute to form a special subdivision within the Design Bureau dedicated to the creation of a new design technology. Working together with a smallish group of operatives from my Academic CC, they began penetrating to the crux of the problem. And with such enthusiasm—enough and to spare!

I think that, despite our enthusiasm and the General Designer's diligence, the whole enterprise might have remained at the level of mere good intentions, if O. S. Samoilovich, one of Sukhoy's deputies, had not so devotedly believed in it. But even he—why pretend otherwise?—failed at first to grasp the essence of the undertaking, thinking that a system of automated design amounted merely to the use of computing methods to facilitate complex engineering computations and organize the technical drawing part of the process. What we were really doing, in fact, was creating a technology for the design and construction of a fighter plane that was new

[31] The Zhukovsky Central Aero- and Hydrodynamic Institute; see e.g. Chap. 1.

[32] To be specific, they were awarded the Prize of the Soviet of Ministers of the USSR.

[33] A Soviet armoured attack plane, first produced in 1975.

[34] Moscow Institute of Physics and Technology.

in principle. For the first little while very few understood this. The horizons of our enterprise did not immediately reveal themselves to everyone.

At first the work went well, till suddenly we were overtaken by a catastrophe: our General Designer died. He was replaced by Pavel Osipovich's first deputy E. A. Ivanov, director of a pilot plant, an energetic fellow, a highly experienced engineer and a competent technologist, but ... neither a designer nor a scientist. Sukhoy hadn't even familiarized him with our enterprise, and at first Ivanov considered all the very costly work that had so far been done to be my personal venture, and Samoilovich's enthusiasm to be motivated by the desire to defend a doctoral dissertation. (I should perhaps mention that he did in fact get his doctorate, only much later.) Evgeny Alexeevich said as much to me eye to eye during our first conversation. It's true that he promised not to terminate our work for the time being, but I nonetheless understood that, given the new General Designer's attitude to it, the perspective was not looking hopeful. Then suddenly the situation was saved ... by my wife!

One day I invited Evgeny Alexeevich and Oleg Sergeevich to my dacha to partake of homemade *pel'meni*.[35] At that time, the wherewithal for a party was affordable for scientific and engineering husbands. Two hundred *pel'meni* cooked by my wife, together with appropriate ancillaries, provided the conditions essential to enlightenment of the mind and mutual understanding. Ivanov, having eaten of the *pel'meni* and quaffed some of the spirituous ancillaries, called our get-together a "grand guzzle," praised my wife and left replete with satisfaction.

My wife and I repeated these "guzzles" a few times in Moscow in order to enlighten the ignorant as to both computer science and the fundamentals of the theory of automated design. I'm unable to judge to what extent exactly our new General Designer took in, after the *pel'meni*, the conception of automated design of warplanes I expounded to him, but at least the boom was raised.

When at last, eight or ten years after that series of "guzzles," we returned to my apartment from the Kremlin and drank champagne (but not only champagne), and dropped the medals we'd been awarded in our wine-glasses, my wife, Antonina Vasilievna,[36] remarked not without justification that she had played a role in our triumph: "Without my *pel'meni*, you wouldn't have succeeded! So I should be awarded half a medal!"

The eventual post-completion fate of the project was unusual.

Following E. A. Ivanov's conversion to our faith, the work really did proceed apace. However, I began gradually removing myself from the project; I was more and more attracted to the field of dynamics of the biosphere, and was thinking about problems of universal evolutionism and other methodological questions. I'll leave that till later, however. So I handed the running of the project over to P. S. Krasnoshchekov, a former graduate student of mine and at the time one of the

[35] Dumplings consisting of a filling of minced meat wrapped in a thin unleavened dough and usually eaten with sour cream or butter.

[36] The author's second wife.

youngest of my current colleagues. He possessed a sharp mind, resourcefulness in solving concrete problems, and exceptional organizational ability. He had suggested several important innovations and carried through their realization brilliantly. The scientific world evaluated his work in automated design accordingly: he was first elected a corresponding member of the Soviet Academy of Sciences and later a full member of the Russian Academy.

In any case, our accomplishments were of undeniable significance. It is even true that without our automated design system, the jet fighters SU-25 and SU-27 would not have rolled off the assembly line so soon. The Sukhoy Design Bureau started a new chapter in the history of Russian technology as it relates to aircraft design. Today the implementation of automated design has become almost routine and is standard in design practice. In the early 1970s, however, it was revolutionary, igniting many disagreements and fierce resistance on the part of administrators, and was, in particular, not understood by the then Minister of the Aviation Industry I. S. Silaev. Although the project had been undertaken at the initiative of P. O. Sukhoy and it could not have been completed without the participation of Samoilovich and other younger design engineers, the decisive contribution both of fundamental ideas and input into the project's concrete realization was nonetheless made by the team from the Computing Centre of the Academy. I am unable to recall this fact without a feeling of pride. Later, when I had the opportunity of acquainting myself with analogous systems produced by the firms SAAB in Sweden and Lockheed in the US, I saw that we had surpassed them in the subtlety of our answers to various questions arising in the course of creating our system at the Sukhoy Design Centre.

From the collaborative work on the project there emerged a mixed, friendly and highly qualified collective, made up of those working under P. S. Krasnoshchekov in the Academy and the design engineers under O. S. Samoilovich. It would have made sense for the collective to be maintained and for its members to continue working together. However, it happened that E. A. Ivanov was uncongenial to the Minister of the Aviation Industry at the time, I. S. Silaev, and he was removed from his post. A short time later Ivanov died of a heart attack. He was replaced by a certain Simonov, Silaev's creature, who, in terms of level of thinking, design qualifications, and purely personal qualities, was incompatible with our collective. The link between the Academy of Sciences and the Design Bureau was broken, the mixed collective fell apart, and Samoilovich together with several of his closest comrades-in-arms transferred to the Mikoyan Design Bureau. In spite of the advent of Simonov, however, the design culture did not die out at the Sukhoy Design Bureau, and subsequent SU-type airplanes continued to reflect something of the quality of our labour.

And that's how the work came to a close. Everything comes to an end sooner or later!

Nikolai Sergeevich Moiseev, the author's father

Elena Nikolaevna Moiseeva, the author's mother

Nikita Moiseev, the author, in 1947

Sergei Moiseev, the author's brother, in 1947

Olga Ivanovna Moiseeva and Sergei Vasilievich Moiseev, the author's grandparents on his father's side, early 1900s

The house at Skhodnya, built by the Moiseevs in 1928

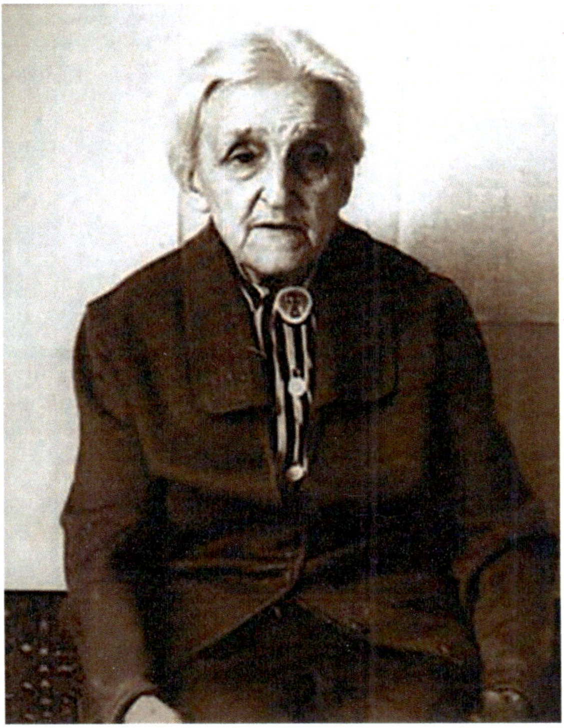

Margarita Vasilievna Moiseeva, the author's stepmother, late 1970s

At Skhodnya in the 1930s (Margarita Vasilievna Moiseeva is on the far left and Olga Ivanovna Moiseeva on the far right)

1948. Volleyball team at the alpine camp of the Central House of the Red Army in the Baksan Gorge (The author is on the right)

The author with his wife Kira Nikolaevna at Skhodnya, 1955

1958. Kira Nikolaevna and the author with their daughters Irina and Alena

Academician I. E. Tamm and the author

A. A. Dorodnitsyn and the author at the Computing Centre of the Academy of Sciences of the USSR in the 1970s

Performing at the seminar

1968, San Remo, Italy. At the Soviet-American Colloquium on Methods of Optimization (The author is second from the left, next to Fanny)

1967. The author with daughter Irina at a Summer School in Shemakha

1977. The author with former students become colleagues, left to right: F. I. Ereshko, Yu. N. Pavlovsky, I. A. Vatel’

1976. Second youth. The author with his second wife Antonina Vasilievna

The author on his 70th birthday, with, left to right, A. N. Nesmeyanov and I. P. Lyubomirov

The author with Antonina Vasilievna at their dacha in Abramtsevo in the late 1990s

In this book I write about my work and my country. It's not a memoir in the standard sense, but rather a record of my reflections against a background of recollections

Chapter 8
Spring of Light

A New Crisis

My new crisis was very personal. But, as always, misfortune brought new perspectives in its train.

The difficulties and troubles of my childhood and youth had not exhausted the share of grief allotted to me by fate. I had yet to go through perhaps the most difficult period of my life: my wife, Kira Nikolaevna, became severely and incurably ill and the disparate elements of my existence became confused: my personal grief, work, health, and the children. This situation lasted well over two years. And then a state of isolation descended on me. It seemed that acquaintances, colleagues and well-meaning people generally are one thing, and friends another thing altogether. Andrey Nesmeyanov was perhaps the only person to whom I could come and ease my soul—simply keep silent, or talk about nothing in particular, or play our version of backgammon. Once in a while down a shot or two. On the night of my wife's tragic end, A. A. Petrov showed sincere concern and had me spend the night at his place. I have remained grateful to him ever since; I could not have remained alone in my apartment that night.

My own health also sharply worsened. During the war my spine had been hurt, but through my sporting activities—skiing in Winter, climbing in Summer, and, most of all, walking with a heavy rucksack—my back had been maintained in such good shape that the injury of December 1942 hardly made itself felt. But now skiing was out of the question and in Summer I was not in a position to be absent for long, so unable to do any of the real climbing of yore. And the pain in my back sometimes took me out of action for several weeks at a time.

At that time also things were not going so smoothly at work. The space exploration theme was clearly beginning to lose momentum. New horizons were called for. From the beginning of the 1960s I had held the position of deputy director of research in the Computing Centre, which in essence meant that I alone bore responsibility for the future of the institute's mathematical subsections. Being

N. N. Moiseev, *How Far It Is to Tomorrow...*,
https://doi.org/10.1007/978-3-030-96651-5_8

much more interested in CAHI,[1] the director, Academician A. A. Dorodnitsyn, paid practically no attention to what was happening on the scientific side of things. Since there was no one to consult with, I had to decide unaided on what new direction the work of the large and talented collective under me should take. A reorientation of the work of the CC was necessary since the situation in our country was changing in crucial ways.

Preoccupied as I was with personal problems, I felt I had insufficient inner strength to formulate new projects and redirect the scientific activities of the CC. I made an attempt, but my thoughts and strivings were in those years diverted towards other things. I had to look for support from outside. I began creating new sections in order to attract energetic, gifted and independent-minded scientists to head them. It seems to me now that this tactic was a good one, since it circumvented the crisis facing the institute and forestalled the promotion of Party or Union functionaries to positions of major influence, an almost ineluctable process at a time when research organizations were encountering thematic difficulties and the immediate demand made of researchers that they carry on in time-honoured fashion was beginning to lose force.

In this way, we created an operations research laboratory headed by Yu. B. Germeier, a programmed control lab run by G. S. Pospelov, and a discrete mathematics lab under Yu. I. Zhuravlev. The appearance of such capable scientists of unimpeachable scientific reputation to a large extent removed the chief onus of my position as scientific overseer, namely the selection of projects, and through this the directions of research of new hirees. Furthermore, during that period our complement of scientists was every year swelled by many bright young people eager to demonstrate their scientific worth. The main, and also most difficult, problem was then that of determining the best use of the skills of such enthusiastic youngsters—much more important, in fact, than the pursuit of my own research activities.

As a result of this expansion, the Computing Centre was transformed into a first-class scientific institution at the international level, acknowledged both at home and abroad.

My home life had also turned problematic. I was having difficulty finding a common language in which to communicate with my children. Instead of uniting around our shared grief, we began living separate lives, each to him- or herself. Of course, I myself was most to blame for this situation. I had always been excessively preoccupied with my own life, work, wife, and mountaineering. I had always loved my daughters and thought a great deal about them, but had not admitted them into my personal life. That was my chief mistake. Sometimes I had taken them along with me to the mountains and we had kayaked together occasionally, but more than this had been required. I should have shown much more affection, should have created a shared spiritual world, and not merely tried to understand their lives, but included them in my own life as my father and grandfather had included me in theirs. Thanks to their ability to talk to me as to an equal, I had then lived in the world of

[1] The Zhukovsky Central Aerohydrodynamic Institute; see e.g. Chap. 1.

their interests and concerns. How I was to benefit from that! But on the other hand I myself almost never spoke to my girls as adult to adult, didn't describe my troubles or satisfactions to them, didn't share my thoughts. . . .

However that may be, following on my wife's tragic end I was left in deep internal isolation from everything around me, yet somehow had to begin to live in a completely new way. I was not up to such an endeavour, however. More than anything I needed compassion and support.

And a helping hand unexpectedly reached out to me.

I met Antonina Vasilievna more or less by accident in the home of mutual acquaintances. We began periodically spending time together. In the case of someone substantially over fifty, a romance develops according to a completely different scenario than in younger days. And his feelings are different in tone from those of his youth. But all the same. . .

> If only there was Esenin's[2] music
> And a sea of light from the window,
> If only my Autumn perspective
> Were always illumined by you. . .

My Encounter with the Humanistic "Intelligentsia"

In early Spring, or, more accurately, at that interlude between Winter and Spring that Prishvin[3] called the "Spring of light," I decided to take a rest. It was a long time since I had a holiday, and the previous year had been onerous, I felt exhausted, and sensed perhaps for the first time that, having reached my sixth decade, I was already far from young. So I decided to go to a sanatorium or "house of rest"[4] near Moscow, with the idea of doing some skiing. Nowadays, in the midst of the present general impoverishment, such an idea seems fantastic, but such spells from work were then possible, even standard, especially for those involved in intellectual kinds of labour.

At a meeting in the Academy of Sciences, I happened to bump into F. M. Burlatsky, at that time deputy director of the Academy's Institute of Sociology. By that time my interests had shifted towards problems closer to the humanities, and I would occasionally show up at the Institute of Sociology. Fedor Mikhailovich queried me about my plans and I told him of my intentions. He said that he himself was the very next day going to the House of Creativity near Ruza:[5] "Why don't you go too? We'll then have someone to talk to." He explained where and how to

[2] Sergey Alexandrovich Esenin (1895–1925) was a Russian lyric poet of peasant background, one of the most popular Russian poets of the twentieth century.

[3] Mikhail Mikhailovich Prishvin (1873–1954) was a Russian and Soviet writer, in particular of children's literature.

[4] In Soviet times there were ample such facilities provided cheaply or even *gratis* for the working populace.

[5] A town about 100 km west of Moscow on the Ruza River, a tributary of the Moscow River.

get hold of a voucher, noting also that the rooms were single with shared bathroom facilities along the corridor, but that in every other respect it was a marvellous place, especially the surroundings.

Heeding his advice, I bought a voucher, got behind the wheel of my "Zhigulenok,"[6] and two hours later I found myself in a picturesque corner of the Moscow environs—or rather Moscow's "further" environs, to use the ridiculous new terminology.

It was early March, when the days are already long and there is ample sunlight, but the snow's whiteness still dazzles, and everything sparkles. Not for nothing did Prishvin call this time of the year the "Spring of light." One might even call it a *riot* of light. And there were birch forests all around where the light was especially bright. I revelled in the weather, the skiing, and the surroundings. I felt that such a combination of light, sun, birch woods, blinding snow and movement through the morning frost was just the medicine I needed.

For the first few days skiing was the whole of my occupation, except for sleeping; I slept endlessly and gradually felt restored to my former self. Then I began to look at the people around me. They were of a peculiar sort, quite outside my ken. This was my first encounter with the "creative intelligentsia" on holiday! The sanatorium's doctor turned out to be the wife of my acquaintance Lieutenant-Colonel Samoilovich, then employed as a scientific research fellow at a military institute in Kalinin. A few years later, Samoilovich, having been promoted to colonel, retired, and they immigrated to Canada, I believe, where his wife had relatives. She had pretensions to literary and musical cultivation. In fact her pretensions were quite general: she knew, for instance, the secret of curing every ailment with aspirin and sleeping pills.

Every evening in *Madame* Samoilovich's apartment there forgathered a company of "intellectuals," as the hostess called them, carefully chosen beforehand. Burlatsky was not invited, while for some reason the honour was bestowed on me. In that season the leading light of our "intellectual society" was the famous Galich.[7] This was indeed a very interesting person: he wrote good verse, which he sang to guitar accompaniment, and said clever and mordant things. However, overall he made an extremely unpleasant impression on me. And this wasn't just a matter of the sardonic manner in which he expounded what were in essence correct thoughts. I am highly sensitive to innuendo.

Russia is both my eternal hope and a permanent source of pain. If I sense the same pain in someone I'm talking to, I can be completely open about everything with him. But it's enough for me to sense condescension, even in negligible amounts, towards "this country" or "this people," for him to become absolutely alien to me. And that is how I reacted to Galich, in spite of all his undeniable talents and the fact that

[6] Diminutive form of "Zhiguli," a car of Soviet make.

[7] Alexander Arkadievich Ginsburg (pseudonym "Galich") 1918–1977, was a Russian poet, scenarist, playwright, writer of prose and singer of his own songs to guitar accompaniment.

it was always interesting to listen to him. Although I was never to encounter him again, in memory he remains deeply inimical to me.

In 1990, during a visit with my wife to Paris, we went to the Russian cemetery outside the city. There, quite close to the modest grave of the great Bunin,[8] was the sumptuous burial place of Galich, to my surprise marked with a Russian orthodox cross.

I didn't like Madam Samoilovich's little society; it was not to my taste. In fact, I find the dissident ambience generally not much to my liking. While I have always been receptive to any discussion of our troubles—our country's troubles—I find it difficult to countenance my fellow citizens' speaking in a too detached way, almost gloatingly, about our common misfortunes. And it has always seemed to me unfitting when they count Sakharov and Solzhenitsyn of their company. For both of these men Russia was an open wound.

To cut a long story short, I made various excuses and stopped attending those evening gatherings of "intellectuals."

I was also not very partial to the company in which F. M. Burlatsky spent his evenings—although it was more agreeable to me than that of the "intellectuals." There I met altogether surprising types, and the mix of these was beguiling, especially to an outsider like me.

Of an evening we usually watched a movie, after which we went to the dining hall to drink kefir. (*Madame* Samoilovich's "intellectuals," on the other hand, drank vodka around that time. This was furnished by her guests, she herself providing tea. Not much vodka was drunk, however, since after-dinner snacks were frowned on.) Thus it was that over kefir the real life of the House of Creativity began. Someone called these get-togethers "kefir orgies," which was in fact a very apposite name for them. They would begin with a discussion of the movie just seen, which would then stray far off-topic. People became extremely worked up. It was very interesting listening to them; these "kefir discussions" were new to me for both the manner in which they were conducted and the topics debated. It was as if I had entered a different world!

Their thirst for kefir slated, the whole company would set off on a walk. The glowing moon, the air fresh from the Spring frost, and the enveloping nighttime stillness was completely at odds with the themes of the "kefir clashes," so that the back and forth arguments gradually subsided and in dribs and drabs the pacified members of the company returned to their rooms without conveniences.

Sometimes during these "kefir orgies" unexpected things would happen causing subsequent revelations. One evening the film *Chapaev*[9] was shown, that marvellous

---

[8] Ivan Alexeevich Bunin (1870–1953) was a Russian writer and poet. He was bitterly opposed to the Bolsheviks, and in 1920 fled to Paris with his second wife Vera Muromtseva. Nobel laureate for literature in 1933.

[9] Vasily Ivanovich Chapaev (1887–1919) was a Red Army commander who became a hero of the Russian Civil War. The eponymous 1934 film was based on a novel by Dmitri Furmanov, a Bolshevik commissar who had fought with Chapaev. One (unhistorical) scene in the film has White army soldiers under the heroic White Army general Vladimir Oskarovich Kappel′ charging

film by the Vasiliev brothers, one of the favourite films of my youth. Later, over kefir, everyone praised the film, and then suddenly someone asked which scene made the greatest impression. There was unanimous agreement that it was that of the psychological attack. And indeed, as one watches Kappel′'s men charging directly at the machine guns, chills run up and down one's spine! Everyone began talking at once about this scene, until suddenly a falsetto voice pronounced: "How we mowed them down! Truly amazing!" For a moment an awkward general silence reigned, until a calm voice said with the characteristic Volga way of rounding ah's to oh's: "Why be pleased about that!? After all they were shooting Russia. And when will their like be seen again!?" No one spoke up in support of this reply, however. The speaker turned out to be Soloukhin,[10] and that's how I came to meet him.

We later conversed a few times. Then, when I had returned to Moscow, someone came to me with a message from him: they were planning to restore a cathedral on the outskirts of Moscow. Of course, I agreed to contribute to such a charitable enterprise. Soloukhin himself made an agreeable impression. I liked the way he talked of Russia and our outlooks on life were close. I think that, like me, he sensed Russia's coming destruction and gave much thought to the historical course it might take up to the end of the millennium.... Nevertheless, a real heart-to-heart and true friendship escaped us. There was too much difference between us as people and moreover we moved in different circles. I was put on my guard by his emphatic claim to be a man of the Russian soil. Though myself certainly Russian, I find all demonstrativeness in that respect repellent, while Soloukhin, although raised somewhere among the unpaved country roads of Vladimir Province,[11] was still able when necessary to speak without rounding ah's to oh's!

That company also boasted a "great poet of the Russian earth," as Kolya Dorizo[12] called him, namely Ostrovoy,[13] of first name Sergey, if I'm not mistaken. Soloukhin also apostrophized him thus. I hadn't read any of his poems, nor did I read him subsequently. But I recall very well how he looked, the reasons for which are as follows.

He was rather short, but sported a very tall fur hat, like those worn by the boyars[14] of the time of one of the Tsar Ivans. He went about with a cane and a dignified air. On the sparkling white snow he used his cane to write a woman's name—Zina perhaps,

Chapaev's position in a "psychological attack," that is, without firing their weapons, only to be mown down by Red Army machine guns.

[10] Vladimir Alexeevich Soloukhin (1924–1997) was a Russian poet and writer of peasant background.

[11] Located west of Moscow, just south of the Volga. Its main river is the Klyazma, a tributary of the Volga.

[12] Nikolai Konstantinovich Dorizo (1923–2011) was a Russian-Soviet poet writing on themes of socialist realism.

[13] Sergey Grigorievich Ostrovoy (1911–2005) was a Russian-Soviet poet, likewise of the school of socialist realism.

[14] Highest-ranking members of the Russian feudal aristocracy from the tenth to the seventeenth centuries.

or Liza. Wicked tongues wagged that it was the name of his new lady-love. Thus on every big pile of snow around the sanatorium the name Zina (or Lyuba or Liza) was carved in metre-high letters. But then someone from among the friends of the "great poet" (or perhaps no friend) poured tea in the grooves forming the letters, lending the inscriptions a highly ambiguous meaning thereby. The days continued sunny, the snow was very white, and the name (Zina or Liza) etched in the snow with tea was there to be seen from every window of the sanatorium. The residents of the sanatorium—the House of Creativity, that is—watched, perhaps with sneaking pleasure, while the "poet of the Russian earth" strove to erase the tea-coloured letters with his cane. This was a considerable labour, and it had to be done expeditiously so that prior to Zina's (or Liza's) arrival nothing remained of her name or of the tea. Otherwise, who knows what she might have thought! And day after day it refused to snow.

No less interesting (for me) were the women who participated in the "kefir orgies." In particular, the wonderful actress Bystritskaya[15] was passing the time there, the marvellous exponent of the leading role in *And quiet flows the Don*. A beautiful, so I felt, intelligent, refined and educated woman, only in real life not much resembling the Russian beauty she played in *And quiet flows the Don*. It was Burlatsky who introduced me to her; he seemed to know everyone and even be on intimate terms with them. The three of us went for walks in the parkland around the sanatorium, during which *Madame* Bystritskaya peppered me with questions concerning the details of my work, my army service, and my family. She even asked me about my scientific publications, but there, in Maleevka, I so little wanted to think of such things that I couldn't recall the titles of some of the books I'd published. She thought I was merely being cute, however. Answering Bystritskaya's questions, I imagined myself some aging scrapper, indifferent to everything. I fibbed as never before or since—in a word, had as much fun as I could.

A few days later Bystritskaya left for Moscow. A car had already come for her, and she, exiting the sanatorium accompanied by two of her cavaliers, suddenly noticed me. Leaving her cavaliers, she came up to me and rather acrimoniously said the following, more or less: "I have a part in a play (or film) about scientific researchers, and have to groom myself for the part. But all you could do was play the fool with me." But then, changing her hurt tone to a mollifying one, she permitted me to kiss her hand in farewell. I did not regret what I had done, however: those strolls in the park had been very pleasant, as had been the role-playing I had indulged in—alas, only in Maleevka and only in words.

There was also a sort of literary lady there, styling herself a writer—not too bad a writer, though strictly of the feminine sort, according to Burlatsky. Again it was he who introduced me to her. Unlike Bystritskaya, conversation with her was

---

[15] Elina Avraamova Bystritskaya (born 1928) is a former Soviet actress best known for playing the role of Axinia in the 1958 film of Mikhail Sholokhov's novel *And quiet flows the Don*. Of Jewish ethnicity and a native of Kiev, she has in her later years shown support for Putin's aggressive stance towards Ukraine.

not very interesting, and she was intrusive and unceremonious. Tonya[16] came to visit me twice and each time I drove my "zhigulenok" to Dorokhovo to meet her. We went for walks, and tried to avoid prying eyes and keep to ourselves as much as possible. But it was precisely on those occasions that the literary lady became especially active. Each time I set off to take Tonya to the station, she used some pretext to force us to take her along, even though the presence of a third person was especially inappropriate. In brief, she irritated me, and I wasn't pleased that Fedor Mikhailovich had introduced me to her.

In the dining hall I sat at the same table as Kolya Dorizo, who, when sober, was a very amiable and pleasant fellow. He once asked the literary lady: "Why do you pester Moiseev so? Don't you understand that it's his wife, a young, delightful woman, who's coming to visit him? Why does he need you along?" According to Dorizo, she replied somewhat as follows: "Well, then I can say I slept with Moiseev himself,“ to which Dorizo replied in turn: "Just make the claim anyway." Yes, that milieu was absolutely alien to me. . . .

The time I spent in Maleevka was my introduction to Soviet humanistic society, or, as they used to say, the society of the "creative intelligentsia." I observed them, trying not to be drawn in to their conversations, mostly just listening, and felt saddened. Alas, they were indeed representatives of Russian culture!

How the whole of that Maleevka public was unlike those people amongst whom I usually spent my time! Both the dissident "intellectuals" and the well-to-do Soviet "creative intelligentsia" were so very different from the intelligentsia I knew from my childhood memories or from encounters with the first wave of Russian *émigrés*.

In the Soviet Union we had individual writers, individual artists and composers, but lacked a genuine humanistic society. There were no (or at least very few) genuine historians or literary critics capable of sincerely propagating their own point of view—heretical maybe, but at least their *own*! It will indeed be necessary to re-establish, to resurrect Russian culture, or even to re-create it all over from the beginning, but not by means of that Maleevka public! Without humanism, without a humanistic consciousness, our nation cannot survive. Even as I speak it's experiencing death rattles. But to expect that it will once more take heart with the aid of exiles like Galich or secure literary ladies, darlings of the Party and the government, is more than ridiculous! The genuine Russian humanistic intelligentsia was exiled, dispersed, or liquidated in the 1920s and 1930s, and any survivors were finished off at the various fronts of the Great Patriotic War. The second wave, the "sixties generation," failed to rise to the requisite standard.

In the light of these observations and the thoughts they prompted, I felt all the more esteem for my engineering and scientific colleagues and the former climbing companions with whom I used to spend my leisure time. I reflected that it would in all likelihood be from among such people that a new Russian intelligentsia would be formed—including a humanistic substratum. It must arise! But more than one generation may pass before a new flowering occurs.

[16] Antonina Vasilievna, the author's second wife.

On the Eve of a New Metamorphosis

However, I didn't allow such unhappy thoughts to burden me too long. I mostly tried to hold myself apart and give myself over to the splendidly sunny month of March. The morning frost lasted almost till lunchtime, the surface was perfect for skiing, and all alone I skied several long trails. On returning, I would take a hot shower, lunch, and then retire to my room to read popular stuff, mainly detective novels, of which the library had rather too many: they should have had reading material more suited to writers! In short, I was on holiday—in earnest and for an extended period.

On the bank of the Ruza there was a trade union "house of rest" where skiing competitions were organized. Once I wandered over there, they gave me a number and I skied ten kilometres. My time was 48 minutes and some seconds, just about ten minutes slower than my usual time in my student days. I was nonetheless very pleased with my effort. I was proud of my achievement and took a bottle of Gurdzhaani[17] to drink with lunch. Burlatsky and Dorizo brought along something more significant, and joined me in a celebratory quaffing. But our company was not exclusively male: that same literary lady came over from a neighbouring table to assist with the bottle.

Leisure, marvellous weather, skiing, and of course Tonya's visits did their work: I began to revive and almost feel young again. In any case, the desire to work returned and I even made certain plans. But most surprising was my return to writing verse, as I had done in my youth and in the first postwar years. They were short poems, each of just a few lines, which I have all but forgotten. A few lines have remained in my memory, however. Here are some I wrote down in anticipation of Tonya's imminent arrival:

> Today she[18] comes headlong to us
> With the blush of early dawns
> And lights a blinding blue fire above us.
> Let the blizzards continue howling,
> And the village track be powdered over:
> With the sun's first ray
> Spring has thrown down the gauntlet.

With complete aptness, I titled these eight lines "Spring." I read them to Tonya and, although she had barely been able to make it back after our skiing expedition, she approved. I also was pleased with them. It had been worthwhile committing them to memory, for they very much reflected the state of my soul at the time.

On the occasion of that visit Tonya left rather late since she needed to rest after the skiing. We strolled along the platform of Dorokhovo Station for a good while, our talk consonant with the aura of Spring.

Towards the end of my stay at the sanatorium the weather suddenly changed: a gusty wind got up, bringing clouds, the temperature went up sharply, and it began

---

[17] A well known Georgian white wine.

[18] Spring—or Tonya?

to rain. This was another, quite different, face of Spring, but splendid too, in its own way.

The evening before my leave-taking, I sat for a long time on my open balcony and listened to the rain. As I listened I again began composing lines, which, unfortunately, I have forgotten. However, I do recall four lines I had composed earlier, in Rostov, where I had also been very happy:

> The wind tore at everything, and frolicsome spray
> Burst in with a wet, melting warmth.
> The night stood moist and lively,
> And one had no wish to go back inside.

That also had been a Springtime, and had likewise been windy and had marked the turning of a new page of my life.

In all probability, I was once again reverting to that state of inner uplift which has always given me the strength to carry on. It seemed to me that I was happy once again, and I felt a renewed desire to work. Moscow beckoned.

Next day there were puddles everywhere and the skis were all put away. I had two days left, but the weather had broken and Moscow summoned me peremptorily. Burlatsky had departed the day before, and Kolya Dorizo even earlier, so I had no one to say goodbye to. At the dinner table I saw only unfamiliar faces. I checked out, got in my car, and set off for home. The roads were extremely slippery: under the surface water lay unthawed ice. But this hardly bothered me: I drove to Moscow feeling I'd undergone a metamorphosis. I was sure that my life would now take a new turn. There would be new problems and new people.

By that very summer Alexandrov[19] and I had completed the first variant of a climatic model, and I had established friendly contact with Professor Burley,[20] who was involved in the American climate control program of the time. Despite the Cold War being then at its height, he promised to help me in any way he could.

A year later Volodya[21] went to Boulder, Colorado for a full eight months in order to carry out computations relating to our climate model, on the first American supercomputer "Cray-1."[22] Thanks to that trip, our model was perfected and its first computer implementation realized. I shall have more to say on this below.

[19] Vladimir Valentinovich Alexandrov (born 1938, disappeared 1985) was a Soviet-Russian applied mathematician. One of the creators of a mathematical model of a possible "nuclear winter." He disappeared while attending the Second International Conference on Nuclear Free Zones in Cordoba, Spain in 1985, and his subsequent fate remains unknown. One of his last papers was titled "Man and Biosphere," co-authored with A. M. Tarko and N. N. Moiseev. See also Chap. 10.

[20] Possibly Marvin W. Burley of the University of Wisconsin, Madison, State Weather Bureau Climatologist for Illinois at approximately that time.

[21] That is, Vladimir Valentinovich Alexandrov.

[22] Produced by Seymour R. Cray's company Cray Research Inc. in 1976.

Chapter 9
On God, Philosophy, and Science

Traditions and Doubts

As soon as the researcher ventures beyond the narrow confines of his specialty and begins working on problems arising at the junctures of the sciences, he finds himself confronted by a multitude of general questions of a methodological and philosophical character. And it is impossible to just wave these aside since they determine the choice of a way forward, a list of priorities, and the value of the researcher's activity and, thence even his eventual fate. For it is precisely here that what lies at the core of a person, his or her original endowment, stands revealed. This may serve him or her as a support or function like blinkers limiting the perspective. I myself am grateful for my good fortune and to the education I received at home as a child for inculcating in me from the very beginning an inclination to avoid all dogmatism.

My family didn't count itself very religious. Like all Russian orthodox families, however, we celebrated Christmas and Easter, and my parents sometimes went to church. I vaguely recall Mama trying to teach me to pray, but I no longer have clear recollections on that score. However, Mama's elder sister, Aunt Manya—Maria Alexandrovna Petrova, that is—did have clear memories of this, and subsequently told me how, before putting me to bed, Mama used to try making me kneel before an icon and pray for the health of our nearest and dearest and also for my own. Recalling these episodes with pleasure, Aunt Manya told me that I used to repeat the simple words of the prayer after Mama, but when it came at last to mentioning my own health, would protest categorically. My reasoning was completely logical: Nikitka is already absolutely healthy and one should free dear God from the necessity of thinking of his health on top of that of all the others. He has enough to do without that. It would appear that back then, at the misty dawning of youth, I demonstrated a propensity for methods of optimization: don't make unnecessary work! As Aunt Manya told it, Mama did not especially insist on my mentioning my

N. N. Moiseev, *How Far It Is to Tomorrow...*,
https://doi.org/10.1007/978-3-030-96651-5_9

own name in the prayer, as a result of which it became shorter! And it was at that time also that there developed in me a decided aversion for the standard dogma.

Following my mother's death, my grandmother Ol′ga Ivanovna took over my religious and moral education. Although of the Lutheran persuasion, she nevertheless faithfully attended the Russian Orthodox church every Sunday, and endured standing for the whole of the lengthy sermon. And each time she would try to make me go with her. When I did go, I found those Sunday services very wearisome. I would shift my weight from one foot to the other and think of things of interest to me. Most often I would fantasize, concocting an adventurous scenario of soldiers, shooting, killings and pursuits. Thus attendance at church completely failed to guide me in the ways of peace.

Nor did it imbue me with piety. Nothing of church attendance remains with me except for a memory of long, tiring, and, worst of all, compulsory periods of standing upright.

However, that was not all there was to my religious enlightenment. Grandmother possessed a marvellous book that was much easier to understand than liturgies. It was called *Sacred History* and contained a brief exposition of the Old Testament for children. However, for me the significance of this book lay more in its illustrations than the text. It contained a few dozen engravings by Gustave Doré. Although Grandmother would sometimes read to me from this book, nothing registered on my consciousness; I related to bible stories as to fairy tales, and these I could make up on my own. But when I was allowed to have the book to myself, I would spend hours poring over the illustrations. Doré was a superb artist and his take on the world cannot but enter into the soul of the viewer, especially if a child. In this way the book left its mark on my life.

Sometimes when looking at Doré's engravings, a tremor of terror would run through me at God's pitilessness and brutality. How cruel He is even to his own chosen people. What then might He do to us whom He has no need of?! To us who are not of his people, who are his enemies?! I came thus to believe in a God with the form of a remorseless and relentless force. I even began sleeping badly: I dreamt of a Jehovah given to punishing me just because I was not Jewish. What should I do, how save myself? Might I become a Jew?

These thoughts not only worried me but actually made me anxious. I complained to Grandmother and she clarified matters for me: the bible tells us about Jehovah, but He is the God of the Jews, and we have nothing to fear from him. We have a Christian God, the one we pray to, and He will always come to our defence. The upshot of her explanations was that I understood nothing at all, and my mind was turned to mush. What's the real truth of the matter? Who is God? Why is there more than one of them? I put such questions to others as well as Grandmother but reasonable answers were not forthcoming. The adults tried to persuade me not to ask silly questions, that I would understand for myself when I grew up, that I would understand that there's only one God.

The conclusion of my religious education was negotiated safely: the book with Doré's illustrations was taken from me, and obligatory attendance at church waived except on special religious holidays. Jehovah gradually receded from my mind, along with the Christian God, whom the Jews had crucified even though He himself was Jewish.... I ceased pestering adults with silly questions to which they were unable to give intelligible answers, and everything went along smoothly once again.

But soon I began asking them "clever" questions again, only this time about the theory of relativity. When I heard that each individual has his only personal time, I asked: "Could it be that God is time?" And of course I had no more success than before in getting a satisfactory answer. But that's another story.

A small spike in my religious feeling occurred following the death of my father. That year Easter came very early and throughout Lent we had splendid sunny weather. In the early morning there would be a light frost and later in the day a steep thaw. My stepmother was grief-stricken, and went to the church whenever she had the time to spare. Since she taught at the school, her fervent religiosity was considered reprehensible. She even had to endure unpleasantness there, which, however, did nothing to prevent her spending as much time as possible in the church. I occasionally accompanied her thither. I was then in my fourteenth year, and I had a very conscious wish to discover what prayer does for a person. I observed the beneficial effect the church service exercised on my stepmother: she came home from the church with a lightened burden of grief, somewhat solaced, and sat down to mark her students' exercise books. I myself, however, never learned how to pray. Although the family was stricken by grief—Grandfather's death soon followed that of my father—and descended thereafter into an abyss of poverty and misfortune, I never learned how to ask anything of God. "If You exist, then help me believe in You"—that was the only thing I felt like asking of Him back then. A very sincere desire to believe in God has remained with me throughout my life, remaining fruitless, however.

And also when at the front, in moments of danger I, unlike many, never resorted to prayer nor called on God for help.

But now that I think of it, I recall a single occasion when I *did* ask Him for help. As a result of an absurd convergence of circumstances, I, together with a technician from the airport service battalion called Benenson—Lieutenant Benenson, a funny little Jew from Belorussia—blundered into a minefield. We were going through a snow-covered field when suddenly a mine exploded, leaving us both unharmed except for a splinter that had penetrated my felt boot and lightly scratched an ankle bone. Nothing serious, although there was plenty of blood. Then when we examined our surroundings closely, we saw that there were mines all around us. By sheer good fortune we had walked 10 or 20 m into the minefield without triggering a mine, or rather just one mine, which was still rather lucky. We froze: a step forward or backward could mean death. My felt boot oozed blood and my vision rippled. Benenson kept repeating: "Just don't pass out; eat some snow." It was at that instant that I pronounced under my breath the words "God, help me if You can, and I will merit it." During those terrifying moments I several times repeated that blasphemous prayer to myself.

I should indeed have been grateful to God, but more for something else, namely the fact of my having been taught some of the elements of mine technology in the Faculty of Aircraft Armaments. Kneeling down and clearing away the snow, I soon saw that the mines were of the same 1905 or 1910 vintage used to demonstrate to us what mines are and how to set and remove them. Another amazing bit of luck or one more blessing from on high!

To put it briefly, in the space of a few minutes which seemed like hours to Benenson and myself, we managed to extricate ourselves from that perilous minefield. I was sent to the medical unit, where I was immediately inoculated against tetanus and given a change of underwear. But that likeable, homely Benenson, who had been so afraid I would pass out, was killed the very next day, either by one of those random shells that sometimes came near or else in one of those absolutely non-random bombardments we were quite often subjected to.

Laplace's Principle

Although I never became a believer, neither, on the other hand, did I embrace atheism. It seemed to me that in that field of enquiry, lying as it does beyond the limits of our intellect, any categorical assertion is out of place. Nothing is provable. There is no logic that might help us answer the eternal question. Everyone answers it for him- or herself independently.

With this attitude I lived at peace with myself for many years. I considered myself Russian Orthodox not according to any religious belief, but in the sense of belonging to the family tradition in which I'd been brought up. Whenever I happened to be abroad on a religious holiday, I would never pass up the opportunity of going to a Russian orthodox church if there was one nearby, but not so much from a desire to attend divine service as out of curiosity: I wanted to observe the people gathered there in the church, eavesdrop on their conversation, and sense the spirit of their world. Occasionally I would meet interesting people in this way.

I understood that I am a person outside the confession, however, that I'm not a "theist." The following anecdote concerning Laplace is well known. At the beginning of the nineteenth century Laplace wrote his famous book[1] containing the first cosmogonic hypothesis,[2] now known as the Kant-Laplace hypothesis. He presented his masterwork to Napoleon, who is said to have actually read it. (Strange times they must have been. Was it that there were few books worth reading, or did the emperor have nothing better to do? Perhaps emperors were different in those far-off days.) Not only did Napoleon read the book, but is said to have reported back to its author in approximately the following terms: "My dear Marquis, I have read your book [on the system of the universe], and found it interesting and clever. However, I found no mention of God its creator." To this Laplace is supposed to have answered

[1] The five-volume work *Mécanique Céleste (Celestial Mechanics)* was composed over the years 1799–1825, earning Laplace the sobriquet "the French Newton."

[2] The so-called "nebular hypothesis," according to which the planets of the solar system were formed by condensation out of a rotating cloud of particles and gas.

bluntly: "I had no need of that hypothesis, my Emperor." Now although Laplace was no theist, neither was he an atheist. It was simply that the concrete details of his exposition required no religious concepts. This is precisely the stance towards religion I consider absolutely appropriate for a mathematician, as well as convenient enough for one living in our disorderly epoch. I adopted it and stopped ruminating on God and religious themes generally.

There came a time, however, when I had to re-think many things, in particular, to make my position on religion more precise for my own sake.

My studies of the biosphere and evolutionism led me to problems concerning humanity. Each man or woman is made of flesh and blood, and, like other organisms, arose as the culmination of an unimaginably complex evolutionary process acting on living matter. Unlike all other organisms, however, a human being has access to a spiritual world, and, responding to its call, is able to frame questions, consciously set goals and strive to achieve them. A human being's consciousness is largely preoccupied with two very different cardinal questions. The first is the question "How?" How does this or the other event take place, how does an arrow fly when released from the bow, how do separate atoms combine to form this or the other substance? And so on and so forth. And from attempts to answer the infinitude of such questions continually confronting humankind, there emerged science, its greatest creation.

It's true that this question also appears, albeit in rudimentary form, among the "higher" animals. In knocking a banana down with a stick, a chimpanzee is solving a certain complex question "How?" In finding a means of achieving the desired goal, the chimp brings to bear not just intuition but also an elementary kind of reasoning. This, however, is far from indicating the presence of a spiritual world.

The emergence of a spiritual world is also, it seems to me, the result of a universal evolutionary process. Elemental matter has nothing spiritual about it. Only at a certain stage of development of sensate and intellectual organization did the phenomenon emerge, perhaps unique to humankind, of what we call a person's spiritual world. This is a synthesis of the sensate and the rational exceeding a certain threshold of developmental complexity. From the viewpoint of his or her spirituality, each person has a certain picture of the world apprehended as a whole. In contradistinction to logical constructs, however, the images born out of a person's spiritual world obey no final standard. The great variety of human beings' subjective representations of the surrounding world, differing so much from the reflexive single-valuedness of the reactions of other species, is one of the characteristic features of the phenomenon of man.

And then, beyond a certain stage of development of sensate and rational apprehension, at last the second question occurred to mankind: "Why?" Why does every existing thing—the sky, the earth, water, and, finally, oneself—exist? There is no logic or science capable of giving any kind of satisfactory answer to this question. Even though it may seem contentless, this question inevitably occurs to every human being at some stage in his or her life. Sooner or later it percolates up to everyone's consciousness.

The indefiniteness of the pictures conjured up by people's spiritual worlds, their intrinsic multiple-valuedness, the absence of any logical answer to the burning question, and the role of the sensate element in our representations of the surrounding world—all of these considerations taken together lead to some sort of idea of a supernatural power. And the emergence of such an idea is at the same time the source of belief in that power. Different established religions are the outcomes of disparate historical developments, but religious feeling, the soil in which belief in a supernatural power grows, is apparently intrinsic to all mankind, and will remain so as long as mankind exists in the universe. But whether a given particular person possesses this feeling is another question altogether.

I have already mentioned that, as a result of my upbringing, or perhaps of certain aspects of my nature, I became neither an atheist nor a genuine believer. Most likely my mathematical education at the university also played a significant part in yielding this outcome: I lacked the necessary conviction of the truth of either belief or atheism. On the one hand, I was unable to deny the existence of a Higher Power, that is, take as not existing something I knew to be beyond my powers of reasoning to conceive of, while on the other hand, I had no internal need of its existence. All the same, I have always harboured an element of doubt. It's likely that many view the problem from the same angle as me, adhering to one or the other religious confession while perhaps even considering themselves atheists. In my case, the religious confession in question is, of course, Russian Orthodox Christianity. This explains why I am never surprised by the peculiarities of other spiritual worlds. I have learnt to have the deepest respect for others' spiritual worlds, as indeed for any sincerely held conviction, and never allow myself to interfere in them or judge them.

Thus I am absolutely convinced that science is compatible with religious belief, and even more so with religious feelings. These two domains are not at all in contradiction with one another, but simply reflect the essence of the above two cardinal questions, the first of which arose from humankind's practical activity and the requirements programmed into our genetic code pertaining to the preservation of the human species, while the second arose in concert with the emergence of human beings' spiritual worlds and reflects certain special properties of the phenomenon of man not yet accessible to us. It is my belief that a person who combines a talent for scientific creation with a belief in something transcending us cannot but be genuinely happy. There are very few, however, who have this happy gift. People like Goethe or Pavlov are few and far between. As for me, fortune (or God) saw fit, alas, to confine me to a life in the world of reason.

I have known people who lived primarily in the world of feelings and belief, and observed how immeasurably happier they were than myself and all those kindred souls who dwell on a different, non-intersecting, plane. For this reason, sometimes when I am very downcast I offer up the blasphemous prayer composed in my early youth: "God, if You exist, help me to believe in You!"

My Picture of the World

In studying the biosphere, in particular its evolution as a single whole (or as a "system," as they say nowadays), I was led inevitably to the need to sketch a sort of "picture of the world" for myself, containing both the biosphere and humankind. I felt compelled to work out my attitude to such fundamental principles as, for example, reductionism, whereby the complex is reduced to its simple components or otherwise simplified, and I eventually became convinced that the world is far more complex and incomprehensible than is usually believed by physicalists and the purveyors of the natural sciences such as myself and my fellow scientists.

Describing my worldview in detail would take me too far afield, and would in any case be out of place in such a book as this. I will, however, describe some parts of that worldview, while continuing to hope that someday the means might be found for publishing my lectures on "universal evolutionism," where my views are expounded in much greater detail.

The sketch of my picture of the world I offer here will also be needed, albeit considered from a different standpoint, when I return once more to the theme of a Higher Intellect.

Fundamental to my world picture are "empirical generalizations," to use the very broad term invented by V. I. Vernadsky[3] to denote assertions not in contradiction with our empirical knowledge: Whatever is, is!—from the point of view of the physicalist, at least. Empirical generalizations allow us to distinguish what is unknown from what is known and provide a foundation for various logical constructions. As a concept they may sometimes seem jejune, but on closer consideration one discovers that with their aid a different way of viewing an object can be inferred and thence a fresh interpretation of it.

I take as the first assumption built into my picture of the world the following: that the whole of the world in which we find ourselves, the whole universe, or, as Teilhard de Chardin put it, the "Universum," is some kind of system. This would seem to be a completely trivial assertion, even banal, since every material object is indeed connected to every other, for instance by the gravitational attraction between them. This empirical generalization in no way contradicts our experience; in fact it's impossible for it to do so for the simple reason that if it were false we could never discover that this was so!

The non-triviality of the foregoing assumption appears only when we take note of the fact that the connections in question should only be such as are accessible to empirical observation. This single empirical generalization thus entails the delineation of that region of my world picture consisting of everything that can become an object of scientific investigation. It is then natural to attempt to trace the possible consequences on the boundary of this region resulting from the postulated situation in the interior of the region, as well as interpretations of these.

[3] See Chap. 6.

Thus for instance if the Universum is a single whole, then it must develop or evolve as a single whole, and everything accessible to observation by us, including us ourselves, is just a component part of the Universum. From this it follows, in particular, that at a certain stage in the development of the Universum there appeared component parts capable of knowing the Universum within certain limits. We shall call the totality of such component parts "the [combined] instrument of the Universum's self-awareness." The aforementioned limits will then depend on the degree of perfection of the instrument of the Universum's self-awareness achieved over the course of the evolutionary process. Humankind is such an instrument of self-awareness, possibly merely one among many. The question then arises of the extent of humankind's ability to know the Universum, an ability which emerged during the evolution of the Universum as one of its features, and continues to evolve. Speculation about this question leads to some of the deepest problems of philosophy, and not only of philosophy but also of practice. For if man is capable of recognizing certain special features of the evolutionary development of the world, then he is capable of influencing that development, thereby extending his knowledge.

That knowledge may, however, be circumscribed by certain clearly defined peculiarities of the evolutionary process intrinsic to the particular "instrument of self-awareness" we call man. The following analogy may clarify what I have in mind.

The members of a certain species of octopus turn out to have brains of complexity comparable to that of human brains. It follows that these our fellow creatures may also form part of the Universum's "instrument of self-awareness," a part arising by means of a process of self-organization of matter different from that of humankind. However, that process happened to endow these particular octopuses with a strong proclivity for cannibalism, resulting in a high probability of such an octopus being eaten by its offspring.[4] In any case, this biological peculiarity renders such octopuses incapable of forming a collective memory since each new generation has to begin from scratch. Might there be an analogous threshold which, by his very nature, man is unable to cross? Might his aggressiveness, for instance, be such a limiting characteristic, the tendency to aggression that man has inherited from those of his ancestors living in pre-ice age epochs, when it was a prerequisite of survival?

The biological process of improving mankind came to an end in the epoch of the sabre-toothed tiger and cave-dwelling bears, when the struggle for survival was unrelenting. Perhaps the real tragedy of humankind consists in the fact that its morphological evolution came to a halt too soon![5]

Conceiving the Universum as a single system leads us to a new way of looking also at other features of the world. We are used to speaking of this or the other "object under investigation." However, this implies that we have the ability to single out the object in question from our system, to somehow sever the connections

[4] Presumably there is some evolutionary advantage to this trait that would explain how it arose.

[5] Relatively recent research based on an examination of the human genome has shown that in fact significant human evolution continues to this day. See also Chap. 6.

linking it to the rest of the world. The multitude of actions on the object stemming from that part of the Universum outside it are then considered "external" to the object of investigation at hand. However this inevitably involves neglecting the reactions of the object on the rest of the system, that is, on its elementary components, which in turn entails neglecting the change in the "external actions" on the object resulting from those reactions. Can the resulting recursive feedback loop always be terminated?

The questions then arise as to when this recursion can be truncated and under what circumstances it is fundamentally impossible to do so. When is the object under investigation crucially just a conditional conception? These are complex questions which, throughout history, have simply been ignored, since in practice the associated difficulties were negligible and moreover most people adhered to a narrow form of rationalism. For most of history human beings have been mere observers, capable of registering only certain special features of the world brought into being by the stupendous mechanism of world-creation, and the problem associated with delineating an object of investigation was left out of consideration. Even today most scientists scarcely recognize its significance. The entomologist, for example, once he has captured a butterfly, places it under a magnifying glass and studies only what interests him, and that's how it is with most other branches of science. Scientists in most fields consider this the only feasible approach to getting results, and in any case it has always been assumed, essentially axiomatically, that influencing the way the grand mechanism of world-creation works is far beyond mankind's capabilities.

But then suddenly this turned out not to be exactly the case. Experiments carried out in the twentieth century unexpectedly demonstrated that there are practical situations where problems associated with the above question definitely need to be taken into account. I'm referring here, of course, to certain experiments of quantum mechanics.

When at university, I took a reasonably good course in physics. Those parts of the course taught by I. E. Tamm was especially noteworthy. The course was not obligatory for mathematics students, and of the thirty or so mathematics students of my year only Oleg Sorokin and I were enrolled in it. For that matter neither were there very many physics students in the course. It wasn't like a standard course of lectures, but of a domestic character, as it were, more like a seminar. We were often distracted from the main themes by certain side issues, and useful and memorable discussions would then ensue.

On one occasion Oleg brought along a paper of Heisenberg containing the assertion that the investigator could never be separated from the object under investigation. This came as a great shock to all present. We students argued heatedly, without, of course, understanding what we were arguing about. And our teacher,[6] who had just finished expounding the formalism of Schrödinger's equation to us, or something of the sort, also found himself at a loss to explain such a statement,

[6] Apparently not I. E. Tamm on this occasion.

coming as it did from one of the fathers of quantum mechanics. All he could do was repeat the then standard dogma to the effect that an electron is always an electron, with properties independent of any observer, and that what Heisenberg was claiming was out-and-out idealism not worth further thought. Oleg and I both tried to protest, but in vain. So the question was left hanging, and, what was especially galling, I subsequently even found I had difficulty formulating it clearly.

Many years later, with the alarum of war already behind me, I found myself having to give lectures on certain topics of theoretical physics, and was thus compelled to return to the problem evoked by Heisenberg's provocative statement. It was then that I came to an understanding of the situation as I described it above. The isolation of any feature of the world is always conditional. Said feature is always a part of the whole, part of a different, more complex system, from which, in some cases, it cannot be isolated under any conditions whatsoever. That is why an object can never really be spoken of except in the subjunctive case. There is no such thing as an electron by itself, a self-contained entity. (Incidentally, the same goes for a person—as the great Sechenov[7] well understood: a person exists as part of a single whole of flesh, spirit, and surrounding nature, as he put it.)

In pronouncing the word "electron," we can only really have in mind a certain definite interpretation of a certain phenomenon, that is, only a certain description of the phenomenon in words, and the discovery in a given context of the relevant meaning of the word "electron" is what science is concerned with. The meaning depends on the context, which includes the observer. An electron together with an observer equipped with a Wilson cloud chamber where the track of the "electron" is visible, is, as indeed for every material particle, one system, while an electron taken together with a diffraction grating, where it behaves like a wave packet, is another, different, one. But separating out, or isolating, the electron as something self-contained is impossible. And a person observing—just observing—what is happening, is already interfering in the process unfolding before him, thereby changing how it proceeds. Even if only by an infinitesimal amount, change nonetheless occurs! It is impossible to understand the forces of nature hidden in the cores of atoms without taking this into account.

From this we infer that the basic paradigms of rationalism, primarily the principle of the independent observer, need revising. Quantum mechanical experiments have demonstrated this unequivocally, and without quantum mechanics there would be no atomic bomb. It follows that we have to investigate every event "from within," that is, from the viewpoint of a participant in the event, taking into account our influence on it and our limitations, themselves formed according to the "laws" of self-development of the Universum. But this rejection of eighteenth century rationalism—or, more precisely, transition to a new rationalism—does not in the least entail a loss of scientific rigour. It's simply a matter of approaching science in

[7] Ivan Mikhailovich Sechenov (1829–1905) was a Russian physiologist, called by Pavlov "the father of Russian physiology." Through his classic *Reflexes of the brain* he introduced electrophysiology and neurophysiology into research and teaching laboratories.

a new way. Niels Bohr, a giant of twentieth century thought, asserted that no really complex phenomenon could be fully described in words alone: the characterization of a single phenomenon requires its examination from a multitude of perspectives. I would like to express this idea a little differently, however. For a person to have an appropriate understanding—which is not at all the same as knowledge—of a phenomenon, he must possess a certain "holographic" representation of it, formed from a variety of interpretations. The task fundamental to a modernized version of physics, and not just quantum mechanics, then becomes that of assembling such interpretations from raw experimental data with which they are in agreement. Note that this approach in no way resembles the process of successive approximation to a mythical "absolute reality" postulated by Hegel and, in his wake, the theorists of dialectical materialism.

The discoveries of twentieth century physics are, moreover, also of relevance to the humanistic sciences. New knowledge, and with it the associated new dogma, changes people's consciousness and hence their behaviour, which in turn leads to a change in historical development. Heisenberg's statement concerning the inseparability of observer and observed thus becomes a fundamental principle of universal import. It's then up to us to learn how to live in this strange world where relativism reigns, and try to extract from that world information facilitating our existence in it, information without which we cannot survive.

Thus it was that one day I began to see science in a new light.

It would appear, therefore, that the rather innocuous empirical generalization concerning the wholeness of the universe and its systemic character entails a revision of many of the basic views that earlier had seemed to me elementary truths given once and for all. The very concept of truth had now to be reconsidered: truth for whom? There is no such thing as absolute truth, after all! We find ourselves in the midst of the chaos of Creation, where we are able to register something of what takes place in our "immediate vicinity" and analyze observed dependencies, but must not take these or our knowledge of them as absolutes. Nevertheless, that negligible particle of Creation called "man" has shown himself capable of extracting an improbably large amount of information from the chaos of creation and, by putting this knowledge to use, changing the conditions of life, and thereby himself and history ... and perhaps even noticeably influencing the entire course of development of the Universum. Who knows?

This represents a new anthropocentrism, no less grand than the conceptions of our forebears. Like the ancient Greeks, we gaze on Olympus, hoping there is a place awaiting us there—in principle, at least.... The way to get there eludes us. We ask in vain whether mankind even has the requisite stature to be worthy of occupying such a place. And doubtless the mystery will remain as long as humankind exists.

As the second empirical generalization basic to my world picture, I postulate that at the fundamental level of creation, at the root of all processes going on in the Universum, there is randomness and uncertainty.

It is indeed an experimental fact that we have no absolutely certain knowledge. While in quantum mechanics the outcomes are explicitly probabilistic, the randomness and indeterminacy at that submicroscopic level make themselves felt at the macroscopic level no less compellingly. The intensity of mutagenesis, for example, is such that it determines many of the characteristics of the newly born individuals of a species. These owe their genetic differences to the contemporary population of members of the species, whose success in continuing and improving the species in the face of changing external conditions depends on their ability to adapt to those conditions through selection among mutations.

Similarly, different people have very different mentalities and different spiritual worlds, as a result of which they construe one and the same situation in different ways, and under identical conditions come to different decisions as to what to do. Thanks to this great variety of human individual types, our species has been able to transform the whole globe into its own ecological niche and survive the great cataclysms of its history.

There are various attitudes one may take to the fact of the underlying randomness of the universe's processes. One may attempt to explain it, or even refuse to accept it, as did the great Albert Einstein, averring that "God does not play dice." But a fact is a fact, and there's nothing one can do about it. God *does*, after all, "play dice!" Without that "playful" element nothing would have turned out as it has!

All the same, we remain indebted to determinism since it was fundamental to the new mode of thought that began its precipitous career at the time of the Renaissance. It is to determinism that science owes its greatest successes and to its potency that we owe civilization. Even today a great many outstanding thinkers and scientists continue to adhere to determinism's precepts. The whole of the theory of dynamical systems, for instance, in particular catastrophe theory, rests on the classical notion of determinism. The outstanding French mathematician and philosopher René Thom even went so far as to equate scientificity with determinism, in the spirit of the nineteenth century.

Life is more complex than any scheme, however, and today life shows us that if we wish to explain what's going on around us, in which we figure as direct participants, it is simply impossible to avoid using probabilistic concepts, i.e., endowing our laws of nature with stochastic interpretations. It is my belief that it will always be thus. Randomness lies at the heart of things: this is precisely the assertion as I originally expressed it when I apprehended it as one of the fundamental empirical generalizations germane to the picture of the world used in my investigations.

To repeat: the probabilistic nature of the world may be interpreted in various ways. It may, for example, be interpreted as a measure of our ignorance of reality or our inability to carry out sufficiently fine analyses. Contemporary science cannot remain satisfied with such answers, however. Ten years ago the American mathematician Mitchell J. Feigenbaum, using a computer implementing the algorithm of successive approximations to investigate the limiting behaviour of successive, fully deterministic, solutions of a rather simple equation, discovered that serial iterations exhibit behaviour very close, in fact indistinguishable from, that of a certain random process. There were analogous such phenomena known earlier, of course, arising in connection with random number generators: by means of an appropriate deterministic computer program, one could generate sequences of numbers having all significant properties of sequences of random numbers. Such considerations prompt further reflections, about which I shall have something to say below.

Although there are other ways of interpreting and explaining away the emergence of randomness in physics, as far as I myself am concerned—someone who during his university years was given an unusually large dose of probability theory by the mathematics faculty and who in his youth felt a rush of delight at becoming acquainted with quantum mechanics, the greatest of all mankind's scientific achievements—it seems more natural to accept the stochasticity of nature as inherent. Thus I take the randomness we have detected in nature to really be there "in fact."

I took this view dogmatically, without thinking about it overmuch, quite some time ago, when I was in the Zhukovsky Academy working on the theory of the scattering of projectiles. It was reinforced when I began teaching at Rostov University, especially after being given the task of leading a seminar on the methodology of physics featuring, in particular, a critique of the Copenhagen School. The upshot of the "Party directive" I was given to criticize this as a "bourgeois perversion of physics," was that I became a staunch adherent of the ideas of the Copenhagen School and counted Niels Bohr among the most influential of my teachers. Back then my conviction of the truth of the position adopted by that school and my belief that that position could be justified by means of correct philosophical and physical arguments, almost cost me by Party membership.

The Secret of the Question "Why?"

It has always seemed to me that the most astonishing and puzzling feature of our world is the existence of what exists. I spoke of this above in connection with the question "Why?" If we take this question to be an insoluble mystery, then we are tacitly admitting the possibility of randomness in the world. For in addressing the question, we affirm the existence of randomness in the world and thereby allow for the possibility that the laws of nature have a statistical character requiring a suitable probabilistic language for their description.

The question "Why?" is concerned not with the random chaos of chaoses, however, but with the observed laws of nature. Science originated, after all, as a means for humankind to foresee the consequences of its actions, and in that connection it would seem that determinism was crucial: if A causes B and B causes C, then A causes C, for instance. If there were nothing but chaos and unpredictability, there could be no science. The reconciliation of stochasticity and determinism, the idea that the difference between them might be small, long defeated me, until Feigenbaum's examples provided me with the needed evidence for how things actually stand. On the other hand, if we were to reject unpredictability in principle, we would thereby also be rejecting the possibility of anything qualitatively new taking place in the world, and this would so limit our horizons that we would perhaps lose interest in pursuing science. This is another reason, based purely on emotion, for my never being able to accept classical determinism: living without the possibility of anything unexpected occurring would doubtless be uninteresting, even boring!

Furthermore, science makes sense only if we take what we are investigating—which might be anything that exists—as really existing. We are thus brought back to the empirical generalization from which we derived the fundamental fact of the randomness at the root of what exists and so also of what we can investigate.

The anthropic principle made a very great impression on me when I first heard of it. In essence it goes as follows: If the universal physical constants, such as the speed of light in a vacuum, the universal gravitational constant, etc., were different from what they in fact are by even so much as a tenth of a percent, the world would be altogether different from what it is. Stable forms of matter could not arise, so there would be no evolutionary process like the one that in our world gave rise to the stars, planets and living creatures, including human beings. The universe would evolve, but in a totally different way, and, what is most important, without observers, without witnesses. Physicists—in the present world they are the chief trouble-makers—formulate the anthropic principle slightly differently: the universe is as it is because we, i.e. people, exist in it!

As I have formulated it, the anthropic principle shows that the development of the Universum proceeds along a razor's edge. For a system as complex as ours, the danger of structural change or catastrophe looms large.

Although the literature on the anthropic principle is substantial, it has been of no practical significance. It is, however, of very great general philosophical importance. For me personally it has been crucial as an influence on my deliberations and investigations over the past two decades.

In studying the stability of complex systems, I repeatedly encountered one and the same peculiarity: while it's true that increased complexity of a system brings with it diminished stability, all the same at some stage of increasing complexity there occurs a reduction in the level of instability, when within the framework of the system a new mechanism arises stabilizing its development. Take, for instance, the system represented by a population of some species of animal, which on the face of it might not be expected to be stable. Here, however, the process of replication, that is, of self-reproduction, is made imprecise through random mutations, and it is this very imprecision, which might superficially seem like a flaw of the system, that gives rise to a peculiar mechanism in the form of a feedback loop, thanks to which the population preserves its systemic character and its ability to preserve its integrity in the face of the difficult conditions of a changing environment.

The appearance of man, in particular, may be considered, insofar as it represents the realization of a collective intellect, as a specialized mechanism with the potential of introducing stabilizing forces into the system of the world. According to the anthropic principle the universe should be a completely unstable system. But is it possible that the collection of observed values of the fundamental parameters is precisely what is needed to eventually lead to the formation of a stabilizing mechanism?

Over the course of development of the Universum, instruments of self-awareness have arisen. The brain of a certain species of octopus is one potential such instrument. This potential has remained unrealized, however, because that species was unable to form a collective memory, with the result that it came to an evolutionary dead end. But we have another such instrument, namely humankind. Although the brains of individual members of this species already stopped developing some tens of thousands of years ago,[8] it has nonetheless proven able to create a collective memory and a collective intellect, which continues to develop at an ever-increasing tempo. How far will this development take us? To this we can give no answer. Perhaps in some other part of the Universum a similar process has gone much further, to the extent that it begins to make sense to speak of the "Intellect of the Universe." Then, however ... the problem of a universe-wide "deterministic program generating random numbers" or some sort of "universal computer" will look quite different.

When one ponders such ideas, one involuntarily finds oneself falling under the spell of such dreams as are spun by the likes of Stanislaw Lem or Ray Bradbury.

But above and beyond all of that there is only overarching doubt and the two questions "How?" and "Why?" I've been discussing, and my childhood prayer.

[8] See earlier footnotes in this Chapter and Chap. 6 concerning recent evidence for the continuing evolution of humankind.

Chapter 10
The "Nuclear Winter" Epic and My Subsequent Retirement

Another Metamorphosis: The Biosphere and Society

Although my research concerning a "nuclear winter" was at most of secondary significance for the large work I had planned and begun writing at the turn of the 1960s into the 1970s, it was precisely my description of a nuclear winter which, originally of little interest to me, achieved general renown, thereby providing significant publicity for the kind of research towards which I was at that time re-orienting the Computing Centre of the Academy of Sciences of the USSR. Up till then the most interesting of the scientific results in the general area of the relationship of human beings to the biosphere, or concerning special features of the self-organization of the material world, had simply remained unnoticed, if not misunderstood. This situation is rather typical in science: far from everything the researcher regards as most important in what he or she has done is so regarded by others since the point of view of the researcher tends to be very different from that of his reader.

The period covering the end of the 1960s and the following few years was perhaps the most intensive and productive of my life. By then I no longer felt the attraction, so irresistible to younger mathematicians, of overcoming the purely technical difficulties involved in proving one or another theorem. I already had a sense of the conventionality of "rigorous science" and claims of "absolute knowledge." I was feeling a greater affinity for the content-rich sciences of nature and the humanistic sciences, and syntheses thereof. Something of the sort happens to all scientists past their prime, perhaps, when their sporting ardour has abated, yielding to an aspiration to discover the "essence of things," to gain clarity, to reach a deeper understanding of some mystery straddling the border between reason and feeling. Perhaps it was a similar desire for clarity—clarity for one's own sake—that stimulated the meditations leading finally to the lucid apprehension of the world attained by the church fathers. It was precisely this lucidity, acquired for personal reasons, for the sake of their interior world, of their

N. N. Moiseev, *How Far It Is to Tomorrow...*,
https://doi.org/10.1007/978-3-030-96651-5_10

internal equilibrium, that gave them the strength to continue living, and attracted and continues to attract to them—even now!—people caught in the toils of everyday living.

When I began working on problems to do with the evolution of the biosphere, and the interrelationships between its developmental processes and those of human society, I soon sensed that I was touching on the holiest of the holy and on the verge of glimpsing something that had earlier been quite inaccessible to me. These new concerns filled my life to the exclusion of all else, and my considerable administrative duties as deputy director of the Computing Centre, which I had borne for the previous quarter century, began to seem onerous. Although the Centre's director, Academician Dorodnitsyn, was very able and a talented researcher, on a personal level he tended to be extraordinarily supercilious and even unkind, and his view of the world was totally alien to me. I began looking for a way to change my social status. This was all the more appropriate inasmuch as the new interests that were reconstituting my life were distancing me from my former activities and from the people I had formerly been close to.

I was getting on, approaching the end of my sixth decade, and understood that I was setting out on a completely new stage in my life, with new values which had yet to be made clear. Back then, in the 1970s, unlike the 1990s, I was materially secure, and with my thousand-ruble monthly salary didn't need to moonlight. I looked for an excuse to be released from my administrative duties, which were preventing me from thinking about the things of interest to me and taking up time needed for investigating the many questions that had become urgent to me. It was not till the mid-1980s, however, that I managed to extricate myself from my official position, thanks to a decree that members of the Academy should have the right to retire from their positions at full salary and with the honorary title of counsellor. It seems I was the first Academician to so retire, at my own initiative, long before normal retirement age. Once quit of the Computing Centre and the department, I embarked on a new period in my life, replete with fresh interests.

I cannot say, however, that this period was a very fortunate one for me since, in the first place, most of the projects I planned over those years remained just that—merely plans. And my parting from the Computing Centre proceeded not entirely to my liking since I could brook no delay in my leave-taking!

But there were many other reasons for not being able to consider this period of my life a lucky one. There were objective reasons, such as the unknown fate of Volodya Alexandrov,[1] the loss of contact with certain of my colleagues and students, the beginning of *perestroika*,[2] along with the associated ending

[1] See Chap. 8.

[2] The period of *"perestroika"* (restructuring) in the USSR was initiated by Mikhail Gorbachov in the mid-1980s, involving, in particular, the decentralization of economic controls and the encouragement of self-financing free enterprise.

of budgetary allocations of resources to research on the modelling of processes of the biosphere, etc., but there were also more basic subjective reasons. The work I had planned would require the collective efforts of a large group of researchers, each of whom, while contributing something of his own to the collective, would need to have an understanding of the project's overall aim and be able to act in synchronicity with the other team members and with me personally. It was precisely this kind of distribution of work that I proved unable to organize, however. And the reason for this lay predominantly with me personally.

The fact is that I coped well with administrative work when it was a matter of giving direction to the activities of a team and then choosing an appropriate person to lead an investigation on a chosen theme or problem. Once this was done, I would usually step aside, fully entrusting the work to the chosen delegate, observing progress from a discreet distance and trying not to interfere. I continue to hope that my junior colleagues hardly noticed my monitoring of their work. I would say that in a goal-oriented organization, this would, generally speaking, be the right approach, and all the more so when its output is of an engineering character, when its nature is substantially determined beforehand, or when the goal is a very concrete one. My choice of this strategy, however, was made largely for personal reasons, for it was in line with the way I liked to work: I valued independence, detesting all interference or intervention, without which I felt I worked much more effectively.

That I was able to endure working as A. A. Dorodnitsyn's second-in-charge for more than twenty years—something my friends had difficulty understanding—can be explained quite simply: in all that time the director never once meddled in the work I was engaged in. Doubtless it suited him that I fulfilled my own duties effectively without encroaching on what he considered his personal prerogatives: foreign connections, the overall strategy for the development of computer technology, and his role as representative of the Centre. He always regarded himself as a director of the old Stalinist guard: everything that went on in his institution was to be considered his personal achievement. And when at last he did change tactics and began interfering in my work, I put in a request to retire.

I believe that the way I organized the research activity at both the MPTI[3] and CC was effective enough. On the one hand, I set the goals of each collective's investigations, thereby giving direction to their work, while on the other I yielded pride of place to the initiatives of my younger colleagues as soon as there was clear evidence that new results were imminent. Although such a format for the activity of a research institute is, of course, not standard, it would seem to be rational enough, even if it means that the supervisor has to be perpetually on

[3] The Moscow Institute of Physics and Technology. See Chap. 5.

the lookout for appropriate projects instead of acting merely as finger-pointing director.

But when you yourself are deeply involved in the problems under investigation, when these are tending at the time in question to even become one's *raison d'être*—however infrequent such a situation may be—that organizational style becomes unworkable. In such cases you need to be able to work with others as a partner rather than a boss. This, however, I proved unable to do. I didn't seem to have the requisite psychological makeup allowing me to turn myself into a mere partner. I was, in essence, a very solitary person. Apart from a very few exceptions, all of my books and scientific articles are single-authored. I had a great many graduate students, several of whom have become not just Candidates but doctors of science and in some cases academicians, yet, despite my being on good terms with them, I wasn't like a fellow searcher after projects to them. Although I was aware of this as a psychological defect, there was nothing I could do about it.

I began my narrative with a description of a certain few hours I spent alone with Lake Ladoga. My mental state then, akin to meditative, was one which I have since childhood often submitted to and which has throughout my life been a great boon to me. Although it is hard to say what exactly I'm thinking of during these episodes, nonetheless a vague outline of the object of my thoughts usually emerges—so vague, in fact, that I always shy away from telling anyone about it. Nevertheless, I have always trusted and obeyed such interior summonses. I am accustomed to trusting my intuition; although it has never given me any reasonable arguments supporting its various positions, it has never let me down. For this reason I find long debates and critical discussions impossible to bear. There have been occasions when I have admitted the correctness of certain criticisms but ... still did it my way. This idiosyncratic mode of thought has made collaboration with others difficult. Attempts at collaborative work, or co-authorship of articles or books has in my case usually ended in failure, and sometimes even in quarrelling.

I can recall only two occasions when I was able to work with someone on an equal footing. The first was at the end of the 1950s, when V. N. Lebedev and I together attempted to find numerical methods for stable computation of the trajectories of space vehicles, and the second was when V. V. Alexandrov and I together worked out the original version of a model imitating the dynamics of the biosphere.

My investigation of the biosphere as a single system, including inanimate and organic nature together with human society, and the interactions of these with the cosmos, would require a reliable company of competent, like-minded people who fully understood and shared the aims of the investigation and were equally engrossed by it. I was unable to assemble such a company, however.

Nikolai Vladimirovich Timofeev-Resovsky [4]

Among my students there was one highly non-standard individual: Yury Mikhailovich Svirezhev. He had graduated from Phystech[5] and done his graduate work in Lavrentiev's department, productive of so many talented scientists. After successfully defending his Candidate's dissertation, under me as supervisor, Svirezhev went to Obninsk[6] to work with the famous biologist and geneticist Nikolai Vladimirovich Timofeev-Resovsky, where he immersed himself in biology and gradually transformed himself, I would say, into a mathematical biologist. It is difficult for me to judge just how highly qualified he was in this his new calling, but he threw out all sorts of terminology so casually that I, among many others, could not help feeling envious of his erudition. In any case, in Pushchino[7] he successfully defended his doctoral dissertation on mathematical problems arising in biology. It was through him that I got to know Nikolai Vladimirovich.

Rather a lot has been written about Timofeev-Resovsky in recent years—about his unusual activities, his disquieting ill-fortune, his very odd personality, his scientific achievements, and so on. Although these writings give the impression of a highly exotic personage, my own impressions of him are somewhat otherwise. He was certainly a titan: strong, intelligent, and charismatic in much the same way as other truly great scientists such as P. L. Kapitsa[8] or A. N. Krylov.[9] There was nothing exotic about him, however, that I could see. He was very Russian, was troubled by the lack of demand for our scientific talent, and understood our potential. He cursed just like the rest of us technological types, which endeared him to us. He tried to work for the good of our country and, like us, averred that Brezhnevs would come and go but Russia would remain and in the meantime we must set to work!

---

[4] N. V. Timofeev-Resovsky (1900–1981) was an eminent Soviet biologist working in direct opposition to Lysenkoism. In the 1930s he directed the Genetics Division at the Kaiser-Wilhelm Institute for Brain Research in Berlin. At the end of the war he was arrested and returned to the USSR, where he was at first rehabilitated, but then re-arrested and sentenced to ten years in the Gulag. In 1964 he organized and headed the Department of Radiology and Genetics at the Institute of Medical Radiology of the Academy of Sciences of the USSR in Obninsk. He was blocked by Lysenkoists from becoming a member of the Academy and remained an un-rehabilitated ex-prisoner for the rest of his life.

[5] The Moscow Institute of Physics and Technology.

[6] A town 100 km southwest of Moscow in Kaluga Province.

[7] A town 100 km south of Moscow in Moscow Province. In the 1950s, the village of Pushchino was chosen as the site of a biological research centre. The Institute of Biochemistry and Physiology of Microorganisms was opened there in 1969, and remains an important scientific centre of the Russian Academy of Sciences.

[8] Pyotr Leonidovich Kapitsa (1894–1984) was a leading Soviet physicist. Nobel laureate for physics in 1978. In the 1920s and 1930s he worked for over ten years with Ernest Rutherford in Cambridge University's Cavendish Laboratory.

[9] Alexey Nikolaevich Krylov (1863–1945) was a Russian and Soviet naval engineer, applied mathematician and memoirist.

Thus he was never a dissident. He was a normal, contemplative, and bold researcher and thinker. In a word, he was very much one of ours!

Sometimes when he was in Moscow, Timofeev would phone to ask me if I would organize a smallish seminar. I would invite a few people, and that very evening my office at the Computing Centre would be the scene of very substantive discussions. But perhaps the word "discussions" does not reflect precisely what went on there. Timofeev did most of the talking. He talked about the Russian sciences of nature[10]—their history, ideas, and philosophy—but most of all about the people who had made them what they were. He was especially well disposed towards Vernadsky[11] and Sukachev,[12] but spoke highly also of Vavilov,[13] Shmal′gauzen,[14] and Chetverikov,[15] among various other representatives of Russian science. He was able to convey to us the great distinction of the Russian sciences of nature, making us feel that we were not mere Ivans ignorant of their birthright, but the heirs of a great scientific culture, for which, moreover, we remained responsible.

I don't recall if we at any time discussed Lysenko[16] or Lysenkoism. I think Timofeev would have considered that topic unworthy of discussion by serious scientists—even by serious people generally.

In fact these seminars—there weren't many of them—constituted a rather well thought-out lesson in natural history for the ignorant, intentionally aimed at us computer-oriented mathematical types. He had an ulterior motive: He had long been persuaded that the time had come when, for the sake of their further development, the sciences of nature, like physics long before, needed to harness the great

---

[10] That is, the life sciences, geology, etc. as opposed to physics, chemistry, and mathematics.

[11] See Chap. 6.

[12] Vladimir Nikolaevich Sukachev (1880–1967) was a Russian and Soviet geobotanist, engineer and geographer.

[13] Nikolai Ivanovich Vavilov (1887–1943) was a prominent Russian and Soviet agronomist, botanist and geneticist, devoting his life mainly to the study and improvement of wheat, corn and other cereal crops. His work was criticized by Trofim Lysenko, whose anti-Mendelian concepts of plant biology had found favour with Stalin.

[14] Ivan Ivanovich Shmal′gauzen (1884–1963) was a Russian and Soviet zoologist and evolutionary biologist of German descent.

[15] Sergey Sergeevich Chetverikov (1880–1959) was a Russian and Soviet geneticist. He was arrested by the OGPU in 1929 and exiled to Ekaterinburg for five years. He later moved to Nizhny Novgorod, where he organized a department of genetics at Gor'ky University. At the behest of Lysenko he was dismissed from his post there in 1948.

[16] Trofim Denisovich Lysenko (1898–1976) was a Soviet agronomist and biologist of peasant background who rejected Mendelian genetics in favour of the idea of environmentally acquired inheritance. His experimental research on "vernalization" of crops earned him the support of Joseph Stalin, following the famine resulting from the latter's forced collectivization of agriculture in the USSR. Soviet scientists refusing to abandon Mendelian genetics were dismissed from their posts and left destitute. Dissent from Lysenko's theories was formally outlawed in 1948. Although Lysenko remained in his position at the Institute of Genetics in Moscow till 1965, his influence on Soviet agricultural practice had declined by the 1950s.

power of contemporary mathematics. He understood that biologists and allied scientists would not be able to do this on their own, and was therefore trying to attract the attention and interest of professional mathematicians, most of all those knowledgeable about methods of computer modelling. Thus neither his choice of our company nor the subject-matter for discussion were in the least accidental. I would say that he was successful in his aim: we began to study Vernadsky's works seriously, befriended the very interesting soil scientist Viktor Abramovich Kovda,[17] and set off on our own journey through the sciences of nature.

Twice I was unable to gather together enough people for a seminar, and then the discussion involved just the two of us. Both such episodes were to have significant consequences for me.

During the first of these one-on-one meetings, Nikolai Vladimirovich asked me to estimate how many human beings our planet could accommodate at the current level of technological development, under conditions of natural recycling of materials. I laboured over this task for a considerable time, perhaps three or four months. One day Timofeev phoned me to ask if I had an answer for him. I said that although the degree of indeterminacy was too high for a precise estimate to be possible, my calculations had produced a figure somewhere between 200 and 800 million people. He laughed and said: "Almost correct, 500 million!—and no calculation needed." His reasoning was as follows: One observes first that only ten percent of the energy used by human beings is renewable, that is, participates in the natural recycling process, the balance coming from the storehouse of remnants of earlier biospheres or from deposits of radioactive materials produced during the Earth's birth pangs. It follows that in order to conserve the Earth's non-renewable resources so as not to upset its natural recycling process and, like every other living species of organism, live in harmony with nature, humanity must curb its appetite or find new technological means of satisfying that appetite or reduce its population to one tenth of its present number. It turned out that Timofeev-Resovsky had the answer before asking me the question but wanted to see how I would go about finding an answer—an elementary leg-pull in his style of humour.

This prompted me to think about a different question. The human species interacts with nature as part of a single whole. What conclusions can be drawn from this basic fact? A great many, it seemed to me: a fresh understanding of the past, present and future of human society, for instance. My attempts to see what consequences follow from this fact led me to a complete revision of my former understanding of the dialectic of societal development.

I had a further discussion with Timofeev-Resovsky much later, this time in Obninsk, when I had already begun reading Vernadsky and thinking seriously about problems of the biosphere. I had, in particular, begun thinking how one might go about studying the interaction of humanity, considered as a single biological species, with the biosphere, of which it is at the same time an inseparable part. Taking

[17] Soviet and Russian soil scientist. Wrote standard works on land reclamation and the evolution of soils and their role in the biosphere. Lived from 1904 to 1991.

into account the enormous variety of human cultures, ways of life, and economic systems, I looked for an initial position which might serve as the starting point for an investigation of this interaction. But how can one investigate a system as a whole from the inside when one has only local information at one's disposal? An experiment on the whole biosphere is impossible if only for the simple reason that it would be far too dangerous to carry out!

Then, one evening following a meeting of some sort at the Obninsk House of Culture, while waiting for the electric train to take me back to Moscow, we talked once more about this problem. I laid out in detail my doubts as to the possibility of an effective scientific analysis of the problem, noting also that I didn't know of any other problem so important for humanity. I said that we could make no progress at all on a solution without computer models of the global processes of the biosphere. But where was one to begin?

Thus, in contrast with our usual discussions, this one began with a monologue from me. Timofeev listened to me without interrupting. When I had finished, he said something like: "I see that you are now up-to-date. Without computer models it's hopeless, and even with them the problem is an extremely difficult one. The game is worth the candle, however. At present, you are the only one in a position to work on the problem or intending to do so, and it's essential that the problem be solved." That's the blessing I received back then, for me of the utmost importance.

That was my last serious discussion with Nikolai Vladimirovich. His wife died and then he himself began very quickly to fade. When sometime later Yu. M. Svirezhev and I travelled to Obninsk to visit him, we saw a different person: the sparkle was gone from his eyes, and he had lost his former wide-ranging curiosity. He asked us questions as if by inertia and barely listened to our answers. I had the impression that his memory had faded.

A little while later I heard that he had died.

Global Problems: Forrester, Meadows, and Others

At the beginning of the 1970s, the phrase "global problems" began turning up more and more often in the language of scientists and politicians, and there appeared several imposing works devoted to such problems. Around that time the International Life Sciences Institute, founded by the noted histologist Professor Maurice Marois,[18] began its work on problems of this ilk, and the Club of Rome[19] was founded. Among the relevant publications I was familiar with at

[18] Actually founded in 1978 by various US-based companies, including Coca Cola, Heinz and Kraft, with the mission of "improving public health and well-being by [bringing together] academic, government and industry scientists in a neutral forum to advance understanding in areas related to nutrition, food safety, risk assessment, and the environment."

[19] Founded in 1968 as "an organization of individuals who share a common concern for the future of humanity and strive to make a difference, [with a] mission to promote understanding

the time, the most important was the truly pioneering work *World Dynamics*, by Professor Jay Forrester.[20] There Forrester attempted to describe, on a planetary scale and in terms of just five variables, the fundamental processes at work in economics, demographics, and the growth of environmental pollution, as well as their interrelationships. To this end he had created a special computer programming language called "dynamo" for analyzing output. I was much impressed by this work, although not by its scientific depth, since most of the functional dependencies it used seemed to have been plucked out of the air and couldn't withstand even the most well-meaning criticism, nor by the methods employed, since these reminded me very much of the old method of "plus-minus factors" used by engineers in calculations relating to electric circuits back in the 1920s. What did impress me was the author's boldness in making an attempt to include human activity in a comprehensive description of the processes going on in the biosphere. Forrester's book echoed the conversations I had had with Timofeev-Resovsky, and represented a response of a kind to my doubts.

The Forrester book was translated into Russian under my editorship, and, by way of an Afterword, I wrote a long article in which, for the first time, perhaps, I described some of my own approaches to the construction of a computer model imitating the functioning of the biosphere. Three years later, while at Yale University on a visiting professorship, I managed to break through Forrester's secretarial cordon and arrange a meeting with him in person. Despite my almost complete lack of English and his very poor French, each of us quickly understood where the other was coming from. Forrester turned out to be a congenial Quaker—there are such amongst that sect—utterly dedicated to his preoccupation, living very modestly, and ... very ignorant of everything outside his profession, that of electronic engineer, with all its pluses and minuses. Everything he did seemed to him the work of a genius and unique of its type. He was actually very little interested in the biosphere itself. What did interest him greatly, however, was the potential use of computers to predict the results of an activity, and machine imitations of real-life processes wherever they might be found, and he used all possible means of discovering new applications for his computer simulation software—precisely *his* software, that is. He knew nothing at all about alternative approaches to numerical analysis of complex dynamical systems.

I told him of certain modelling experiments undertaken by us. I think his tendency to be sceptical had to do with the fact that what we had done was a degree more complex and interesting than what he himself had achieved. Although his technique was remarkable for its simplicity and accessibility to physicists and mathematicians, he himself was not up to working at the level of real mathematicians and physicists.

---

of the global challenges facing humanity and propose solutions through scientific analysis, communication and advocacy."

[20] Jay Wright Forrester (1918–2016), American computer engineer and systems analyst.

Global problems, i.e., problems on a planetary scale, soon began to feature among the concerns of international organizations, in particular UNESCO. In 1971, at the initiative of Professor Forti,[21] an official with UNESCO, the first conference on global problems took place in Venice. The main event, the focus, if you like, of the conference was a talk by Forrester's graduate student Dennis Meadows titled "The Limits to Growth." The book version[22] published subsequently became a bestseller: it was printed in huge numbers and translated into many languages. The whole of the conference was centred on Meadows' exposition, demonstrations of the computer language "dynamo," and computer experiments using the suggested models.

I would say that Meadows' talk had a "deafening" effect, somewhat like a heavy metal concert, only modified to suit a sedate society of scientists. Every speaker praised Meadows, some even singing dithyrambs to him, his supervisor and his colleagues. My own talk was the only one where a dissonant note was sounded.

I liked Meadows very much personally for the audacious, sanguine young fellow he was in those days. There was, furthermore, a great need of his computer models at the time, and it was essential they be widely publicized: it was important that the general public be informed of the fact that our contemporary way of life brought with it a multitude of dangers. As many people as possible had to be made to heed the warning that the looming catastrophe might unfold without warning, that there would be no time to change the ship's course into the maelstrom at the last minute. On the other hand, however, I felt absolutely unable to grant real scientific status to Meadows' work. An example to be used in student demonstrations was the best that could be said of it! The trouble with Meadows (and his teacher Jay Forrester) was that they took their approach to investigating the actual dynamics of what was unfolding on our planet too seriously, to the extent that they actually believed it adequate to solve the problem. Meadows failed to understand the basis of my doubts. He even seemed to believe that the phenomenon of exponential growth following a loss of equilibrium was a new discovery of his.

In fact, the problem is infinitely more complex and difficult than is indicated in Meadows' and others' work on modelling the planet's development using Forrester's technique. A more authentic description of the interactive processes occurring both within nature and with nature is essential to the problem. As a first step, I proposed constructing a computer model imitating the interactions of the oceans, the atmosphere, and the biota,[23] and treating man's activities, such as those relating to economics, in various separate scenarios. And in fact I had attempted a sketch of the plan of such a computer simulation, albeit at the level of an improvisation.

[21] Paolo Forti is at present a professor at the University of Bologna. He has expertise in hydrology, geology, and geochemistry.

[22] A report of the same name published in 1972 dealt with a computer simulation of exponential economic and population growth with a finite supply of resources. The findings (of a team of 17 researchers led by Meadows and three others) were first reported at international meetings in Moscow and Rio de Janiero in the summer of 1971.

[23] A taxonomic superdomain comprising all (nonhuman) life on Earth.

My talk on that theme was received without enthusiasm, and the comments following it were highly critical, suggesting that the very possibility of realizing such a system of models and the software to underpin them was doubtful. Nevertheless, I returned to Moscow with a fresh understanding of the problem and a deeply held conviction that there is simply no other way to investigate global problems! Of course, by themselves, models of interactions within the biosphere are not adequate to a profound analysis of the developmental perspectives of the human species, but they are nonetheless essential to such an analysis. A new direction of fundamental scientific research was emerging of inestimable importance for humankind, and mathematical modelling was central to it.

When, back in Moscow, I gave the same address at a seminar in the Computing Centre, although the audience was unusually large, once again I met with little support for my ideas. Dorodnitsyn, in particular, was harshly critical. "We have enough problems of our own, so why bother with those about which we know nothing"—that was the leitmotif of his commentary.

But then I unexpectedly got support from the Earth sciences section of the Academy. Following my talk, Sidorenko, at that time vice-president of the Academy, was active in supporting my proposals, and I was granted the equivalent of six salaried positions, which was pretty good even for those times, when science was still well funded. When supplemented with resources from the Computing Centre, this sufficed for the organization of two new laboratories, one to be devoted to problems of modelling the processes of the biota, and the other to modelling the dynamical system consisting of the oceans and the atmosphere. Yu. M. Svirezhev was put in charge of the first and V. V. Alexandrov the second. In order to ensure the integrity of the program, they were required to maintain close contact.

And so began a happy phase in my life.

Under Svirezhev's guidance I devoted myself to a serious study of the biota, since it soon became clear that I knew extraordinarily little about it. Eventually I began to understand something of it, and even wrote a few joint papers with Yury Mikhailovich. By the end of the 1970s, however, the smooth surface of our collaboration had begun to develop a few cracks. This had to do partly with Svirezhev's inclination to do, not what needed doing, but what he already knew how to do, as the easier alternative. The work I wanted him to do required concentrated effort in order to bring to light new aspects of the problem, instead of which he began busying himself with problems of his own devising. The upshot was that I lost one of the laboratories I had set up. I lacked the resources for making up the resulting shortfall of resources, which was disastrous as far as the integrity of the project was concerned.

But, on the other hand, things went going better than expected in the other laboratory. Here I had a much easier time of it since I was a "professor of hydrodynamics," so in a position to understand what was being investigated rather than having to assume the role of student. Furthermore, Alexandrov was without the ambitious streak which in Svirezhev had hindered collaboration. We worked together in the evenings, discussing the mass of details of our project, and organized seminars, inviting a wide variety of specialists, some from abroad. Of the problems we encountered, the most difficult was that of arranging equality of accuracy among the component models of our system: the degree of computational accuracy had, of course, to be the same for every component. There was also the difficulty of deciding amongst the wealth of details which to neglect and which to include. And then there was the enormous number of computations required. To cut a long story short, by the end of the 1970s we had worked out the first version of our system of models, the result of a collaboration in both labour and learning. Dorodnitsyn had been right about one thing: we had a lot to learn.

Next was perhaps the most difficult step: the development of appropriate computer software. To achieve this, first the right algorithms had to be chosen, then appropriate finite difference schemes for implementing them, and finally the computer programs written. This Alexandrov did all by himself! My participation was minimal, confined, more or less, to that of critic. But then at the end of it all we were unable to do any actual computing. The Computing Centre had at its disposal only a BESM-6 machine,[24] clearly not up to the task.

The Americans came to our aid. Professor Burley,[25] director of the American climatological program, provided funds for the testing and debugging of our program at the Centre for Climatological Research in Boulder, Colorado, and Alexandrov duly went off to the US on an eight-month field trip. Not only did he debug the program and carry out a great number of computer experiments, but also made a film of the display on the computer monitor, which he brought back to Moscow. The film showed the changing isobars—curves of constant atmospheric pressure—calculated by our program using as initial condition the weather on December 31, 1970.

I took Alexandrov's film and flew to Novosibirsk, where G. I. Marchuk, the then Chair of the Siberian branch of the Academy, assembled a group of meteorologists. When what looked like one of those infamous Siberian anticyclones appeared on the screen, recognition was instantaneous, and in one voice they exclaimed: "Typical January weather!" This represented, in effect, a very positive affirmation of the quality of our model, indicating that it correctly reproduced the basic peculiarities of the dynamics of the atmosphere and hydrosphere. The only true test of our model was that it correspond qualitatively to reality, so the approval accorded it by the meteorologists was for us proof of the possibility of modelling mathematically the extremely complex processes taking place in the biosphere—the movement of air masses, the formation of clouds, the motion of the oceans' waters, etc.—thereby providing a good idea of how these actually interact. This was of the greatest importance to us.

We understood, however, that we had made only a first and very tentative step towards a solution of the problem: that we had yet to understand how to describe the dynamics of the biota within the framework of the system we had created, and then how to include a model of humankind's activity, not to mention a multitude of

[24] The first second-generation Soviet computer, built using transistors.

[25] See Chap. 8.

other extremely difficult questions demanding answers. But as far as I personally was concerned, that memorable Novosibirsk seminar at which specialist Siberian weathermen immediately recognized their prevailing January climate had at least dispelled the doubt that had been tormenting me: was it possible that the enterprise of modelling the biosphere, in which I had involved so many people and on which substantial funds had been spent, was frivolous? This doubt had been all the more real in that there were no precedents for the enterprise, neither in the US nor in Europe. Our work was of a truly pioneering character.

After testing the ground, we could now move cautiously forwards. Years of hard, professionally very demanding, work lay before us. Despite Svirezhev's not wanting to join our company, that work went along reasonably well.

Carl Sagan and the First Fall-Out Scenarios of Nuclear War

The rise to world-wide prominence of the concept of a "nuclear night" was unanticipated. Prior to March 1983 none of us had thought of it. All our efforts, in particular, had gone into creating the means of analyzing the character of possible interactions between humankind and the biosphere—as a real instrument of scientific research, however, and not merely a variant of Forrester's *World dynamics*, useful only for exposition in student seminars as a demonstration of potential dangers. We had as yet not contemplated any practical application of such an instrument, apart from the attempt by A. M. Tarko, an active member of our team, to estimate the possible change in productivity of the Earth's biota resulting from a doubling of the concentration of carbon dioxide in the atmosphere. I believe this work was the first of its kind.

But then in March 1983, the well-known American astronomer Carl Sagan published descriptions of various scenarios of the aftermath of a putative nuclear war in which nuclear strikes are exchanged of energy equivalent to thousands of megatonnes of TNT. Such a calamity must clearly have serious planet-wide climatic consequences. Using a purely speculative approach, Sagan and his colleagues advanced a scenario of the "nuclear night" that must follow from vast urban conflagrations and the pall of soot that would envelope the planet following the nuclear explosions .... Then, as a consequence of the "nuclear night," there would begin a "nuclear winter," since the sun's rays would for a considerable time be prevented from penetrating to the Earth's surface, which would therefore quickly cool.

As soon as we received detailed information as to what Sagan and his colleagues had done, I said to V. V. Alexandrov and his co-worker in the laboratory G. L. Stenchikov: "Here's a chance we mustn't miss. As of today, ours is the only program capable of checking the truth of the astronomer's conjecture!" By that time, thanks to the efforts of Alexandrov and especially Stenchikov, the computational system and associated software had been brought up to such a degree of efficiency that we could carry out the appropriate computations on our old BESM-6.

By June or July 1983, all our computations had been completed and accompanying illustrative graphs prepared. When Stenchikov showed me the results of the initial computations, I found the numbers hard to believe and made him re-do the computations several times with various initial conditions. The output always told the same story, however. Even in the case where the warring parties use altogether only between thirty and forty percent of their nuclear arsenals in strikes carried out on urban centres, the quantity of ash and smoke that would as a result accumulate in the upper atmospheric strata would be such as to obscure the Sun for many months. Except for the smaller oceanic islands, partially shielded by the oceans' action as an efficient thermostat, the temperature would descend to below freezing everywhere on the Earth's surface, and in certain regions of the globe—Saudi Arabia, for instance—the temperature would drop to thirty or more below zero Celsius. Only after about a year had passed would the temperature gradually begin to rise.

But the planet would not then return to its former state. The biota could not withstand such a blow. The tropical forests would perish along with every organism living in them. The fate of northern forests would depend on the time of year when the nuclear catastrophe occurred: if during the Winter, they might survive, but if during the Summer, the taiga would share the fate of the tropical jungles. The oceanic biota would more readily withstand the blow, although it would nevertheless be subject to radical re-adaptation.

Thus our analysis of the scenarios following immediately on a possible nuclear war predicted a complete restructuring of the whole biosphere, which, although not entirely wiped out, would metamorphose into a new state qualitatively different from its current one. But in this new biosphere there would be no place for humanity, even leaving out of account the resulting omnipresent lethal level of radiation at the Earth's surface!

On October 31, 1983, a grand two-day conference dedicated to the consequences of a possible nuclear war opened in Washington, D.C. The first day was for the press, and Carl Sagan gave a brilliant address. The second day was devoted to professional analyses, and the main address on behalf of our Computing Centre, including an explanation of our model, the techniques used to analyze it, and the results of our computations, was given by Alexandrov. It was a triumph.

The Americans had been able to analyze the possible dynamics of atmospheric changes only for the first month following a nuclear exchange, while we were able to paint a picture of the whole of the first year. I asked the head of the American research centre that had carried out their computations why their prognosis was restricted to just the first month. The answer came that although the Americans had a significantly more complete model of the dynamics of the atmosphere than we did, that model neglected oceanic dynamics. Although our models were rather primitive by comparison, they had the advantage that they had been integrated into a single system. The oceans, after all, act as a thermostat which, in the last analysis, determines the fate of the biosphere. Computations relating exclusively to the atmosphere are capable of giving a more or less correct prognosis of the sequence of events only at the very beginning, when the heat inertia of the oceans can be neglected, whence the limit of a month on the relevance of the American computations.

For both the Americans and us it was highly significant that the qualitative results of the respective computations predicting the sequence of events over the first post-catastrophe month were practically identical. These predictions were subsequently put to the test by a great many researchers using a range of models, on various computers, and utilizing a variety of numerical algorithms. The upshot was that no serious corrections of our results were needed, that our very simple model sufficed to bring to light the fundamental fact that in the aftermath of a nuclear war the biosphere would be qualitatively changed to such an extent as to exclude the possibility of life for human beings.

The success of our presentation owed much to the personality of Volodya Alexandrov: truly gifted, he was genuine in everything he undertook, even in his frivolousness.

He and Yura Svirezhev had been undergraduates together in the same group in Academician Lavrientiev's department in Phystech, and both of them had passed through my hands in my roles as professor and dean. They were both given to laziness and playing up, but in very different ways. Volodya tended to skip lectures and obligatory seminars but always managed to come good at the last moment. I was his examiner three times and, although very exacting, had no choice but to give him fives![26] As far as the esteemed Professor Svirezhev was concerned, however, I always found myself unable to award a good grade, even on the final exam for his Candidate's degree.[27]

Yet I nonetheless took him on as a graduate student. I was not mistaken in so doing, I would say, for I sensed in Yurka Svirezhev a kernel of what some might call "God's gift," notwithstanding his egotism and his liking to feel himself boss. "My lads" is how he tended to patronize his co-workers. How much effort I put into trying to wean him of this habit of relating to people and work! It was precisely this trait which prevented him from becoming the really great scientist that he should have become given the correctness of my estimation of his creative potential. If the time has come for a summing up not just of my own career, but also the careers of those of my students who have passed the fifty mark, then I would have to say, albeit regretfully, that I had expected more of Yurka Svirezhev.

Alexandrov was a different case altogether. He was a team player by nature and without a hint of protest was always ready to pitch in merely because he could! He acquired foreign languages without making any special effort to learn them. He spoke English with the Texan accent beloved of Americans. He was always ready to communicate frankly, and around him a circle of friends would usually very quickly gather. In the end, perhaps, this quality may have led to his death—but I shall talk about that a little later on.

---

[26] The top grade in the Russian and Soviet system of grading. See also Chap. 2.

[27] Approximately equivalent to the Ph. D. degree.

Our address at the Washington conference owed its success in no small part to Volodya—to his way with language, his temperament, and, I would say, his lecturer's charisma. He subsequently performed before the US Senate and the Pope, and he met with success on these as on other occasions.

Following the conference, our little group found itself in a rather tight spot. The tragedy with the Korean Boeing[28] had occurred, and the whole world was baring its teeth fiercely at the Soviet Union and its citizens. No airline company would sell us tickets. Each day a representative of the US State Department would show up at our hotel and note with a smirk the number of days left till our visas expired. Then, just one day before the date of expiry, we managed to book a flight to ... Mexico! Visas were issued at once, and next day we flew to the land of the ancient Aztecs.

There was a stopover in New Orleans, where by reason of bad weather we were able to spend a few hours downtown, even though strictly off-limits to Soviet citizens, but by evening we were already enjoying the hospitality of the Soviet ambassador to Mexico and a selection of his colleagues. There was much in the newspapers about our success at the Washington conference and our difficulties booking a flight, and about each of us individually. In view of this publicity, it would have been miserly not to allow us a few days' relaxation in Mexico while awaiting a Soviet airliner.

On the day following our arrival, we travelled to the plateau on which the old Aztec city with its famous pyramids is situated. There I saw what form a city might take in a country lacking the concept of the wheel! Not only the pyramids, but the whole city was made up of steps. It was just this that astounded me most of all.

The following day was the anniversary of the October Revolution, and we were all invited to a sumptuous reception at the Soviet embassy. There were performing artistes and marvellous food and drink, but what I most remember was the uneasy conversation I had with an American military attaché, evidently a ranking general. He had been briefed about our success in Washington, and was extremely disapproving of what we had done, which, as far as he was concerned, was simply harmful! And not only to America but also to the USSR. "You'll see. They won't be patting you on the head for what you've done"—such was the general tenor of his remarks. Of course, everything he said had to be understood relative to the sizeable quantities of superlative vodka and black caviar consumed. All the same, I believed in his sincerity: we had delivered a serious blow to the military, its budget, and its worldview, and in due time this would come back to haunt us. And, indeed, back home I was met with a cool reception from many sides. During a break in a meeting of officials of the Soviet Military-Industrial Complex at which I happened to be present, Smirnov, director of the Complex,[29] let fly with the following words:

[28] Korean Air Lines Flight 007 was a scheduled flight from New York City to Seoul via Anchorage on September 1, 1983. It was shot down by a Soviet Su-15 interceptor en route from Anchorage when it deviated from its planned route into Soviet airspace. All 269 passengers and crew were killed. The incident precipitated one of the most tense periods of the Cold War and led to an escalation of anti-Soviet sentiment, especially in the USA.

[29] And deputy chairman of the Council of Ministers of the USSR at that time.

"What do you think you're doing crawling off like that. You'd have been better quietly busying yourself with your scientific pursuits here at home instead of making such a din abroad. If not for you, no one would have believed that astronomer." This last was in fact true. Our computations, carried out independently, provided a basis for the most powerful arguments of opponents of nuclear war. At General Military Headquarters, and elsewhere, I was received with similar coolness. And my remarks to the effect that it was essential to begin looking for alternatives to our long-held principles—alternatives compatible with the new scientific discoveries—met with sour irony: "Just what one might expect of those civilian types."

Not liking dissidents, I have always shied away from them. But now I received an unexpected blow from their opponents at the round table of the journal *Our Contemporary*:[30] I stood accused of anti-patriotism. How much hidden meaning there is in that single word!

Our marvellous sojourn in Mexico culminated in the following amusing episode.

We were staying in a charming two-storey hotel with an interior Spanish courtyard, and every morning an engaging company of parrots was brought out into this courtyard; since the nights were cold, they spent them indoors. All day long the courtyard was filled with their chatter in what was, most likely, a parrot version of Spanish, although occasionally something more or less Anglo-Saxon might be heard. One of their company apparently took a shine to me, and began making overtures of friendship. Whenever I went into the courtyard, he would begin flapping his wings and uttering sounds similar to Russian swear words. This association strongly appealed to me, so I took to approaching his cage and giving him something tasty, while repeating the phrase: "Beastie, beautiful green beastie," whereupon the parrot would quieten down and listen attentively to me, watching my mouth.

Sometime following my return to Moscow it happened that a high-ranking Soviet public servant or perhaps even a minister was booked to stay at that same hotel. This guest's mien must have reminded the parrot of mine. In any case, he welcomed him in his characteristic parrotty way, flapping his wings and enunciating clearly in Russian: "Beastie, beautiful green beastie." Alas, our high-ranking personage failed to appreciate the genuine fellow-feeling expressed by my parrot friend, and even went so far as to complain to the ambassador. I was told that they moved him to another hotel: re-educating parrots is no easy task!

Alexandrov's Fate and the End of the Story

For never was a story of more woe
    Than this of Juliet and her Romeo ....

The finale of our ever so promising project was clouded over by a series of sad and even tragic events.

[30] A Russian-language, Soviet literary and socio-political monthly.

Svirezhev's refusal to collaborate in the project had already begun severely limiting our horizons. He was replaced by A. M. Tarko, but, despite the considerable respect and empathy I felt towards Alexander Mikhailovich, and my gratitude to him, this replacement did not fully make up for the loss of Svirezhev.

The main blow, however, came from outside. It would appear that in 1985 Volodya Alexandrov perished. It is a tragic and mysterious story. On the eve of his departure from a conference he was attending in Madrid, he left his hotel to take a walk and ... disappeared. His things and his money (of which there was a significant sum) were discovered intact in his hotel room. Several organizations participated in the search for him. In particular, the Spanish secret service was, it would seem, especially assiduous in its efforts to find him. A year after his disappearance, the Spaniards officially announced their finding that he had definitely not crossed the Spanish border. And on that note the investigation was ended! The Alexandrov affair attracted journalists from Belgium, America, and other countries. Our American colleagues also made a considerable effort to find out what had happened to him, but the mystery of Volodya's disappearance remains a mystery to this day.

I have my own version of what happened to him. I think—am almost certain, in fact—that a leading role in the tragedy was played by one or more of the special services. However, which one of these was in at the end, the KGB or the CIA, I do not know.

Certain of my co-workers, like me, went abroad quite often, and it was inevitable that we should occasionally find ourselves in the company of certain persons, or in certain places, of interest to our special services. Representatives of the latter would quite often approach us with a request of one kind or another, usually trifling. Nonetheless, I always avoided fulfilling such requests, since after "A" there almost inevitably follows "B." And in any case, ever since my experience of "friendship," with the regimental "special type,"[31] I had tried to keep as great a distance as possible between myself and the special services, even though appreciative of the fact that every government needs such services. And I had several times warned certain of my colleagues about taking on commitments beyond those falling within the purview of their official duty. In Volodya's case, I had made a special point of this. He was our official representative in our work with the Americans, and back then we were collaborating with the Livermore Laboratory,[32] considered a closed institution. I tried to organize matters so that the necessary seminars were held in Moscow, but sometimes found it necessary to travel to Livermore to communicate with "the best of the best:" Teller[33] himself was working there.

And Volodya, thanks to his relaxed nature, his openness, and his Texas slang, was always counted as a friend by our overseas colleagues. Thus he never stayed at

---

[31] See Chap. 1.

[32] The Lawrence Livermore National Laboratory was founded in 1952 in California as an American federal facility devoted to research in science and technology applicable to national security.

[33] Edward Teller (1908–2003), Hungarian-American theoretical physicist. Provided, along with Stanislaw Ulam and others, the theoretical basis for the hydrogen bomb.

the assigned hotel, but lived with one or another of his American friends, sometimes for weeks at a time. For this reason he had access to much that was out of bounds to other visitors. Knowing our system, I was absolutely certain that our spy agencies had their eyes on him. That's why I warned him. But, knowing his thoughtlessness and insouciance, I imagine that he would not have refused to do as they asked.

An American professor who was very well-disposed towards me, once told me that there were people in the US highly displeased with our being allowed to have contacts within the Livermore Laboratory, although not so much with the collaboration as with the character of the personal contacts established. This was not long before the tragedy. I passed on the professor's remarks to Volodya, laying out the facts explicitly. Clearly this was all too late. Shortly after Alexandrov's disappearance, E. P. Velikhov[34] asked me how our contact with Livermore was going. I answered: "It's finished," to which he replied: "You should continue with the work in order to re-establish the contact that's been broken. It isn't as if everything depended on Alexandrov." But in fact everything *did* depend on him, since the contacts he had made were very personal and were now broken once and for all. Although this might be seen as implicating the CIA, I think it would be overhasty to lay the blame on them, since it's not difficult to imagine situations where Alexandrov might have become an expendable obstruction to our own intelligence agencies. Volodya's disappearance in a city's downtown had all the earmarks of a professional job. Who apart from security services had the means to do this? Who else could possibly benefit from his demise?

Following this tragedy, Alexandrov's laboratory began to fall into disarray. It's true that we managed a few further successes. Both A. M. Tarko and I gave addresses at an international conference in Helsinki, and some of our later computations achieved international resonance. A computation carried out by V. P. Parkhomenko and A. A. Mochalov using a modified version of our system turned out to be especially apposite. In the early 1950s, American generals were in all seriousness considering the feasibility of a preventative strike on urban centres in the USSR. The proposal involved dropping an atomic bomb of the same type as that dropped on Hiroshima, on each of five hundred or seven hundred—according to which of various scenarios was under consideration—Soviet towns and cities. Although the effect on our towns and cities was more or less obvious, it was interesting to try to understand what the planet-wide consequences of such a strike might be. It turned out that a real "nuclear winter" would not occur in such circumstances, although of course certain significant climatic shifts would certainly take place. But the most interesting conclusion of our computations was the following one: when the distribution of radioactive fall-out (of iodine and strontium) was estimated, it turned out that on the territory of the USA—of the aggressor, that is—there would fall no

[34] Evgeny Pavlovich Velikhov (born 1935), influential former Soviet, now Russian, physicist. Vice-president of the Academy of Sciences of the USSR 1978–1991, and of the Russian Academy of Sciences 1991–1996.

less than twenty times the dose of these radioactive materials that had irradiated Chernobyl.

Apart from sporadic successes, however, there was less and less work being done on global problems, and funding for such work was also drying up. At the same time, my relations with Dorodnitsyn were becoming insupportable and I understood that I simply had to leave the institute I had worked in for over thirty years. The edict concerning counsellor status, according to which retired academicians would retain their salaries, meant that I would be spared financial problems: sitting at home and not having to worry about working on the side, I would have the leisure to think about the scientific questions of interest to me.

I understood that I was entering on a new stage of my life, which would be organized completely differently from anything that had gone before. And, to be frank, I was a little afraid—afraid of the change in my work and social status. So it was not without some anxiety that I submitted my request to retire to G. I. Marchuk, the then president of the Academy. I am very grateful to my wife for supporting my decision to take this step. I now understand that it was the only way out of the dead end I found myself in 1985.

I was in my 68th year, but still felt fully capable of working. Furthermore, I had a large project to work on whose general outlines I had configured as far back as the early 1970s. I was compelled to introduce crucial changes to my plans for that project, however, since I had then reckoned on having a whole team to work with and also on the possibility of carrying out a great many computer experiments, whereas now I had neither a collective to call on nor access to a powerful computer. I had only myself to depend on.

In Retirement

Becoming a "homebody" turned out to be less painful than I'd expected. I was more or less ready for the change. I had lived a sort of double life over the preceding several years. On the one hand, there was the large collective I had been involved in, which had given me much satisfaction, and all the more so in that its scientific work had been going along quite well. And, on the other hand, there was my more intimate scientific life, which few had knowledge of and which had its own value and inner logic.

The latter interior scientific life evolved according to its own rules—according, indeed, to a logic of its own over which I had limited control. The most unexpected circumstances might influence its development, but only those that fit a certain definite though unforeseeable profile. A special sort of spiritual world emerged, a kind of mental attitude sometimes hard for me to understand. I shall now describe a few perhaps disconnected fragments relating to the preoccupation that in retirement kept my life taut and active—no less so, in fact, than when I had been working busily in the Computing Centre.

I don't know exactly why, but even when still quite young I had very good relations with my former university professor A. G. Kurosh. We even spent the 1958 Summer break together at a tourist camp in the Carpathian Mountains. Alexander Gennadievich came from Smolensk, so knew my family background, and was always solicitous towards me, even though my mathematical interests were very far from his algebraic ones. When already myself a professor, I was introduced in his office to one of his doctoral students, the academician-to-be V. M. Glushkov, who, following the successful defence of his doctoral dissertation, went off to Kiev to head the Institute of Cybernetics there.

Glushkov and I found that our views on many things coincided. We took to going on walks together, at first in Neskuchny Garden[35] and later in Goloseevsky Forest in Kiev. We talked of many things, in particular our various plans for the future. He introduced me to his young colleagues Mikhalevich, Pshenichny, Ermoliev and several others. Subsequently, aided by Viktor Mikhailovich's light touch, I was to act as opponent at the defences of the doctoral dissertations of practically all members of this Kievan company. Glushkov was highly intelligent, and, I would say, a splendid fellow. I found his company interesting. Our attitudes to mathematics were also similar. On one point, however, our views were very different, and in this case the difference reflected a profound divergence of principle.

Viktor Mikhailovich thought in the style of a vividly expressive technocrat. In one of his papers he had even expressed the opinion that once the number of electronic computers in the USSR had risen to a thousand or ten thousand—the precise number was not essential to his argument—all problems of economic management and social order could be solved. He suggested that then the problem of greatest importance would be that of optimal decision-making in economics and the control of production. In his institute there grew up a rather talented group of specialists in methods of optimization, under the supervision of Mikhalevich. I had very cordial relations with the members of this collective.

At that time, I likewise was working hard on methods for finding optimal solutions to problems arising in technology. We ran joint seminars and Summer schools with the Kievan group, to our mutual benefit. It seemed to me completely inappropriate, however, to identify economics with technology, since the leading role in economics is played by the human being, and the role of the electronic computer, no matter how highly developed and important it may be, is still really only ancillary.

Back then I thought about these matters a great deal, in particular about the management of the economy, and sometimes even felt inclined towards making assertions in a technocratic vein. I now occasionally feel discomfited when I recall some of my publications of that time along those lines, especially the primitive treatment of the programmed method of control attributable in equal shares to Glushkov, Pospelov, and me. By way of self-justification, I might mention that this sort of "computer euphoria" was at that time characteristic not only of us Soviet

[35] The oldest park in Moscow.

specialists: the thoughts of our colleagues abroad were stuck in the same groove. However, then, in the mid-1960s, I was already beginning to appreciate that the world, with society embedded in it, is of a far more complex structure than it seemed to us specialists in applications of computing technology.

It is sometimes difficult to understand why one particular event, often quite trivial in itself, provides the impetus to a complete revision of one's thinking.

The event I have in mind occurred at a small consultative meeting held in one of the guesthouses on the Klyazma[36] in 1967 or 1968. There were present the organizer, V. M. Glushkov, together with G. S. Pospelov and various persons representing the Military-Industrial Complex, in all no more than 15 or 20 people. The aim of the meeting was to agree on general principles for the implementation of contemporary data processing methods in the management of the "people's economy," in politics, and in military affairs. Unfortunately, nothing whatever resulted from this meeting—even worse, we all ended up quarrelling. The blame for our disagreement could not be laid solely on our different interpretations of the phrase "the national economy." I, for one, took a rather extreme point of view, reckoning that at any level of centralization of the economy and any level of efficiency of computing power, the economy could still not be controlled rigidly, and in any case even if it were possible, rigid control would be mortally dangerous! On this Glushkov begged to differ.

As far as I was concerned the arguments in support of my view were obvious. As someone who had spent years using numerical methods to find optimal solutions of problems, I knew very well how unstable such solutions could be. A small inaccuracy in the initial conditions, a hitch in the optimization procedure, and one obtains a result very far from optimal, perhaps even worse than one chosen at random! Computation of an optimum is useful not so much in providing an "optimal solution" to be applied in practice, or, even less, a directive to action, but in giving an idea of the limits of the possible, so as to curb utopian ambitions. It seems that it was right there and then, in the midst of the discussions in which Viktor Mikhailovich Glushkov had invited me to participate, that I began to formulate the worldview that I later came to call "universal evolutionism," a worldview allowing me not only to draw certain philosophical conclusions, but also to make inferences of a purely practical character.

Back then, however, in that memorable guesthouse on the Klyazma, what had been a purely scientific discussion degenerated into a quarrel. I understood only later what was behind it. Glushkov wanted to organize a "gang," as he termed it: "Nowadays people live through gangs!" By this he meant a group of like-minded persons capable of energetic activity on several fronts, aimed at fulfilling the desires of their ataman. In the role of ataman, of course, he saw only himself. I saw the virtue of strength in numbers, but only provided the team's captain works for all, rather than acting like a gang leader. And I could certainly visualize Viktor Mikhailovich as captain of such a computing team: he was younger, more energetic, had access

[36] A river not far from Moscow, a tributary of the Oka.

to the highest authorities, and was of a capacious intelligence. Furthermore, he actually wanted to be at the head of such a group of the like-minded. The difference between a gang and a team was so great, however, as to make mutual understanding impossible for us, even assuming unanimity of scientific views! But in any case there was no such unanimity.

I understood all this somewhat later, and also that with my theoretical arguments I had become an obstruction to the realization of his ambitions. We parted acrimoniously and for a whole year only nodded curtly to each other when we happened to meet. Our relations eventually normalized, but there were to be, alas, no more walks together in Goloseevsky Forest, nor joint field trips abroad, nor, worst of all, those conversations so full of substance that we used to have. Such is life's logic, however: although paths through life intersect with one another profusely, it is extremely rare for two of them to coincide! This fact may in part account for the delight in being alive, providing as it does a variety of possibilities for the development of the society of those creatures latterly become conscious of their humanness!

Yes, the discussion at the guesthouse on the Klyazma jolted me into completely changing my way of thinking. Little by little I ceased working purely on mathematical and information-processing problems. Humanity is a part of Nature; it developed as a single whole via extremely complex evolutionary processes. This fact should be considered as truly fundamental, as a crucial factor determining the features peculiar to human society's development. Marxism is largely concerned with the socio-economic factors characterizing this development, contrasting these with "natural" processes. But perhaps one should approach the question from the opposite direction: try to see what the processes that have moulded our social nature have in common with Nature herself, i.e., ask ourselves why everything has proceeded as it has, and not otherwise, and look to Nature for the answers. One should not divorce Intellect and its possessor, Humankind, from Nature, but study how to view them in the context of the overall development of Nature, as a Natural phenomenon.

Once this was done, we would be in a position to appreciate why there are no optimal solutions in Nature. More precisely, we should understand the real reasons underlying the fact that the trajectory of every natural process may be regarded as a minimum value of a suitable functional.[37] This purely physico-mathematical fact was to be of very great importance to my reflections: Nature appeared to have a set of goals, albeit infinite, rather than behaving completely randomly and chaotically.

Once I had set out on the path whither such reflections led, I unexpectedly found myself completely alone. My circle of acquaintance, my former students and colleagues had all been trained up to entertain only fully concrete technological

[37] Perhaps the author had in mind the fact that this "Principle of Least Action," as it is sometimes called, is an emergent probabilistic property of the random processes operating at the deepest level of the physical world. See e.g. Richard Feynman's *QED: The Strange Theory of Light and Matter*. Princeton University Press, 2006.

or mathematical themes. They simply refused to accept my "new wisdom." My dear friend Alexey Andreevich Lyapunov,[38] with whom I had enjoyed friendly relations since before the war, also failed to understand what I was getting at. Yu. M. Svirezhev, then already gone over to biology, showed indifference when I attempted to tell him of the new scientific interests preoccupying me. But, as mentioned earlier, it was he who had introduced me to Timofeev-Resovsky, and the latter, unlike everyone else, took my spiel seriously, very empathetically and without irony. Unfortunately, my meetings with Nikolai Vladimirovich were brief and sporadic. Nonetheless, they helped me to draw two conclusions: first, that the questions preoccupying me should not be abandoned as the fruit of amateurish thinking, but were sufficiently serious and, furthermore, substantiated, to be worthy of investigation by appropriate professional means; and second, in order to lend credence to them as arising from an integral construct of reality, they should be viewed through the prism, so to speak, of some concrete work utilizing my professional skills. The questions of interest to me would in this way be given the badge of respectability as part of a genuine "scientific program."

So I began studying the biosphere as an integrated system. The problems of mathematical modelling that at the first stage of this investigation made up the bulk of the work, gradually yielded priority to methodological problems. I returned once more to pondering questions that had come up in discussions with Glushkov, but now on an altogether different plane. My new starting point was the fact that in Nature all processes are subject to stochastic variation, and proceed with an enormous degree of uncertainty. The development of human society, in particular, cannot then be otherwise!

In this way the system of views which I was later to call "universal evolutionism" began to take shape. This turned out to be, in essence, an alternative treatment of the principle of self-organization, or "synergetics," as it is called, which at that time began to be investigated the whole world over. My treatment, however, was not at all a mere reformulation, but an original, distinct construct, an independent take on the problems of self-organization. My approach had the advantage that it lent itself to further development, to a variety of mathematical interpretations, and, more importantly, to a multitude of applications.

An exposition of the entire system would take me too far afield. Such an exposition exists, however, in the form of notes from lectures delivered at Moscow State University and the Russian-American University,[39] and recorded on the hard disc of my computer. If some day funds for the publication of these lecture notes should be found, they would make a sizeable book ....

[38] Alexey Andreevich Lyapunov (1911–1973) was a Soviet mathematician. He was one of the founders of cybernetics, and also worked on real analysis, set theory, programming, mathematical linguistics, and mathematical biology.

[39] Possibly the Moscow International University, founded in 1991 by presidents Gorbachov and Bush.

To summarize: work on computer models of global processes in the biosphere and on problems of evaluating the outcome of large-scale nuclear war, culminated in my retirement. I tolerated the change quite well, however, since I had work on several fronts, of both a methodical and methodological character, ready and waiting.

All's for the best in the best of all possible worlds. Finding myself alone with my computer, I felt as if I had attained to a freedom of thought I had never before experienced. And I set about taking advantage of it. But that's another story.

Chapter 11
My Agricultural Career

Remembering Mark Twain

There is a delightful anecdote by Mark Twain about his editorship of an agricultural newspaper,[1] and the sequel. The episode the great writer describes is not typically American, however. How many unqualified persons, for whatever spurious reasons, have been appointed editors of our own Soviet newspapers, for instance! Are the competence and propriety of an editor so important, when all is said and done? There are far weightier qualifications rendering an ignoramus suitable for an important position in the mass media. We know the consequences of such postings from experience, most often bitter. It's hardly necessary to give examples. Although Mark Twain's story strikes a chord with us, we fail to be surprised by the assumed unusualness of the situation it describes. With us the incongruity might be much greater!

I wish to describe an extremely improbable event which could only happen in *our* country, however: how one Nikita Nikolaevich Moiseev, despite being unable to distinguish parsley from celery except by taste, became, if you can believe it, a full member of the Lenin All-Union Academy of Agricultural Sciences of the USSR.

Notwithstanding the fact that it partakes of a large element of the humorous, this tale had significance for my work activities and fortunes generally. The story of my transformation into an agricultural academician has its origin in events occurring some time ago, which helped me to understand many of the principal ways of my country and were crucial to me when it came time to think about ways and means for us to get out of the bog fate had led us into.

But I shall relate all that in the right order.

[1] "How I edited an agricultural paper," published in July 1870 in *The Galaxy*.

N. N. Moiseev, *How Far It Is to Tomorrow...*,
https://doi.org/10.1007/978-3-030-96651-5_11

Concerning Ivan Nikolov and the Utility of a Holiday at the Golden Sands Resort[2]

At the time of writing, a very interesting person by the name of Ivan Nikolov—in Russian, Ivan Nikolaevich—continues to live and work in Bulgaria. A professor of political economy, in the 1970s and 1980s he served as director of the Institute of Management under the aegis of the Central Committee of the Bulgarian Communist Party. For those times Ivan Nikolaevich was rather free-thinking, with an opinion of his own on each and every topic, and for this reason the Bulgarian powers-that-be did not pay him much. His record as citizen and Party member was brilliant. Before the war he had been a member of the underground and a partisan, and when Bulgaria went over to Hitler, had fought against the fascists. He was well known in his specialty, and not only within Bulgaria. He was the first economist among those I knew in the Soviet Union to openly criticize the legal interpretation of the phrase "public property" adopted in all socialist countries. If one notes also the fact that, apart from certain minor quirks, he was a decent, open-hearted person, then it becomes clear why he had the respect and fellow-feeling of the Bulgarian intelligentsia but was not liked by the authorities. In the Soviet Union, in certain, in particular, Party, milieux he was also well known and respected as an authority, and this would prove to be of some importance to me.

In a word, the top brass didn't like him, but left him alone in the prestigious position of director of an institute linked to the Central Committee. However, although at the time it would normally have been difficult to remove him, Zhivkov[3] eventually found a way to do so, appointing in his place Ognian Panov, a person of very different calibre, mediocre but obedient. This occurred somewhat later on, however, and has no immediate bearing on my story.

Nikolov and I were friends, with much in common. There was our collaborative work, especially from the time when I became head of the Department of Computer Science of the Academy of the People's Economy.[4] We held the same views, seditious back then, on many political and economic questions. And it was of no small importance that we were of the same age and so both belonged to the war generation. But of all that united us in friendship, most important was our mutual love of nature, travel, and sport. In the 1960s and 1970s, I often travelled to Bulgaria, where I hiked with him on the high meadows of the Pirin Mountains,[5] and visited his birthplace Petrich, the capital of Bulgarian Macedonia. There we

[2] A resort on the Bulgarian Black Sea coast.

[3] Todor Hristov Zhivkov (1911–1998) was the communist leader of the People's Republic of Bulgaria from 1954 to 1989.

[4] Now known as the Russian Presidential Academy of National Economy and Public Administration.

[5] A mountain range in southwestern Bulgaria.

were once entertained by a blind clairvoyant[6] who described my interior world with astonishing accuracy, and from Petrich climbed up to the place where the borders of Greece, Bulgaria, and Yugoslavia met.[7]

One Winter, Ivan Nikolov and I attempted to climb Musala, the highest peak of the Rila Mountains.[8] We were both fifty years old, but in excellent alpine form. However, a fog descended on us, a heavy snowfall set in, and we were unable to locate the desired peak along the crest. Moreover, we missed by a mere 10 m or so being swept away by a terrifying avalanche that rumbled by. We spent the night in a mountain *khizha*, as the Bulgarians call the huts scattered in considerable numbers about the slopes of the crest of the Rila mountain range, warming ourselves with Bulgarian *rakia*[9] while recalling the dangerous ordeal of our unsuccessful climb. Ivan raised his glass to the health of the splendid and eternally young Musala, who had rejected the advances of us two old men.

There were thus abundant reasons for our mutual interest and empathy.

Then one fine day Ivan Nikolov invited me to come to Bulgaria, not to work, but to visit just as a friend, and with my wife. Naturally, as was the custom in those days, the invitation had to be sanctioned by the appropriate Bulgarian higher authorities. Since he had invited me over the telephone, I had to wait a week or two for the official invitation: "The Central Committee of the Bulgarian Communist party formally invites you to come with your wife to spend 18 days at our Central Committee's sanatorium on the Black Sea."

Once I had received this high-level document, I took it to the Presidium of our Academy in order to obtain the appropriate authorization. But there a difficulty emerged. It turned out that although I had a perfect right to travel to Bulgaria on a work-oriented field trip, moreover at the expense of the Academy, to go there just to bathe in the sea was a no-no, even at the invitation of our sister communist party! (The invitation had been signed by the secretary of the Bulgarian Communist Party.) The problem boiled down to the fact that, by virtue of my work in the Computing Centre, I had access to the most highly classified documents, and such secret personages were permitted to holiday and be entertained only at home. So there!

At first I felt discouraged. My leave had already been formalized and I had paid the airfare. But then I had a brain wave—a not unusual occurrence among your scientific types.

[6] Most probably Vangelia Pandeva Dimitrova (1911–1996), or "Baba Vanga," famous clairvoyant and herbalist living in a region of the Kozhuh Mountains of Bulgaria.

[7] The place in question is now where the borders of Greece, Bulgaria and the Republic of (North) Macedonia meet.

[8] A mountain range in southwestern Bulgaria.

[9] A fruit brandy popular in the Balkans.

I phoned Ivan in Bulgaria and explained the situation to him in the Aesopian language.[10] Ivan Nikolaevich understood me at once ("twigged," as my younger daughter used to say then), and a few days later I got another telegram inviting me to visit his institute for a whole month to give a course of lectures and engage in collaborative research. This was enough to clear the way to having my visit officially sanctioned, and what's more since I was no longer going on vacation but on an official field trip, the Academy would pay my way. Thus all that remained was for me to buy a plane ticket for my wife!

On landing in Sofia, we saw a black stretch limousine approaching our aircraft. My wife said: "Look, they've come to meet some big shot Moscow swindler." It turned out, however, that the "swindler" they'd come to greet was me, for I was an official guest of the Bulgarian Communist Party and guests of such high standing had to be greeted with appropriate ceremony, even to the extent of plying them with cognac.

There followed 19 sunny September days on the shore of the warm, caressing waters of the Black Sea, a holiday free of all worry and free of charge to boot. Then back to Sofia and a week of intensive work: I had to meet the terms of the invitation, after all. And then back to Moscow, where the event I've been leading up to took place.

When I entered my office in the Computing Centre, my secretary said to me, not without a certain malicious satisfaction: "While you were swimming in the Black Sea, I had you elected to full membership in the LAUAAS."[11]

At first I thought it was a stupid joke. My wife had always laughed at my inability to identify certain vegetables and berries except by taste. It soon became clear, however, that what I had been told was precisely the truth: I had been elected, moreover unanimously, to full membership in the V. I. Lenin All-Union Academy of the Agricultural Sciences, and moreover this had been done by my own secretary! But how could this be, and in my absence?!

As I shall now relate it, my tale may contain certain inaccuracies, some errors of detail arising from incomplete knowledge. I can vouch for the general trend of events as I describe them, however. All the same, I shall begin with a conjecture based on a rumour, albeit highly plausible.

Events had unfolded as follows. The highest agricultural authority in the land (then M. S. Gorbachov, Secretary for Agriculture of the Central Committee of the Communist Party of the USSR) had decided to introduce into the ranks of the membership of the LAUAAS a few full members of the big Academy—the All-Union Academy of Sciences—specializing in fields of possible relevance to agriculture, but not represented appropriately in the Agricultural Academy. In particular, it was decided to appoint to the LAUAAS an expert in computer science,

---

[10] A way of speaking conveying an innocent meaning to an outsider but a hidden meaning to someone in the know.

[11] The V. I. Lenin All-Union Academy of the Agricultural Sciences.

or, more precisely, in agricultural computer science: the time had come for even agriculture to be computerized.

There was, naturally, only one possible candidate for the position: E. P. Velikhov[12]—"omnipresent Velikhov"—had an almost exclusive claim to it. He was then academician-secretary of the Computer Science Section of the Academy, and his credentials for the position would then have seemed impeccable. He had, furthermore, always been one for cultivating intimacy with the throne, in particular with the future General Secretary. And in any case, who besides Velikhov was there? Well ... here we have another conjecture, but one founded not on rumour but on the word of A. A. Nikonov, President of the Agricultural Academy at the time.

Velikhov's here, Velikhov's there, Velikhov's everywhere, and this put the powers-that-be on their guard. Then, to add to this, it came out that E. P. Velikhov had no experience in processing large masses of data nor, what's more, in working on computational problems arising in agriculture. When these circumstances were communicated cautiously to Gorbachov, he is supposed to have said: "Look for some other academician, one that has at some time worked on problems involving the processing of agricultural data." It was at this juncture that my name came to the top—I shall give the reason for this below[13]—and turned out to be in some sense the only alternative, since apart from me there were simply no active members of the All-Union Academy satisfying the condition of having "at some time worked on problems involving the processing of agricultural data."

So my name was put forward and approved. But then it emerged that I was at that moment at an unknown address somewhere outside the Soviet Union. I was at the Golden Sands Resort blithely swimming while the deadlines for submission of the requisite documents were passing by. A scandal was brewing.

The situation was saved by my secretary, a plucky lady indeed. Without giving it too much thought, she assembled the requisite documents and ... signed on my behalf everywhere where necessary, in particular, at the foot of the document announcing my putative intention to apply for election to full membership of the Agricultural Academy. From that point on everything went swimmingly since I was the only candidate for a lone computer science position, and, being far removed from the internal machinations of the LAUAAS and, moreover, having no intention of interfering in them, I wasn't putting anyone's nose out of joint. So everyone was happy, and all the more so in that I clearly had no intention of introducing computing methods into the work of the Agricultural Academy. The result was predictable: the appropriate box was checked, my election was approved unanimously, and the authorities were satisfied.

Thus it was that a holiday free of charge—a "freebie," as they say nowadays—moreover at the nicest time of year, got me yet another scientific title. Although this was perhaps neither needed by me, nor, to be frank, deserved, all the same ....

---

[12] See also Chap. 10.

[13] That is, how the author had gained experience "processing agricultural data."

This amusing tale serves primarily as an example of the sort of thing that could happen "over here" as compared to "over there," since over there such a situation could never have arisen because ... it was ruled out. And they criticize socialism *à la* Gorbachov and how it was implemented! Well, *that's* how! But I interpreted the affair in my own way, namely as the Soviet system making up for its former treatment of me as an outcast!

A. A. Nikonov and My Friendship with Stavropol′[14]

My appointment as Academician in Agricultural Computing—the word "election" seems inappropriate in this case—had, strange to relate, a certain rational basis after all.

In Stavropol′ there is a Regional Agricultural Institute which was, back then, one of the scientific institutions affiliated to the LAUAAS. It was a reputable institute, with reputable traditions, and a reputable research faculty—at least such was the case back in the 1970s. Over the period of its existence, the institute's staff had accumulated a vast amount of data, an analysis of which might conceivably be of help in formulating rational schemes for the extension of agriculture and the rational use of land in the Northern Caucasus. The institute was then headed by A. A. Nikonov, an agreeable and clever fellow, and furthermore, like me, a war veteran; he had fought well, been wounded, and awarded the Order of the Fighting Red Banner, the decoration most esteemed by front-line soldiers. A few years after my arrival he would become president of the LAUAAS.

There was indeed a great quantity of material piled up at the institute, so great, in fact, that it had become difficult to access. It had become a matter of urgency to bring order to this mass of information.

This could be achieved, naturally, only with the help of modern computing technology, and that is why Alexander Alexandrovich Nikonov turned for help to the Computing Centre of the Academy of Sciences—to me, in other words.

Our institute was then going through a rather difficult time. For many years we'd been working chiefly on problems in rocket and space technology thrown up by the military-industrial complex. The challenging nonstandard computational problems that arose required a developed mathematical culture for their solution and were often beyond the competence of the engineers working in the design bureaux. The Academy of Sciences thus played a leading role in establishing Soviet space technology in its early phase.

But as time went on, the various industrial organizations began assembling their own collectives of highly qualified mathematicians. Furthermore, over the first two postwar decades the military-industrial complex became thoroughly bureaucratized

[14] Capital of a region of southern Russia in the northern Caucasus. A notable segment of the population is made up of Don Cossacks.

and developed a sense of self-sufficiency, especially from the time when parity with the West in missile and nuclear arms was attained. There was less willingness to engage in the risky technological innovation that was so evident in the first wave of design bureaux and which accounted for the upsurge of our defence industry. The functionaries gradually brought in to replace people like Korolev,[15] Yangel′[16] and Chelomey[17] found it more convenient to begin copying Western models than plunge into unknown waters. This was the beginning of our lag behind the West, the start of a gradual diminishment of our technological and intellectual resources.

This trend was first felt by certain institutes affiliated with the Academy—in particular, by our Computing Centre—in the form of a sharp reduction in the number of industrial contracts. The question then arose of finding a new overall strategy and new problems to work on.

M. V. Keldysh, the president of the Academy at the time, insisted repeatedly on the need to find a way to utilize, anywhere where it had a chance of being useful, the experience acquired by applied mathematicians over years of working on projects in armaments, economics, and the management of production. This, the so-called "Keldysh principle," was of great importance, and I supported it whole-heartedly. Applying the principle in practice, however, turned out to be very far from simple.

Despite the fact that the scientific institutions affiliated with the military complex bestirred themselves and began producing problems for the "ancillary sciences" to solve, and were so successful in this that they no longer had much use for the Academic institutes, our Military-Industrial Complex was still unable to get by without good fundamental research. This we understood clearly, far more so, perhaps, than the new wave of functionaries with their heroic stars[18] and Academic titles. And that is why we tried our best to support our research initiatives; Keldysh did what he could in that regard. The resources of the Academy were limited, however, and we had to look for support on the side.

If in the mid-1960s our defence branches were operating more-or-less at the same level as the West insofar as their utilization of the resources of applied mathematics, numerical methods and computer science were concerned, in other areas of productive activity we were in this respect lagging far behind. A process was under way reminiscent in certain respects of what happened in Russia following her victory over Napoleon. Back then the system created by Peter the Great[19] proved unable to match the tempo of the industrial revolution, and the effects of this lag on its armaments capability became evident in the Crimean War: in the Crimea the rifled barrels of the Allies' artillery were what secured their victory, despite the fact that the relevant theory had been developed first in Russia. Although General

[15] See Chap. 1.

[16] See Chap. 7.

[17] See Chap. 1.

[18] Such as those awarded to "Heroes of Socialist Labour."

[19] At the beginning of the eighteenth century.

Mayevsky[20] had worked out the basic theory of the ballistics of rotating shells, and was the first scientist to do so, Russian industry proved unable to furnish the army with rifled artillery.[21]

The process described above as beginning in the USSR in the early 1960s was extremely dangerous to our country. In the initial stages it wasn't so much the military sphere that was threatened, since it had something in reserve. The real danger lay in the lag occurring in areas of activity crucial to the well-being of society at large. If a System is unable to deal with such a falling behind, then in the matter of a few years stagnation must ensue, followed by societal collapse. The underlying problem, it seemed to us, was the pervasive lack of understanding of the implications of the computer revolution, and ignorance of the trend towards higher, energy-efficient, precision technologies requiring new kinds of technical training, new attitudes to work, and new kinds of organization.

As early as the 1960s, we, the natural scientists and engineers, seeing our country's situation clearly and understanding how it had come about, began to be concerned for its welfare. I would say that, instead of being preoccupied with questions of freedom of speech, the dissidents, with few exceptions (such as the physicist Sakharov and the mathematician Solzhenitsyn), would have been better off showing real concern for their country by bringing attention to this trend.

We were frank about the dangers and tried to explain how the latest progress in the scientific-technological revolution had opened fresh perspectives. However, our attempts to stimulate the interest of the various relevant branches of government in the new possibilities opened up by developments in computer science were not very successful. The chief obstacle to such attempts at persuasion was the notion of self-sufficiency that had embedded itself in the consciousness of the monopolistically inclined executive bureaucracy. We were told that "no one needs anything." I went from one office to the next with one and the same message: "Take us on at no cost, take our knowledge and find a good use for it, whatever use you prefer, it's mostly up to you." We were ready, at our own expense—at the expense of the Academy, that is—to refine the appropriate data-processing algorithms, implement new computational systems, and devise routines of automated design of anything at all, from airplanes to seed drills. Unfortunately, however, the System rebuffed practically all of our proposals, and seldom turned to us with requests for help. We were all in the same boat: from the young and energetic V. M. Glushkov, who in Kiev had built up the Soviet Union's largest cybernetics institute, to General G. S. Pospelov, head of the Academy's Defence Section, who had been trying to get contemporary analytic methods applied in the military sphere, and the many others in between, all striving to hold the country back from tumbling into a technological crevasse, or from fatal stagnation.

[20] Possibly Sergey Ivanovich Mayevsky (1779–1848), a Russian general who served in the Napoleonic wars.

[21] According to other sources the British and French artillery used in the Crimean War was unrifled except for heavy cannon used in the siege of Sevastopol.

At the time in question, it was not, in fact, the dissidents who were of most danger to the bureaucrats embedded in our System, but us, since we were nudging them towards the necessity of re-training, of their being forced out of their sinecures, and of the destruction of the old order. Our enlightened actions carried the implicit message, terrifying to many: "If you can't do the job, then get out!"

Back in the 1970s, while pondering the inevitability of catastrophe and the ruin of our System, and thence our country, through its inability to negotiate the latest bends in the path of scientific-technological progress, I nonetheless thought that it would take place somewhere over the horizon, in the twenty-first century, and that it would be up to later generations to deal with the consequences of the collapse. I saw the future of my own generation as a quiet sinking ever deeper into the bog. At other times I—always the incurable optimist!—hoped that a gradual, more gentle transformation to a society organized along more rational lines might yet be possible. I envisaged in this connection a new upswing in technological progress in our country; in fact, such a turnaround had begun, so one didn't have to be a seer in that respect. What I didn't foresee, however, was that the speed at which the "technological revolution" was expanding would be so great as to bring us to the brink of disaster almost immediately. The looming catastrophe was evident even to our ruling *élite*, who were forced to begin *perestroika*. They had at last got the message: Things could only get worse!

But that's a theme for another day.

Let's return to the story of my agricultural career.

As already mentioned, in the mid-1970s we received an unexpected request from A. A. Nikonov to come to the aid of the Stavropol′ Agricultural Institute. It was understood that we would work *gratis*. In any case, by then it was rewarding enough for us that someone had a use for our knowledge. Furthermore, the request had originated with M. S. Gorbachov, then First Secretary of the Regional Committee of the Communist Party,[22] and a rising star in the Party establishment. Naturally, we were happy to oblige. A team of moderate size was assembled, and off we went to Stavropol′.

A homely, pleasant, well-run, southern town with limitless horizons opening to the view from the town's hills, a receptive, hard-working populace, and, it must be noted, cheap fruit of good quality, a boon to Muscovites.... In short, we liked the town. Gorbachov impressed me as a liberal, well-intentioned, intelligent person, although with a penchant for talking too much and turning every dialogue into a monologue. We were also pleased with what we saw at the institute, and especially with its director. Although, it's true, Gorbachov and then Nikonov left for Moscow shortly after our arrival, they both kept up an interest in Stavropol′ while attending to the duties of the important positions they subsequently occupied. This meant that we could always depend on their help in settling organizational questions, a good enough reason for us to feel able to set to work.

[22] Officially residing in the town of Stavropol′.

Gorbachov was replaced by Vsevolod Serafimovich Murakhovsky, a thick-headed fatty who eyed everyone and everything with suspicion. Things went along much as before, however, for the simple reason that an institution's work routine depends not so much on its boss or even his second-in-charge as on its executive apparatus, so that replacement of one chief by another is hardly noticed, at least for the first little while.

So we took pleasure in our work in Stavropol′, and evidently that work bore fruit. My colleagues' chief contribution, perhaps, was the introduction of an appropriate kind of computer culture into the work of the institute, providing its scientific members with the necessary connection to the computing world with its hardware and software. For me personally, however, a different aspect of our work was of greater significance.

Vsevolod Serafimovich Murakhovsky, the new secretary of the Regional Committee, asked me if we could create a personal information system allowing him to see at a glance how the harvest was proceeding, that is, providing him with a daily visual tableau of the results of the harvest so that he could monitor its progress. It was important that he be able to read off what he needed at a glance. Nothing whatever along these lines had been done earlier: the information needed by Murakhovsky was regularly tabulated by the regional computing centre, but the files were too thick to get the gist of, and in any case those in charge[23] had insufficient leisure to interpret them.

Two of our Academic colleagues, both called Seryozha, at that time both Candidates[24] and both still young, began transforming these tables into graphical form, designing a graphical representation of the data from which it would be easy for Comrade Murakhovsky to read off the information he needed, and providing the software used in the transformational process. They also took a personal computer, in those days a rare and precious object, from the Computing Centre of the Academy and brought it to the Stavropol′ Agricultural Institute. Their work done, each morning the information collected by the regional computing centre was transformed into graphical form and downloaded onto a diskette. Then towards 10 am, this diskette would be brought to the office of the Regional Committee and inserted into the personal computer, now installed in the office of the "micro-general secretary."

My two Seryozhas taught Comrade Murakhovsky how to insert the diskette into the computer and which keys to press in order to bring up the graphical display our shining light had requested. That September, when the harvest was coming to an end, with evident pride Vsevolod Serafimovich showed me a whole bookshelf in his office filled with diskettes, arranged like gramophone records. He could now retrieve at will the most essential information about the harvest for any day of his choosing. He was able to see how things stood at each collective farm and Soviet farm of a very extensive region, how their harvests varied from day to day, etc. etc. In

[23] Probably just Murakhovsky himself.

[24] That is, holding degrees equivalent to the Ph.D.

other words, the bridge of the great agricultural dreadnought was now equipped as her captain desired, so that he might the more effectively discharge his anti-market duty.

I might mention that through this work I found myself in a rather delicate position. The computer that had been brought to Stavropol′ belonged to the Computing Centre of the Academy, and as such was my personal responsibility. It had been agreed that I would have it returned when the harvest was done, but Murakhovsky wouldn't part with it. Try separating the first secretary of a regional committee from anything he would not willingly part with! And in any case he really did need the computer, and learned to use it to appropriate ends. I was saved once again by A. A. Nikonov, who got hold of another computer for Murakhovsky. The old one was returned to its perch at 40 Vavilov St. I have since then taken to calling this computer "grandpa of the Russian Navy" since now no self-respecting programmer would deign to work with it. But what your youthful scientific worker finds unacceptable is entirely satisfactory to at least one academician: the computer stands in my study at home and I am word-processing this book on it!

The completely run-of-the-mill work I have just described played an important part in establishing professional relations: doors were opened to me in Stavropol′ region. I was able to travel all over the region and meet with chairmen of the collective farms, sub-regional committee secretaries, and other personages in authority. I conversed with ordinary collective farm labourers, Cossacks, and people from other towns. I learned a great deal over those months, and gradually formed my own ideas of how Russian agriculture should best be organized. I would later do quite a lot of talking and writing about this, but, as always, to little effect.

I have long been cured of any illusions, understanding now how futile one's efforts to do something useful for our country can be, even those aimed at merely explaining how dangerous certain actions may be. All the same, I continue trying to understand things and talking about them. The stimulus to this most probably comes from my subconscious mind, like the frog in the La Fontaine fable, which, having fallen into a pail of milk, should in theory have drowned, but didn't: it kicked and kicked with its hind legs until the milk had been churned into butter.

In recounting above the rather trivial work carried out over three months by two young Mathematics Candidates from the Computing Centre of the Academy, I was partly prompted by what happened as a result: Murakhovsky told Gorbachov how the personal computer from our Computing Centre had turned up in his office and how from then on a computer had become obligatory furniture for the office of the Secretary of the Regional Committee, and as a result I was officially recognized as a specialist in agricultural computer science and made a full member of the Agricultural Academy.

It was, in other words, a plot worthy of that great American writer, but one than could only be realized in the Soviet Union.

Russia's Fate Will Be Determined in the Depths of the Countryside

I was involved with the Stavropol′ region for over ten years, till 1985, up to the beginning of *perestroika*, more or less. And I'm indebted in a great many ways to my intercourse with the town itself, one of the jewels of my Motherland. I learned a tremendous amount there and, what is no less important, made many friends.

A kind of great primal strength can be sensed in those boundless steppes, in the intelligent folk, in their unhurriedness, love of work, and kindness. From chairmen of collective farms to simple Cossacks, they are easy to converse with. Over a bowl of excellent *borshch* at a camp in the fields or a cup of home-made wine in someone's home, I was able to put the most delicate of questions to them. They always told the truth, even though they were sometimes a little wary of the exacting Muscovite travelling about in an official regional committee car. They, likewise, had questions for me, and what questions they were! Enough to make a party organizer's hair stand on end! But it's the fault of the party organizers, even Gorbachov, that they have long ago forgotten how to talk candidly with people.

The only place where I experienced some difficulty was in the Regional Committee Headquarters, although even there I was received respectfully, even warmly. The difficulty I mention had its source in my own concept of truth, however, and was therefore unavoidable. Otherwise writing this book would have been pointless since I couldn't claim to have been true to myself. Yet one more thing that I eventually came to understand!

The situation I'm referring to was as follows. On my travels about the region, I was usually accompanied by one or another "lad" working as instructor in some section or other of the Regional Committee. This suited me since all the practical bother associated with each foray fell to my current fellow-traveller, and he generally refrained from interfering in the business at hand. My fellow-travellers always had their own agenda and concerns far from mine. Their attitude to me was the same for all: they were all alike astonished by my questions, attributing them to my crankiness—what else could you expect from an academician!—or incompetence. This explained the total lack of interest they showed in the discussions I had with the people we met. They took no part in them whatsoever. This suited me down to the ground, and on my side I tried not to interfere in the doings of my travel-companions. But, alas, I wasn't always able to refrain from interfering.

On one occasion, we arrived at one of the big Cossack settlements around 60 km from Stavropol′. Everything in the town interested me, even though I was in a hurry to get to the field camp where we were expected. We drove along wide streets between solid houses, and suddenly found ourselves in the market place, or, more precisely, in a bazaar or, in local parlance, "collective farm market." My travelling companion stayed in the car and I got out to stroll about and have a look at the life around me. The market was extraordinarily scanty despite the fact that it was grape season. The fruit-sellers had come from far away and were all "Caucasian nationals,"[25] as they say nowadays. There were hardly any locals, and, although of course not as high as in Moscow, the prices were still pretty steep. When I asked my travelling companion how this could be in such a rich region, he just shrugged and said: "I don't go to markets." The canonical answer!

Next I went into a dining hall, which called itself a restaurant, and ate, and made my companion eat, the standard "people's lunch," costing a ruble, which for the time of the year and the money of those times was by no means cheap. The meal turned out to be of the "people's norm" sort, and very average at that. Once again a query, and once again a dumb shrug of the shoulders. I would guess that a great deal of time had passed since my travelling companion had played the role of ordinary mortal in an ordinary eating establishment.

Then, to the latter's annoyance, I went to the town library, which consisted of a few clean and cozy rooms. The chief librarian, a local Cossack of the female variety, recently retired from her position as school teacher of Russian, was a decent sort, intelligent and not without a sense of humour. A few words of conversation sufficed to established mutual empathy. Sensing my interest, she talked for quite a while about how she lived day by day, and about the region and its traditions. And while she talked I was thinking: what pleasant people live in these Cossack towns, how easy it is to have dealings with them, how well they know and understand everything, and how capable they are. How important it is not to get in their way, not to hinder them in their life and work! As I said goodbye, she invited me to return the next day to partake of tea. I promised to return, and was in an excellent frame of mind as we travelled to my *rendez-vous* with the collective farm's chairman.

At the collective farm's administrative office there was no one to greet us, and my companion voiced his irritation: "He was told I was coming!" It turned out the director was at a field camp, some 20 km further on across the boundless, freshly harvested steppe. On arriving, we were surprised to find ourselves in an extraordinary place: a small hamlet in a deep depression, with a babbling brook, and an apple orchard all around. I thought to myself how splendid it must have been there in April or May! But even in August it was pretty good.

The chairman turned out to be of the same age as me, and, seeing on my breast-front the Order of Leningrad, was immediately inclined to look favourably on me. He began talking about the war. It turned out that we had been assigned at the same time to Khvoinaya Station, on the Volkhov front, only then I had been a senior lieutenant and he a senior sergeant. After a few minutes of mutual reminiscence, the chairman began addressing me as "thou"[26] and I had become for him just a rather inquisitive "Nikolayich." For some reason he didn't like the name Nikita.[27]

The chairman took us to lunch, which was welcome after the meagre "people's lunch" we'd had. We dined on perfectly fresh *borshch* with sour cream and some

---

[25] That is, from the Caucasus.

[26] That is, using the familiar form of the second person pronoun.

[27] So he used a standard abbreviated form of the author's patronymic instead. It seems he didn't like the name "Nikita" because it was also Khrushchov's.

kind of meat as second course. There was a jug of milk on the table and watermelon for dessert. And the bread! I hadn't eaten such bread for a very long time. I told our host about the market where Caucasian nationals sold their produce, and about the "people's lunch."... Frowning, he exclaimed: "I know, I know all about it, Nikolayich! If those youngsters didn't prevent me"—he glanced at my travelling companion—"I would flood the whole town with produce. Then mustachioed guests wouldn't need to lunch here. And the restaurants would serve more than just preserves. Do you know how much I charge for the lunch you've just eaten?"—he was now very worked up—"You won't believe it. Half a ruble. And I still make a profit. Sometimes as much as 20 kopecks. Yet even though they charge a ruble for the "people's lunch," they still make a loss!"

In the evening I spent a long time with the chief agronomist discussing how things stood. He also was a Cossack, aged about fifty. He had graduated from the Agricultural Institute by correspondence. He had spent his whole life in the fields. When I expressed an interest in knowing how the recommendations of the Stavropol′ scientists were carried out, he answered: "They aren't. In Spring we receive orders: where, what and how.... There's no taking any initiative."

Next morning, on taking leave of that hospitable hamlet, I came on a revolting scene. My travelling companion was shouting at the agronomist, while that elderly gentleman stood at attention before him saying in self-justification: "Don't you worry, Nikolai Stepanovich, everything'll be tip-top."

We got into the car and set off. But as we drove up out of the hollow onto the wide steppe, unable to control myself I lashed out at my travelling companion: "Nikolai Stepanovich! How could you? You might be his son. What right do you have to address him in the familiar 'thou'?" Where does this boorishness come from? And in any case he understands everything a hundred times better than you do." Although my companion was clearly offended he was wary of responding too sharply. I was apparently a guest of Gorbachov, after all, and lunched with the boss. But, frowning, he said something making an impression much worse than any angry squawk: "But you don't know them. Let them have their will and .... They're ruthless in...." In what, I didn't hear. And till we arrived back in Stavropol′ I observed a gloomy silence, sunk in my own thoughts.

A month or two later, M. S. Gorbachov invited A. A. Nikonov and me to meet with him. By then he had become Secretary for Agriculture in the Central Committee, which meant that he was in charge of the agriculture of the whole country. And Alexander Alexandrovich had been elected president of the LAUAAS. The conversation turned to Stavropol′, in particular, to the interaction of the big Academy with the Stavropol′ Agricultural Institute. The atmosphere was professional but very amicable. Mikhail Sergeevich asked about the achievements of the institute, demonstrating a good grasp of detail, and the discussion went along in a smooth and business-like fashion.

But towards the end of the discussion it hit a snag. Mikhail Sergeevich asked me what my general impressions were: how the farm work stood in the Stavropol′ region, how results issuing from the institute were being put into practice. Lulled by the friendly, reasonable tone of the meeting, I said what had been on my mind for some time: "The region's on the rise, that's clear. So many hard-working and competent people. But there's one problem: the Party apparatus of the Regional Committee. They should stop their meddling once and for all." And I proceeded with instances.

As I went on, Gorbachov's face grew darker and darker, until at last he interrupted my tale with the laconic asseveration: "The Party apparatus is much more complex than you think." No monologue of the usual sort, but a cool leave-taking without any parting good wishes. The office's occupant was clearly angered. Only later did I come to understand my tactlessness: one must never infringe on the "holiest of holies."

In our System, the distribution of power, especially in the Party, was based on an unwritten but well-defined pecking order, obligatory for all: from your rank-and-file instructor, to your secretary of the Central Committee, extending, as we were subsequently to discover, even as high as the General Secretary. Being embedded in the Party System, they all observed the rule of one-sided adherence to the "thou" form of address. Everything belonged to the Party apparatus: its members were the only actual owners of personal property. However, each apparatchik's ranking in the system determined his or her prerogatives: those of relaxation, goods and other necessities of life, and, naturally, the right to address as "thou" all those standing a rung or more lower in the Party hierarchy. It was categorically forbidden, moreover, to discuss anything relating to these entitlements with anyone outside the System, that is, outside the Party *nomenklatura*, or even with colleagues standing on lower rungs. What went on behind the green fences,[28] what was discussed there, and what they ate and drank, was none of our business. These were secrets more rigidly kept than all military ones taken together.

Thus my instructor behaved himself badly at collective farms, yelling at elderly people, not because of anything to do with farm business or any mistakes made, but in order to reinforce by the day or even hour the people's understanding as to the land's actual owners. If they, the *nomenklatura*, should loosen just one of the threads binding Gulliver, they risked losing everything. I believe Gorbachov understood the rules of this game better than anyone. I have come to understand that these "rules of the game" evolved gradually, by inertia, and that they even worked against the true interests of the Party and its power, so that obedience to them was leading the Party on to its demise. However, even if they understood the tragism of their situation, there was nothing anyone could do about it. I think Gorbachov knew this. He nonetheless overestimated the System's stalled potential, and this prevented him from setting appropriate goals and choosing an easier means of switching our society over to a "natural" path of development.

[28] Around the dachas of the *nomenklatura*, or ruling *élite*. See also Chap. 4.

The Blinkers of Urban Thinking and the Liberalization of the Countryside

From my travels about the Stavropol′ region and conversations with locals of varied social status, I was able to get a clear picture of the peasantry of southern Russia and of many of the realities of our agricultural system, including some never reported in newspapers nor mentioned in books. And, towards the end of the 1970s, thanks to this close acquaintance with Stavropol′, out of a variety of ideas for a rational re-organization of the pastoral world and its means of production, I began to assemble a unified system. I understood that it wasn't a question of agricultural science or agronomy, and, even less, of computer science or computerization. We have many first-class agronomists, after all, people whose professional competence would certainly suffice for the introduction of rational farming methods, which are at present underused to an extent that is nothing less than criminal.

The main task confronting us today is the re-organization of agricultural production, in particular the present system of ownership, ensuring an appropriate relationship between people and land. At first my endeavours in this direction consisted in thinking aloud and holding discussions with people whose opinions I valued. Then I began speaking publicly on the theme, and, finally, published my basic views in a book titled *Creative Paths.* I soon saw, however, that this book had few readers—perhaps there weren't any!— so had no appreciable influence on people's thinking.

It took me a considerable time to reach my final judgment of the situation: I was prevented from taking in what was in plain sight by my blinkered urban thinking and the difficulty I had giving up certain principles. The first thing I grasped was that the dead hand of the Party apparatus was a bane to every living thing. It wasn't just that your average apparatchik was ignorant, nor that he lacked ability, nor that he was disinclined to penetrate to the essence of the specific problems arising. It was all much more complicated. An apparatchik had his priorities, and acted and gave orders on the basis of his own corporate self-interest as far as the rules of the game permitted. Even if he were well-educated, his actions were still dictated primarily by those rules, and not by the interests of the farm economy of any particular district or even wider region.

However, a sudden repudiation of the established order would be fatal to the tens of thousands of people in positions of power having access to the "nation's body," i.e., its true owners, that special class whose members, fully absorbed in their immediate concrete concerns, have no thought for the welfare of the land, nation, or party. They understood only too well the importance of the role played by the established order in determining their immediate fate. And in their opposition to the imposing of any kind of limitation on our autocratic regime's powers over everything, including the smallest economic details, they will be desperate—but for death rather than life. And all the more so if the question of a different law regarding ownership of property should be considered.

The imposition of the system of collective farms represents yet one more error committed by urbanites. I belong to a generation that experienced at first hand the nightmare of collectivization. This, the new serfdom, was imposed by fire and sword, accompanied by the intentional liquidation of the best of the peasantry. The most reflective and hard-working peasants were done away with! And, once established, the system of collective farms amazed by its absurdity, irrationality, stupidity and mismanagement. It seemed to me, as someone living in Moscow, that the peasants must hate the collective farm system and be dreaming of its complete destruction. So I was shocked to learn that such was not the case, that the situation was much more complex than I had thought: my oversimplified assessment amounted only to the spare-time rumination of a theoretician—and an urbanite, to boot.

Among the peasants, especially the Cossack ones, with whom I had occasion to hold frank conversations, I encountered a great variety of opinions about the problem of private ownership of land and the system of collective farms. There were those who were straight-out aggressive, all of whom had the same attitude: "If only I had a piece of land, I'd show them." One forty-year-old Cossack said to me: "If the authorities didn't interfere, I'd work with pleasure."[29] He then outlined a plan, a combat tactic, as he called it, as to what and how things should be done, what's useful and what's useless. But there were surprisingly few like him.

What was totally unexpected as far as I was concerned, was the fondness of many for collective farm life and the way it was structured. Recalling the opposition of the peasants during the period of collectivization, I had felt certain that they would consider a return to private ownership as a gift from God. But it wasn't like that at all. Although there was much about the collective farm system that the Cossacks didn't like, and which they berated in the most vivid language—the foremen, the chairman's ignorance, the drunkenness of the bosses (and not only these), and the pilfering—for them none of this indicated that the collective farm system should be abandoned. The leitmotif of all they told me was that: "It's good to live in a commune. But if only the people from the Regional Party Committee would stop interfering! Let us live by our own brains!" And this I heard not recently but in the mid-1970s, twenty years ago!

Subsequently, I tried to see if my impressions would be similar in other regions: in Kaluga Province,[30] in Belarus, and in the Moscow environs. I found that their general tendency was the same, although less robust and clear-cut than in the Northern Caucasus, where the Cossack traditions, still quite strong, provide a special setting. Furthermore, the collective farms of the Stavropol′ region were richer, and the people were more content with their lot so less willing to risk it. There were problems, of course: even an inexperienced eye might observe results of mismanagement and bad organization, and the locals were aware that they should have been much better off than they were, living in such a well-endowed region. In

[29] In the original this sentence has the form of a rhyming couplet.

[30] Kaluga is a city situated about 90 km southwest of Moscow.

fact, one didn't need to be a specialist to grasp this—even a mathematician could work it out.

Such were the peasants' sorrows.

But there was much I failed to understand back then. The richness of the area played a role, of course, and the traditions of the Cossacks, especially their communal way of life, which lulled many of them into believing that life could be arranged similarly on a collective farm, with social ranking preserved.

It wasn't just a matter of tradition, however. On one occasion I conversed with a collective farm worker of very advanced years from out of town. He had originally belonged to the impoverished middle-level peasantry and could still remember how farms were run individually. I put the following direct question to him: Would he like to have his own allotment, to work for himself so as to live independently? His reply was lengthy and ambiguous: "On the one hand ... but on the other hand." The sense of what he said, however, was that he wouldn't, under any conditions, want to go back to being a peasant working just for himself. Yes, his living conditions were worse than before collectivization, even though, unlike the Cossacks, he had been a middle-ranking peasant even among middle-ranking peasants; the Cossacks were all more or less *kulaks*.[31] But back in those days, he worked from sunup to sundown, and if he should be given his own plot of land, he would have to work just as hard again. It wasn't so much having to work hard that he feared, however. At present he was under the protection of the government, which, although it did his thinking for him, still and all fed him. "And if there's a famine? And what if new technology should appear? And how would I market my grain? It's not as if you can get it to the bazaar like grapes. So the way things are now, life is more tranquil." It was then that I grasped the true meaning of Nekrasov's[32] lines:

The great chain has broken,
Broken, recoiled and struck
The master with one end,
The serf with the other!...

I thus acquired the deep conviction that, although the collective farms as they then existed could, of course, not long endure, may God save us from a dismantling of them by decree. It all has to be done gradually and very cautiously; it must never happen that the snapped chain recoil on the producer. In the countryside, one comes up against the age-old contradiction inherent in human society generally: very different sorts of people have somehow to rub along with one another. At one extreme, there are those with boundless energy, aggressors by nature, intent on riches or glory, prepared to work day and night and risk everything, even life itself, sometimes in the pursuit of mythical goals understood only by themselves.

[31] Wealthy land-owning peasants, reputed to exploit farm labourers and the poor.

[32] Nikolai Alexeevich Nekrasov (1821–1878) was a Russian poet, writer, critic, and influential editor of literary journals.

At the other extreme are those satisfied with a modest life, avoiding stressful work and, especially, responsibility. For this sort security and stability of existence are of foremost importance. They are terrified of the unknown brought on by changing circumstances.

Although the variety of human characters and goals is the source of inequality between people, and of their struggles with one another and with the difficulties of living together, it is also a source of human happiness in providing opportunities for overcoming all that humankind encounters on history's thorny path.

Thus it was that back then, at the beginning of the 1980s, I grasped the necessity and inevitability of a reorganization of the whole of our agricultural system, and, insofar as my profession qualified me to do so, began to reflect on possible strategies for such reorganization and the ground rules for the necessary—inevitable, I would say—"revolution from above." I tried to share my thoughts with others, but neither the reformers nor the conservatives would listen to me. Although V. A. Tikhonov's[33] views came closest to mine, he had arrived at them by a completely different line of argument. I then also understood that it was precisely in agriculture that the key to the future development of the nation was to be sought. The fortunes of the agricultural sphere are of far greater importance to the nation than tanks or rockets, and industry must learn to furnish the countryside with the needed technology cheaply and in sufficient quantities. The issue is not just the survival of the countryside, but of the whole country. But how is it to be decided? One thing is clear: the urban centres must become economic partners with the countryside, and the market should be oriented mainly towards the latter. And this cannot be achieved without proactive political will on the part of the government.

One day I communicated something along these lines to M. S. Gorbachov. He listened attentively, and although he didn't respond, it seemed to me that from that time on he was more attentive to me, several times requesting that I think certain things over.

So I understood that we would get nowhere without a revolution from above: decrees would be needed. But the kinds of decrees that would give free reign to individual wills, that would allow the natural inequality among people, to which we owe our progress, to come to the fore. The decrees should also be such as to shelter humankind from the dangers threatening it—dangers yet to appear! There are few people capable of providing for their advent.

At this juncture we inevitably arrive at the question of private ownership, or, more specifically, private ownership of land.

[33] Vladimir Alexandrovich Tikhonov (1927–1994), Soviet and Russian economist and political figure. Academician in the LAUAAS.

Private Ownership of Land: What I Understand by This

It's precisely the countryside where today the fate of the country, of our nation, is being decided. And this is so even though the countryside is at present in a state of degradation, and the peasantry no longer exists as a social force. There are several regions where it is not so much a matter of solving the "problem of the countryside," as bringing the countryside back to life, or, more precisely, of creating a new form of country life. Country life constitutes the essential foundation of every culture and society, providing not only sustenance but also functioning as a natural link between man and the land, whence arise many of humanity's moral values. A sense of the "power of the land," that unchanging and eternal human valuation, represents the most important component of a rational society's culture.

I'm convinced that as soon as the key to the resuscitation of our pastoral world has been found and the appropriate process set in train, urban life will also be diverted into an appropriate new channel. To put it briefly: stabilization of society will come via stabilization of the pastoral world. It might be said that this doctrine was the starting point of all my subsequent reflections. It emerged at the time of my "going to the people"[34] and, as I later understood, it accords well with our national traditions. The doctrine might seem controversial to many, but the more closely I come to know Russian life, the more arguments I find in its support.

I cannot agree with Marx's well known statement as to "the idiocy of country life." Everything comes around in its turn. Ways of life may arise that are misunderstood and considered alien although they have their own logic, their underlying rationale. While it's true that civilization is becoming more and more urbanized, the connection to the land cannot be broken! Cultivation of the land will never be replaced by any kind of hydroponics. A sense of closeness to nature is essential to every healthy civilization, and the source of this sense is the countryside! Anyone divorced from nature loses his essential humanity.

In approaching the problem as to how our agricultural system should be reorganized, we first have to address the question of private property. Note by the way that although this question is the starting point, of course not everything reduces to it. So what is the right answer to the question today? There are many published opinions on that score. The position of V. A. Tikhonov and V. I. Belov is closest to mine, although departing from my views in several respects.

The land represents a public value, a good held by the commonality of humankind—not property but a good! It might seem that it should therefore belong to human society as a whole. But by its very nature, by virtue of the fact that it is individuals who till or otherwise exploit it, land belonging to one subject cannot belong to another, and this is a brute fact, not a matter of local knowledge or culture!

[34] In Russia in the 1860s and 1870s, following the emancipation of the serfs by Tsar Alexander II, students and certain members of the middle class, called at the time "Narodniks" from the word "narod" for "people" or "folk," put the words of their slogan "going to the people" into practise by travelling into the countryside in order to convert the peasantry to a socialist political view.

The inherent contradiction has to be understood and reckoned with. It follows that land should be owned by some person or group of people, whether by individuals, farmers, collective farms, Soviet farms,[35] or, finally, the government. But then we need to somehow reconcile private ownership of land with its overriding importance to society at large. And that is why, although private ownership of land is necessary, the way the land is exploited should be regulated by the citizenry. I would propose setting up a special land court with a mandate to deprive a subject of the right to own land if land currently owned by that subject is being improperly exploited—for instance if it is becoming exhausted.

Under a societal arrangement of this sort, through one or another of the locale's social institutions as intermediary, such as our local councils,[36] should these continue to exist, a citizens' government would play the role of landowner, as it were, and the citizenry would thereby have a stake in the effective use of the land and its conservation. Furthermore, in parallel with increased agricultural yield, the citizenry would receive proportionately more revenue in the form of land taxes, which might be used to fund social programs.

Only in such a context, where the rights and duties of private landowners are circumscribed by those of society as represented by its local councils or other appropriate elected assemblies, do I see a rational resolution of the contradiction inherent in the concept of private ownership of land, a means of averting the conflict between the interests of the private landowner and those of society at large.

If a farmer or other individual owner, or a collective or Soviet farm, should prove incapable of working the land, whether granted or purchased, or if the land in question should lie waste, or its fertility decrease, then, as in the case of bankruptcy, it should be put up for sale—but only after a court trial! There should be no ownership in perpetuity. Apart from this, all relations between landowner and government—one representative of the citizenry, of course—should be purely economic ones. The citizens' government should have no right to interfere in the particular way a landowner exploits his or her land, nor, more generally, to dictate how he or she should behave. But the government should provide assistance to the landowner since this is in the citizenry's interests. And for it to be able to provide such assistance, we would need to have government-run model farms, experimental stations, research institutes, a system of regulations, tax incentives, easy credit, and so on.

I'm convinced of the benefit that a pluralism of forms of farm organization would bring, and of the harm resulting from the prescription of a single "best" way of doing things. Human beings' ability to adapt to circumstances is so great that, given the different kinds of agricultural activity and differences in locales, in each case a rational combination of a variety of forms of ownership and social organization can be expected to arise naturally in interdependence with the local culture and

[35] In Russian *sovkhozy*. In contrast with the collective farms, which were run by collectives of peasants, these were run directly by the Soviet government.

[36] In Russian *soviety*, a hierarchy of which constituted the governing structure of the USSR.

conditions of life. Social engineering is harmful, even dangerous, and especially so when applied to the countryside. Cautious recommendations and demonstrations by example are the most that should be permitted. Although at present society's attitude to collective farms as a form of organization of production is mostly negative, this attitude would seem to be not only nonconstructive, but in fact unjustified. If collectivization had proceeded otherwise than it did, then as a form of management of farm production, in many regions it may well have turned out more competitive, given the fact that feeding a country such as ours requires a large-scale, high-quality farm economy. In our present situation, however, farming will take some time to get up and running. The farmer is a capitalist dependent on the market. When will ours be of that sort again? I note in this connection that the original plan of agricultural cooperation which began to be developed in the distant 1920s[37] was consonant with the social spirit prevailing at that time in the Russian countryside, not entirely extinguished even now.

The centralized command system ruined the free labourer, turning him into an agrarian proletarian. In particular, we must never forget how much harm was done to the collective farm system by my namesake Nikita.[38] His idea was to consolidate the collective farms by liquidating subsidiary farms.[39] In the central belt of Russia, I saw many smallish collective farms grouped into villages, which were quite efficiently farmed. There people knew each other well, they worked with a will, and the generations were in touch with one another. But then one fine day there comes from on high the command to consolidate the collective farms—in connection with the production of corn, in particular—and, worst of all, to liquidate the private plots. Immediate impoverishment followed, in particular of the lives of the farm workers, and the collective farm markets vanished, never to return. In view of this, one should be extremely cautious in carrying out all organizational restructuring. In each region, life as it is lived there will show which organizational structure, under which conditions, is most cost-effective, efficient, and, what is of no less importance, compatible with the traditions and character of the local populace.

By the "liberalization of the economy" I mean allowing in equal measure for the possibility of introducing any of a variety of ways of organizing production, in particular, and, I think, primarily, in the countryside. But of course the idea of a rational organization of labour applies more widely than just to farming. Looking, without prejudice, to the West, we see, of course, the importance of having recourse to various forms of cooperation. For example, in the United States nearly ninety percent of citrus fruits are grown "collectively," that is, in farming

[37] Presumably, this was during the years of the so-called "strategic retreat from socialism" called the "New Economic Policy" (NEP), begun by Lenin in 1921 and ended by Stalin in 1928. See also Chap. 2.

[38] Nikita Khrushchov, who in the mid-1950s instituted reforms in the Soviet agricultural system, including cultivation of large areas of virgin land. Although such measures led to a significant increase in farm production, the Soviet Union remained non-self-supporting in food production.

[39] That is, private family plots. According to some sources, however, these continued to account for up to a quarter of Soviet farm production.

cooperatives, whose regulations remind one very much of what Chayanov[40] wrote along those lines, and which were established for our lands in the 1920s, long before cooperatives made an appearance in California's Salinas Valley.

Farming by individual farmers is an extraordinarily efficient way to organize farm production, as indeed no one would dispute! And this holds true especially when the farmers' cooperative extends to firms engaged in processing their produce. Towards the end of the 1970s, I had the opportunity of seeing this for myself when I was invited to Canada by the firm *Petro-Canada*. On Canada's prairies they grow the world's cheapest wheat, but that didn't interest me so much as what I observed when shown around the dairies of some dairy farmers in Quebec. What struck me at first was the amount of capital needed to make the enterprise cost-effective, and how closely integrated it was with the market, in our case lacking. And in our situation, without government support and a costly program of "farmerization," successful farming operations will remain thinly scattered oases. The second thing I noticed in Quebec was the element of risk involved in determining what to do and how to do it, where money should be invested, how to reckon with the state of the market, and so on. In addition to being trained in advanced agronomy and possessing the technical skills needed to engage with the complex technology they owned, including computers, the prosperous farmers I talked to were qualified businessmen. They knew all the ins and outs of banking and the stock market, and much else that one needed to know if one were not to go belly up. Whence the conclusion: you'll never make a farmer of a peasant. An adequate reserve of wealth and generations with experience of successful farming are what will be required.

As far as I can see, in the meantime we should have recourse to farming operations managed by individuals. Initially, of course, such enterprises will not be very cost-effective or profitable. For the first little while, however, they will not have much competition. Although there will inevitably be some failures, a good many of these enterprises would undoubtedly develop into efficient farms. I anticipate that any advice along these lines that I might give to our government would fall on deaf ears: our rulers suffer from the complex of self-sufficiency, and will long continue to suffer so, despite the shuffling of premiers.[41] I shall, nevertheless, venture a small piece of advice: the government should do all in its power to provide support for those who want to work and have the know-how. No matter how things may turn out, it is just those capable today of working, learning, and taking risks who will give the nation a chance of coming through the present crisis and resuming its place in the vanguard of civilization.

---

[40] Alexander Vasilievich Chayanov (1888–1937), was a Russian and Soviet economist, sociologist, social anthropologist, and writer of utopian fantasy. The term "moral economics" is due to him.

[41] This refers to the political situation starting in 1990, when Yel'tsin began appointing a new prime minister every year.

One more thing I realized in the stagnant nineteen-seventies, years of stagnation indeed: it will be much more difficult now for us to restore our agricultural sector and feed the nation, than it was during the NEP period. In 1921 the peasant villages were still there, and there were *bona fide* peasant landowners who had just regained the right to work their own land. And did they ever work! Two or three years into the NEP and the markets were full to overflowing with bread and meat and all the rest of the rich bounty the countryside could lavish on the starving cities. One should remember also that back then 80% of the populace lived in the countryside, as now in China, and that four peasants fed one city dweller. Now, however, the situation is reversed: farm workers constitute only 20%, so that government aid is essential for the countryside to revive.

But of main concern is the formulation of a strict system of laws making any high-handedness on the part of local authorities of any rank punishable, thereby fostering among the common folk a sense of the law's inviolability, a belief that henceforth it will guarantee the farm labourer his independence, and a conviction that his energy and any initiative taken by him will not merely be taken for granted but proactively welcomed by society.

The Russian peasant's attitude today is one of passivity. Our peasant is weary—weary from all kinds of fuss and bother. The arbitrariness and thievery of his bosses have made him mistrustful of everyone. To give him faith in the future should be society's chief aim!

Such are the free reflections prompted by my work in Stavropol′ Region.

A Modest Conclusion

At the beginning of the 1970s, I began working on problems arising from a study of the biosphere as a single whole, and of its interactions with human society. I have already spoken of that.[42] Around the same time, we received a request to go to the aid of specialists in Stavropol′, thanks to which I was given an invaluable opportunity to experience our agricultural reality "as lived." This coincided with the initial stages of the formation of the worldview which I later called "universal evolutionism." I was given a unique opportunity to study the dynamics of processes at work in the countryside: the evolving organization of the life of the people with its hierarchy, among a host of other features of country life which I interpreted as instances of the mechanisms of natural self-organization.

The young fellows who accompanied me on my journeys about the region were astounded at the questions I asked, and soon showed their indifference to me. And in any case, could I have explained to them what I myself had difficulty grasping?

When setting out to write this chapter, my intention was to tell the story of how I, someone far from being a man of the soil, became, by the merest accident,

[42] See Chap. 9.

an agricultural academician. Although I had wanted it to be a funny story, it has turned out to be a serious one after all. Although my election to membership in the LAUAAS was of little relevance to me or my work, on the other hand, the ten years I spent communing with people of the soil turned out to be very rewarding indeed. I am sincerely grateful to A. A. Nikonov and M. S. Gorbachov for providing me with such an opportunity.

Chapter 12
A "Golden Age" or Thoughts on the Origins of Communism

The Phenomenon of Attractiveness

I've thought a great deal about the reasons for the popularity of the communist doctrine. I myself, after all, was a member of the communist party for almost fifty years, although merely *formal* membership such as mine was, might prompt a somewhat different question. I was accepted for Party membership in the Summer of 1942; a punishing war was under way, and the fate of my country, of my people, was being decided. These are no empty words: we know full well the actual intentions behind the fascist[1] slogans. And even back then we all knew what was going on in the occupied territories. Although I identified our Motherland neither with the government nor the Party, the frontline slogan "Forward, Communists!" struck a chord close to my heart; I wanted to be with those in the vanguard. Both wanted and strove. I strove to recover from my branding as an outcast. I felt myself to be Russian, and here, in Russia, I had to be in the forefront of things.

It's another matter altogether to feel attracted to the communist ideals, however, to believe sincerely in their absoluteness, in their indisputability. It has always astounded me that such strong convictions could be held by the people *en masse*. Frankly speaking, I found such mass belief frightening. What was at its root? Skillful propaganda or something more profound? I should confess that for a long time I myself thought that real communism was coming, that its dawn was inevitable, although, it's true, not today, and not here, but somewhere over the horizon. It was only in the 1970s, when I began my probing of the developmental scheme of the material world, resulting in the theory I call "universal evolutionism," that I began to see these matters, among others, in a new light. Nevertheless, for a long time a resolution of the "ideological paradox" eluded me. Pondering that paradox, I recalled Rousseau's words to the effect that no tyrant is capable of making

[1] Nazi, that is.

N. N. Moiseev, *How Far It Is to Tomorrow...*,
https://doi.org/10.1007/978-3-030-96651-5_12

a people do anything it doesn't want to do, and that is why I still feel that the ideals of communism must be of widespread general appeal.

In reaching an understanding of the attraction communism holds, I was greatly helped by a book I came across by accident, though not so much by the book itself, which is a more or less standard utopian novel, but by the history of its publication: how it became a bestseller and at the turn of the nineteenth into the twentieth century was issued in huge printings. It was for this reason that, in the course of preparing these "reflections," I extracted from my computer an article I had begun writing in those far-off—already infinitely far-off—1980s, when it was still possible to hold a quiet conversation about a future that might hold neither cataclysms nor bloodshed. Perhaps that unpublished article, together with commentaries added later, will allow the reader, if interested, a glimpse of the character of that epoch and the worldview of its author.

Edward Bellamy, Average American[2]

It's 1988 and *perestroika* has begun. I have just finished reading the novel *Looking Backward*, written in 1887 by the Bostonian Edward Bellamy. By means of nothing more than a sort of hypnotic sleep, he transports his hero forward to the last year of the present century.[3] I was curious to see myself through the eyes of a nineteenth century American, and reflect on what might have led him to picture to himself a future so far from the reality.

On publication, Bellamy's novel enjoyed a great success. It was translated into every European language; in Russia it went through five editions. From the late nineteenth into the early twentieth century, Bellamy's book was widely discussed and attracted the interest of Russian *littérateurs* and philosophers. In particular, N. F. Fedorov,[4] philosopher, expert librarian at the Rumyantsev Museum,[5] and one of the founders of Russian cosmism,[6] was moved to write critical essays about the

---

[2] Edward Bellamy (1850–1898) was an American author and socialist influenced by Marxism and Fourierism, and made famous by his utopian novel *Looking Backward: 2000–1887*. The vision of a harmonious future world described in this work, led to the formation of a great number of "Nationalist Clubs" in the US, dedicated to the propagation of Bellamy's political ideas and their realization.

[3] That is, to the year 2000.

[4] Nikolai Fedorovich Fedorov (1829–1903) was a Russian Orthodox Christian philosopher, a founder of the Russian cosmism movement and an early transhumanist. He claimed to use purportedly scientific methods to draw conclusions as to physical immortality and the resurrection of the dead.

[5] Moscow's first public museum, deriving from the personal collection and library of Count Nikolai Rumyantsev (1754–1826), last of his line.

[6] A broad theory of natural philosophy, combining elements of religion and ethics with a philosophy of the origin, evolution and future of the cosmos and humankind.

book. Perhaps it would not be without interest for us also to compare and contrast the prognosis of a late nineteenth century American with the reality in which we find ourselves immersed a hundred years after he wrote it down.

Surprising though it may seem, this book, written by a genuinely "average American of the end of the nineteenth century," as its author describes himself to his readers, sheds more light than many a scholarly essay on certain of the sources of the socialist dogmas that have kept our thoughts and actions in thrall for most of the present century.[7]

From time immemorial—at least since Plato's time but doubtless far earlier in the history of humankind—, it has been natural for people to wonder what lay in store for them, and imagine how, at some future time, in some golden age to come, they would like to have the way of life of the society they belonged to re-organized. Such thoughts arise inevitably in connection with human beings' natural striving to improve their lot. As Goethe says, almost every human being lives with the feeling that "he is born to something higher."

There have always been and always will be utopias, dreams of a golden age; they are part of human nature. I believe that utopian schemes are in fact necessary: they serve to affirm ideals essential to human beings. Man does not live by bread alone, after all.... But these utopias may conceal dangers. Every utopian scheme is, when you come right down to it, merely a social engineering program of one kind or the other, a purposeful reconfiguring of people's lives and the social aspects of their society. And such schemes are sometimes put into practise, and that is precisely where lurks the danger of certain utopian schemes for the fate of humanity. Once the idea of a utopia has penetrated deeply into people's minds, it may breed fanatics, whose actions may be such as to have unpredictable and sometimes tragic consequences. Our own bitter experience must surely convince us of the utter truth of this! That is why social engineering is as dangerous to humankind as genetic engineering. Look before you leap!

In certain exceptional circumstances, however, changing the social system may turn out to be the only way to save a society and perhaps even the lives of the people comprising it. In such cases the needed social reconstruction will inevitably need to be carried out in the context of speculative schemes of one kind or another, and such schemes cannot be other than utopian since we human beings are unable to look far into the future. We find ourselves on the horn of a harsh dilemma about which we can do nothing: change is inevitable yet extremely dangerous. The best we can do is to bring genuine and profound knowledge to bear, in the hope of easing the tragic uncertainty of our situation. The dilemma cannot be fully resolved, however, since it is in the nature of things, and we must therefore take account of it in practice.

---

[7] The 20th.

Utopian Gloom

Although both the authors of utopian works and the times they lived through are many and various, their representations of future societal organizations always have certain features in common. Be it Tommaso Campanella, Thomas More, Fourier, Marx or anyone else, their ruminations on the future are based on common assumptions concerning the developing nature of humankind as a socio-biological phenomenon. These common elements derive from archetypes that have taken shape over thousands of years of reflection: the principles of freedom, equality and justice. The meaning attached to these words has varied considerably over time, of course, but the essence of human striving, as it is given expression in the various utopias, has nonetheless remained practically constant. Your human being has always desired, and will continue to desire, the freedom to choose his actions and way of life, and to be master of his fate. He wants work that is interesting and commensurate with his abilities, and which, within reasonable limits, secures his existence and the future of his children. And, finally, he has always favoured equality, both material and legal. Most people would be prepared to wish that in the world of the future there be neither poor nor rich, and that all be free on the same terms: "the freedom of each guarantees the freedom of all." This, Marx's principle of "real humanism," aptly expresses the general mood.

No matter how much the high-sounding words "freedom" and "equality" are bandied about, however, any contemporary of mine living in this country, and remembering the first famine of the time of "war communism"[8] and all that followed, is bound to read between the lines of utopian writings a summons to repeat what we have just lived through, a period which, I hope, has been thrown forever into the dustbin of history: barracks socialism, unfair equalization, regimentation of private life, and prostration of one's person before the rulers. As Engels is known to have said: "Today a man might be working as an architect, but if need be, tomorrow he'll be pushing a wheelbarrow." Is that not ghastly?!

Our duty to the present government[9] embraces every sphere of human activity, leaving no room for an individual's own personal pursuits. A sort of averaged-down omnivorousness. Recall Mayakovsky's[10] line: "…he'll plow the earth, and write verse." And he wasn't being satirical! For him it was a principle of life, an ideal!

As the utopists would have it, as far as our private life is concerned, we can expect only petty bourgeois security and adherence to the notorious "moral code of communism." A spiritual life, a quest for the truth—for God, if you like—, enquiry into the surrounding world, with all its complexity and ambiguity, controversial problems relating to the inconsistency and ambiguity of people's personalities—

---

[8] The name given by Lenin to the economic system in Russia imposed in part to ease the economic difficulties brought on by the civil war of 1918–1921.

[9] The Soviet government.

[10] Vladimir Vladimirovich Mayakovsky (1893–1930), Russian futurist poet of the Bolshevik Revolution.

all that is to be swept away into the depths of the stage, on which, instead of human passions, suffering, and grief, but also true happiness, even if only fleeting, we are offered a kind of safe but sickly sweet existence.

In all of the ideal schemes conjured up by utopian writers, philosophers, and all those claiming to have laid the foundations of scientific socialism, the desired future of society is pictured as an existence without conflict. Teilhard de Chardin talked of a transcendent future life of humankind, without differences of race or class, when humanity will become merged into a single whole with Nature and God. And the attainment of that final oneness is supposed to represent the culmination of the universal evolutionary process—the end of history, in Hegel's words. Although orthodox Marxists would protest against such visions of history's end, their idol also talked of a certain limiting situation, a certain final state of society, if you like. In one of his best works, *The German Ideology*,[11] he talks of communism as a certain process culminating in the establishment of "real humanism," when, as mentioned earlier, "the freedom of each will secure the freedom of all." He claims that at that final stage of societal evolution, class-distinction will have vanished, and with it all conflict between people. It's true that, unlike Teilhard de Chardin, Marx does not promise the complete absence of conflict in the coming transcendent life: conflict remains, but only between Humankind and Nature, not among people. According to Marx, all that will remain is for humankind to overcome the forces of Nature, rendering her submissive.

Thus all these visionaries of former times saw in the future a sort of "golden age," a bizarre ideal state that human society *must* reach some day, since it's *fated* to do so! And in that ultimate state, in that new communal dwelling-place, a certain general order will have been established—call it a phalanstery, or a barracks commune, or what you will. There all people will be equal in respect of rights, obligations and freedom, and the government will provide all alike with the same delicious food and comfortable quarters. It is precisely this grim society, devoid of interior stimuli to self-improvement and self-organization, that the utopists present to us as an ideal to strive for and one which we shall inevitably reach sometime in the future. At the fork in the road where the Volokolamsk and Leningrad Magistrals[12] split, there stood for a very long time, right up to 1986, a billboard with the slogan: "Communism is inevitable!" So there!

To repeat: the overall tendency common to all the utopian visions of the future organization of society cannot be coincidental. It must surely have its source deep in human nature. A probing of that source would make for a very interesting and important enterprise. A detailed discussion along these lines would, however, take us too far afield.

[11] Co-authored with Friedrich Engels.

[12] In Moscow.

The Search for an Alternative

In the second half of the nineteenth century, the United States was like an active volcano. People were bursting with energy. Industry was developing by leaps and bounds, new cities were being built, financial empires rising, and a railway was abuilding from ocean to ocean. On the one hand, there were the fantastically rich, and on the other, the much more numerous impoverished and disadvantaged. And the number of the latter was on the rise as the result of overt and ruthless exploitation of one group by another. The struggle for rights and a place in the sun was growing ever more desperate, as indicated by the first May Day demonstrations. In other words, in the America of that period an unruly early form of capitalism of the most unregulated kind held sway, that same capitalism that Marx observed and studied, and of which the general trend was seen only by Eduard Bernstein,[13] one of Marx's most gifted disciples.

Bellamy, American to the core, and a resident of Boston, at the time America's intellectual capital, was a witness to all that. To him, as to a great many of his contemporaries, the unfettered capitalistic system that had become established seemed inhuman and irrational. A re-structuring of society seemed essential, and Bellamy addressed himself to the task. Like every utopist, he tried to represent to himself the ideal way of life that he would like to see established in his country. And the picture he sketched turned out to be attractive not only to his compatriots: his book became a bestseller in Europe as well as America. It would seem that Bellamy had caught the prevailing mood, and the "ideal social arrangement" that he described in detail in his book, was very much in tune with the aspirations of a wide swathe of the reading public. People were looking for an alternative, and Bellamy proposed one in a form that answered to their hopes. (Did not much the same thing happen in Russia in 1917?)

So what did that ideal society look like, that coming paradise called by Bellamy "the golden age"?

It's hardly worthwhile paraphrasing the whole of Bellamy's voluminous novel; I shall confine myself to just the key ideas forming the basis of the social organization he proposes.

Here are its main features. In the first place, over everything there reigns thorough-going "equalization." There is no money and no such thing as payment for work. Instead, each person is given a certain amount of credit, the same for all, which is recorded in a special logbook, in which all of that person's expenditures are also noted, along with everything else that society, or, more precisely, the government, has given him. He has the right to use his allotted credit in any way he chooses. There is no trade of any kind, so no shops. Instead, there are distribution warehouses, from which each citizen is free to order whatever takes his fancy,

[13] Eduard Bernstein (1850–1932) was a German social-democratic Marxist theorist and politician who rejected the parts of Marxist theory based on Hegelian metaphysics. He was member of the Reichstag for Lower Silesia 1920–1928.

provided his credit is not thereby exceeded. These orders, made on the basis of samples of what is up for grabs, form the basis of the system of production, which is designed by the highest governmental authorities. Thus in some ways, though not all, this "golden age" resembles a Fourierist phalanstery.

Secondly, the government has ownership of absolutely everything: land, buildings, factories, roads,. . .. Every person, be he peasant, doctor, engineer, or whatever, is nothing more than a worker in the one and only enterprise: the government. The citizens own nothing beyond things used for their immediate personal needs.

Bellamy also describes the process whereby his golden age is to be established. In essence it is one of monopolization, and he goes on a great deal about the blessings this would bring. One should remember that at the time Bellamy was writing his book, monopolies were spreading like wildfire. In his tale, everyone gradually comes to appreciate the benefits brought by the concentration of industry and the establishment of monopolism, so that the process is unhindered, favoured as it is by the whole of society. It culminates in the unification of all enterprises in a single whole, owned by the nation. The real owner, however, is the government. Everything superfluous, in the sense that it has nothing to do directly with the activity of production, disappears, in particular, advertising. And competition, of course, is no more. Market relations are replaced by distributional ones, acting via a precisely functioning system of accounting. (This recalls Lenin's dictum: "Socialism is keeping accounts!" and also the disappearance of the word "shop" from the Russian dictionary of the early 1920s, when in its place the word "distributor" was to be used.)

Bellamy talks a lot about how the switch to the "single enterprise: the government" would simplify the management of the economy. The managerial, i.e., distributional, process he describes seems to him so logical and simple that "only a fool would find it unacceptable." Furthermore, the process would be so well organized as to represent a going concern, so would not require highly qualified managers to run it. Hence employment in the Washington government would be open to even the most mediocre of people: the government might be run by a cook![14] Managerial positions would be prestigious though not onerous.

Bellamy also predicts that towards the end of the twentieth century there will be no further need of prisons in America, for the simple reason that the social sources of criminality will have been completely eliminated: no money, no private property, nothing to buy, and nothing to sell. Hence stealing and burglary, in fact any crime of property, are rendered meaningless. There remain only accidental crimes, and crimes against persons, to combat which it will suffice to set up a citizens' court "to administer justice." The concept of a person's "rights" will have become an archaism.

It would indeed seem that, under these conditions, neither a code of laws nor lawyers or prosecutors would be needed. The only crimes that might occur from time to time in the society of "the golden age," would be those expressive of a sort

[14] This is a veiled reference to Khrushchov's boasting that the USSR could be run by a cook.

of atavism, and the perpetrators would be subject to cures in special hospitals. Thus, instead of prisons there would be hospitals!

I would say, however, that the notion most germane to the social structure described by Bellamy, is that of an army of labour. In "the golden age," work will be compulsory in much the same way that military service was compulsory in the nineteenth century. And the hierarchy in Bellamy's army of labour is also similar to that of an army: new recruit, rank and file, officers of various ranks, etc.

The life of an individual is divided into three periods. Till age twenty-one, his or her life is devoted to study. In this connection, everyone gets the same higher education, *gratis* but compulsory. Then, for the next twenty-four years, each citizen serves out his time at work like a soldier serving out his tour of duty. In accordance with a rigid scheme devised by the government, people are assigned occupations in line with their abilities (but depending also, it's true, on the wishes and needs of society) and slotted into the appropriate kind of work, automatically ensuring their promotion up the social hierarchy from new recruit to general. However, whatever the position in the hierarchy of one or another member of the army of labour, all alike, from private to general, are to receive the same allowance,

Finally, once a person has attained the age of forty-five, he leaves the army of labour, is absolved of the requirement of doing further work of any kind, and spends the rest of his life enjoying "agreeable leisure." His allowance—credit, that is—remains the same as it's always been. However, any person may elect, should he so wish, to undertake one or another social responsibility, chiefly in the management or judicial spheres, where his twenty-four-year-long experience in the army of labour might be of use.

Society, or, more precisely, the government, assigns value to the activities of its citizens not by means of material rewards, but by advancing them in the social hierarchy or conferring other marks of society's esteem. Nevertheless, a person's social ranking is determined not by the final outcome of his labour, but the degree to which he has realized the potential inherent in his abilities. Thus it is not the worker who works more and better who is considered best, but he who has come closest to reaching the limits of his capabilities. In other words, "from each according to his ability." Yet how likely is it that Bellamy was acquainted with the *Communist Manifesto*?!"

Thus in Bellamy's society there are no material stimuli, only moral ones or those related to personal ambitions, to social prestige. Thus here, of all the possible engines of societal development and productivity, only competitiveness over one's ranking in the social hierarchy is operative. And this has to subsist amid a rigid military discipline excluding all laziness or negligence.

The author of this "sociological novel"—such is its subtitle—considers that only a clear understanding of what lies at the root of personal gain will provide enough impetus to set in motion the mechanism of transition from the old "primitive capitalism" to the barracks communism he proposes in *Looking Backward*. He believes that neither revolutions nor any other kind of violent social upheavals will be necessary for this transition to occur, but only a gradually dawning of joy in people's hearts and minds at the prospect of the new social order. Its rationality alone suffices for it to recommend itself.

Bolshevist Ideology, An Average American's Ideals, and N. F. Fedorov's *The Philosophy of the Common Cause*

What I have described above is perhaps enough to show what Bellamy had in mind. But for us his own frame of mind is not really so important. What is of much greater significance is the fact of the tremendous resonance his novel had with the society of the time, that it responded to the mindset of a wide swathe of the educated public of that time, in particular in our own country. The reaction of the public to Bellamy's novel shows that many, perhaps very many, people were just then looking for an alternative to the reigning acquisitiveness, as well as relief from painful anxiety about what the morrow might bring, finding in the novel the portrait of a more peaceful, satisfying, and measured existence oriented towards the concept of "agreeable leisure" at the core of Bellamy's ideal society.

Comparing the ideas expounded in Bellamy's book with the principles expressed in connection with the October Revolution, we come to a better appreciation of the atmosphere in which the doctrines of Bolshevism and its sources were formulated. Although Russia is, of course, very different from America, people are much the same regardless of the continent on which they happen to live. And Marx was far from being the single main preceptor of Bolshevist ideology, no matter what the Bolsheviks themselves may have claimed.

Just as with Bellamy, in the Bolsheviks' ideology private ownership of property is ruled out and the market replaced by a centralized distributive system. And Bellamy's book shows that the idea of an army of labour was in the air long before the October Revolution. However, the attempt to realize these principles led not only to a barracks society and subordination of the individual to the government, but also to bloodshed and genocide.

The fact that Bellamy's book went through five editions in pre-revolutionary Russia is a sure indicator that it had a significant influence on critical thought there. On the whole the response to the book was more than merely sympathetic. Among the more critical responses we find the aforementioned one by N. F. Fedorov, in which, without the least criticism or even commentary, he accepts most of Bellamy's proposals: the idea of an army of labour, the levelling out, and the implied freedom from conflict of such a future society. Evidently Fedorov found it completely natural to expect that these principles would be implemented in the society of the future. He reacted negatively to only two of Bellamy's ideals, namely the lack of a common cause and the ideal of petty bourgeois security: "When the ladies rose from the table [in a general, free dining hall],[15] they [the gentlemen] sat on for a considerable

[15] This and the following two insertions were made by the author.

while, sipping wine and smoking cigars [also provided free of charge]." Was it worthwhile transforming society over a period of a hundred years for the sake of free cigars? That's what worried Fedorov.

The main work published by that philosopher up to that time was his *The Philosophy of the Common Cause*. Like all Russian cosmists, Fedorov regarded humanity as a natural part of the universe, but one endowed with Reason and Will, thanks to which it has tremendous power to act on the surrounding world. According to him, humankind should devote all its resources to the securing of the harmonious development of humankind and nature, transforming the confrontation of man with nature into a relation of benevolent concord. He regarded the establishing of this harmony as *the* common cause of all of humankind. Although he called this process one of "subordination" of nature, the meaning he gave it was the same as that given today to the phrase "the coevolution of humankind and nature": the directed development of nature and human society ensuring general progress.

It should be mentioned, however, that he didn't consider harmony between humankind and nature as an end in itself, but rather as essential to the persistence of the human species on sinful Earth and the resurrection of the dead. By this he doesn't mean the Christian "resurrection of the soul," but actual physical resurrection! And he believed that people should prepare themselves for this ahead of time. However, it's hardly worthwhile our discussing this thesis here, lying as it does beyond the confines not just of science, but even of religion.

From a modern point of view, what is of greatest import in Fedorov's work is the more rational portion, largely having to do with affirmations as to the necessity of achieving harmony in the mutual relations between Nature and Man through appropriate changes in the behaviour of the latter. The sense in which Fedorov intends this assertion is closer to Vernadsky's take on it than to that of Teilhard de Chardin. The latter suggested that the merging of Man with Nature was inevitable, that it would occur regardless of what humankind might desire or how it might act, while Fedorov talks of the tremendous effort needed by Humanity for it to be able to establish "normal" relations with nature. In the absence of such a "common cause," harmony, and consequently also the future growth of the human species, will be impossible to achieve.

Thus in his essay, Fedorov comes down hard on the concept of a mindless and idea-less existence, on the idea of life lived, from the age of forty-five on, merely for the sake of pleasant relaxation, which Bellamy proposes as an ideal to be pursued by the whole of humanity. Instead, Fedorov considers that the chief characteristic of the society of the future will be a striving to further "the common cause," that society's efforts will be dedicated to gradually resolving the crisis in the mutual relations between Man and Nature. Not petty bourgeois invulnerability, then, such as Bellamy describes, when, having completed his enforced tour of duty in the army of labour, a person has earned the right to devote himself to "pleasant leisure" at his own discretion. Rather, as Fedorov sees it, a person's life should henceforth be entirely taken up with the work and struggle involved in aiding "the common cause."

* * *

Thus the ideal social organization described in Bellamy's novel was evidently not far from the concept of an ideal future society very widely espoused at that time. Perhaps his utopia was a sort of "tendential limit" of the ideas of the reflective liberal intelligentsia of his time, since it attracted the attention of the reading public and met with general approval. It wouldn't be too much of a stretch, perhaps, to call Bellamy's ideal society a communistic one, or at least a variant of such. Not merely socialistic, but precisely communistic, since just as with communism so also in the society of "the golden age" is class distinction eliminated, along with private ownership and conflict between people, the freedom of each securing that of all. In both cases, furthermore, production is centrally planned, each person is supplied "according to his needs," and, finally, work is demanded from each "according to his ability." Weren't these precisely the ideals implemented in the first years of the revolution—fated to determine our lives over three generations!—in terms of which the Bolshevik faithful imagined the "golden age" of humankind—although, it's true, only after "setting the world on fire" and razing everything to the ground? In this respect their vision was unlike Bellamy's, where the transition to the new social order was expected to be achieved painlessly, since "only a fool would be incapable of appreciating the good things that would accrue to all of us."

Wasn't complete submission of individual to government also one of our ideals? Not to society, but precisely the government, as represented by that group of persons who had assumed the right to speak on behalf of the people and presumed to know what each citizen needed better than he himself.

We have all seen to what tragedies the putting of such ideals into practice leads. The source of those tragedies is to be found not in particular nations or peoples, but in the structure of the system based on such ideals. Wherever these were adopted into practice, there followed the spilling of blood, oppression, a lessening of human dignity, and a retardation of development, both material and spiritual.

Nowadays we know very well how exceedingly utopian social organizations like Bellamy's "golden age" are: their realization is impossible in principle, since an individual human being will always remain what he is, with all his passions and designs. But we know even more: that social arrangements of this kind inevitably degenerate into totalitarianism and dictatorship, and are accompanied by bloodshed and oppression. That's what experience has shown us—and according to Marxist philosophy, experience is the best guide to reality.

In the West and now even here in Russia, Marx, Lenin and the other fathers of communist ideology are generally vilified, and that ideology's historical sources are also subject to negative criticism. However, all the propagators of communist ideology were children of their time, and so incapable of proposing any other ideals than those they did in fact advance. Back then, in the times of the Klondike gold rush and early capitalism, all educated people, observing the injustices and evils of "savage capitalism," were naturally prompted to imagine a diametrically opposite system. Such a system could only be one of absolute equality, with human relations ordered according to some kind of "barracks" mode of living, since that is the only possible one under the condition of equality. And there could be no room for the market in any form, nor for private ownership nor any other form of competitive striving, and therefore any kind of independent thought.

Among Marx's nineteenth century followers, perhaps only Eduard Bernstein was capable, at the end of that century, of penetrating the mists of time and catching a glimpse of the new horizons the societies of the developed nations would reach. It was his feeling that the principles of social security, the primacy of the individual, and observance of civil rights would gradually prevail against the greed, ignorance, and neanderthalism[16] characteristic, alas, of humankind. And, according to him, this would not be brought about by conscious activity on the part of the better part of humankind capable of logical reasoning, but simply as a matter of course amidst a society of ordinary people not a whit different genetically from ancestral hunters of the woolly mammoth and sabre-toothed tiger. The change was to come about again via a logic of history, one of "iron necessity," or the survivability of people possessing the same qualities as had given rise to the "Klondike era" and "savage capitalism." The transformation of society would not be carried out by means of a process of social engineering managed by a small group of chosen sages or members of a sword-bearing clan, but via the same market mechanisms, based on competitiveness and everything else that one might naturally call the creation of the millions.

Nations, like individuals, choose a path into the future, sometimes by accident and sometimes as the result of a struggle or purposeful attempt to realize new social structures. Then life itself takes over, pitilessly rejecting whatever fails to conform to "nature's strategy." This is precisely that process of self-organization by means of which those who stray from the natural path of development, who wrongly judge the trend of worldwide development, recede from the footlights of the historical stage into its depths. Now, in the late twentieth century, we perceive that the most successful countries are those whose economic systems have departed from the tradition of the "savage market" and gone in quest of a rational symbiosis of various forms of private ownership, i.e., of liberalization, adhering to principles of equality at least as far as initial opportunities, and perhaps even perspectives, open to each person are concerned. Only the first few steps have been taken in this direction so far; there is much yet to be done. But we can already begin to make out the path along which civilized governments will lead their people. We also know, however, that in following that path we shan't reach the "heaven to come" or any "golden age" such as the utopists dreamed of.

One should not underestimate the role of human beings' reason and will in the world process of self-organization. On the basis of an analysis of the past, knowledge of the environment, and the particular ways in which productive forces develop, together with an understanding of the limitations on our ability to foresee the future, we should, rather than devise utopian schemes, concentrate all our efforts on shielding humanity from possible crises. In this connection what is of greatest importance is that we prohibit the social engineering dictated by utopian

[16] An unmerited slur, perhaps, against the character of *homo neanderthalensis*.

constructions and every kind of fanaticism pretending to be in the know as to absolute truth and able to see beyond history's horizon and lead the people into a "golden age." Our history[17] has provided humanity with a harsh but graphic lesson. It's true that we shouldn't overestimate the value of that lesson: Hegel once said that the meaning of history is to be sought in the fact of our never being able to learn anything from her. But one might still say, at least, that the extent to which such a lesson is just depends on how clear it is.

But what of a "golden age"? Will such a thing arrive someday?

It was, is, and always will be: the golden age is just the continuing struggle to overcome the fresh obstacles continually arising on the path of development of the human race, and the pursuit of a gradual consolidation of our collective mind and the formation of a new morality. It is there in our everyday life, in our overcoming of the various difficulties resulting from attempts to realize utopias of one kind or another, in the search for a "natural path of development" and in our conviction that the "strategy of Reason" is approaching ever closer to a "strategy of Nature."

* * *

The fear of competition, where as a rule winners are always far fewer than losers, is one of the reasons for the appeal of the idea of general equality and communism. People aspire to equality, ignoring the fact that competition is perennial. Replacing the market by governmental redistribution will not eliminate competition. The maxim "an earring for each sister" applied in practice yields green fences,[18] Gulags, and new forms of inequality.

Communism hasn't disappeared, since it has never really appeared. However, it always was and always will be, in the sense of being embedded in the spectrum of those biosocial laws, against which a positively developing morality has always fought. Its extreme versions have disappeared, although the danger remains of their return in one form or another, a mortal danger to civilization and the future development of Humankind.

And that is the main lesson I took from the utopian narrative of an "average American," written exactly one hundred years ago.

---

[17] Of the USSR.

[18] Around the dachas of the ruling *élite* in the USSR. See also e.g. Chap. 4.

Chapter 13
Russia's Twilight: Dawn or Dusk? Russia at the Crossroads

And black, earthen blood
Promises us, swelling our veins,
Destroying all boundaries,
Changes unheard of,
Rebellions never yet seen. . .
(Alexander Blok, "Requital")

A Time of Warning

Not since the Time of Troubles,[1] perhaps, have the Russian people found themselves in such a catastrophic situation as the present one.[2] The nation is laid low. We have to start over almost from zero, gathering together those truly able to believe in a better future and work towards it.

However, what is at present happening in Russia is by no means specifically Russian in nature, or purely an outcome of its former socialist system. It is rather just one indication, albeit in hypertrophied form, of a general crisis in human civilization, including, perhaps even primarily, Western civilization in both its European and American variants. The "*laissez faire*" principle, originating in pre-revolutionary France, has lost its force—perhaps even some time ago. And many others among the postulates underpinning our way of life have long ago become mere brakes on civilization's development. The Western world is becoming ever more alienated from Nature, its people are becoming degraded, and the spirituality so necessary to sustain human beings in their most dire trials, has faded. The idea of progress is discredited. We are faced with a crisis of rationalism. It is beginning to dawn on many of us that, as classically understood, equality and freedom are incompatible.

[1] The period of Russian history comprising the years of the interregnum between the death of the last tsar of the Rurik dynasty in 1598 and the accession of the first tsar of the Romanov dynasty in 1613. During that time Russia suffered greatly from famine, occupation by the Polish-Lithuanian Commonwealth, and a series of civil uprisings led by false pretenders to the throne.

[2] Just prior to 1993.

N. N. Moiseev, *How Far It Is to Tomorrow*. . .,
https://doi.org/10.1007/978-3-030-96651-5_13

I believe this to be the main source of the difficulties besetting our civilization. Since the beginning of the nineteenth century, we have continued to identify freedom with private interest. In most people's minds freedom means freedom to consume.

We stand on the threshold of a new century, and a new millennium, indeed, in the course of which we shall have to devise a new "existential paradigm" for humankind, or, more precisely, such a paradigm will have to emerge by itself out of the creativity of billions of individual people. It is urgent that we gauge the form that paradigm will take beforehand, since otherwise Blok's prediction, already having come true once, will once again become our reality.

The order established in the first post-World War II years, the world order of the twentieth century, is undergoing a planet-wide process of breaking down, and not only in the former Soviet Union. Thus our Russian take on the future, on Russia's future in particular, cannot possibly be correct if we regard our internal, national crisis as narrowly peculiar to us, unconnected with the current processes of global development generating a sense of insecurity world-wide.

This feeling of uncertainty has been the subject of much thought, and related themes have been widely discussed and written about, including, among many others, those concerning the mutual relations of nature and human society, already so very inharmonious as to make us aware of an approaching ecological crisis and of the inadequacy of modern society's organizational structures to the ever-increasing power of a technogenic civilization. It is my belief that the crisis of modern civilization is partly a consequence of the conservatism of our ideas about the reality we inhabit. We fail to notice that a transvaluation of values is under way, and continue trying to live by standards established in the nineteenth-century European consciousness. We think that those ideals are invariable, or at least that's what we would like to believe. But this is not so. The concepts of freedom, democracy, equality and many others commonly valued by humanity, are constantly changing and being given greater precision, and their current interpretations, bequeathed to us by the nineteenth century, barely answer to modern needs. We continue trying to rely on past ideals, to speak using a language already become alien to us, yet without acknowledging the situation. We desperately need new ways of making sense of everything. Today the appropriate concept of freedom should be freedom to reject evildoing and no longer freedom to do as one likes!

The twentieth century was extraordinary for its greatness, but also for the tragism of much of what has transpired and continues to transpire. Much has been made of the fantastic upward flight of human thought and energy that has endowed us, in a relatively short time, with technological achievements that, even for the coming generation, might seem fabulous. But, on the other hand, much has also been said and written about the outbreaks of savagery and consequent misfortune, not only during two of the most frightful wars of history, but also in the great number of localized conflicts that flared up and continue to fester in various places around the globe, even in a Europe thought to be beyond all that. The twentieth century has become one of warning: humanity must show itself capable of adumbrating the future's general contours so that it can take preventative measures against possible disasters, and not simply bow to the inevitable.

And it's time that Russia, in particular, began looking for answers to the questions confronting all of civilization.

A Biosocial Interpretation

Although the meanings of ideals, among other human concepts, change with time, there are nonetheless certain human values that are immutable. Human beings are Nature's creations and therefore bear fixed birthmarks acquired over the course of their evolution which can change only with a corresponding alteration of their nature. The values I'm referring to are the laws of the *Sozium*, or social laws. Perhaps a better name would be *biosocial* laws, since they date from the homeostasis of the human individual and human society occurring in the early stone age,[3] with origins to be sought in the distant reaches of the past when our ancestors were still being moulded into human beings. The laws in question were gradually framed over hundreds of thousands of years by the conditions of life prevailing when our forebears lived more like other animals. But now civilization, culture, legal systems, and morality have become barriers to changing those laws, so completely different are our lives from those of our forebears, who feared being eaten by sabre-toothed tigers, among other dangers we are no longer subject to. Nowadays, instead of perils to be dealt with by means of aggressive action, we have nuclear arms and other means of wiping humanity from the face of the Earth. Alone reason and universal consensus are capable of saving our civilization, and that is why today the conflict between the ancient biosocial laws governing us and our present reality has reached such a pitch. And the situation can only get worse.

The more detailed our analysis of a phenomenon, the clearer our picture of it is, and the more various our interpretations, the greater the number of perspectives. There is no such thing as absolute knowledge, just as there is never just one interpretation of a phenomenon. Thus future historical trends might be predicted, for instance, on the basis of the perspective lent by an analysis of the conflict between the biosocial laws and civilization's attempts to limit the range of actions stemming from those laws. And we shouldn't abandon historical interpretations already to hand, such as those of Marx and Kant as well as other past thinkers, but preserve them while at the same time keeping their shortcomings in mind.

The dominant, monopolistic, role of human beings in nature is due not only to the development of their brains, to their intellect, but also to the extraordinary range of their aspirations, desires, and abilities. This is a very important biosocial factor, with a multitude of serious consequences. At one end of the spectrum of

[3] But relatively recent research has shown that marked evolution of human beings *has* occurred and continues to occur right up to the present. See earlier footnotes, e.g. in Chap. 5.

aspirations and desires we have the boundless energy of individuality, intolerant of all restriction, while at the other, there is the herd mentality, giving rise to the ideal of the anthill, where each individual is nothing more than a cog, where even digestion is a guaranteed function of the commonalty. This variety explains on the one hand our amazing adaptivity, our survivability under extreme conditions, but on the other the internal contradictions that arise in human societies, which, although they may intensify social development, may just as well distort and even destroy a society.

By the "twentieth century order" one usually means the ideological opposition between communism and capitalism, and this interpretation of the spirit of the century is in many ways an appropriate one, encapsulating as it does its most important general aspect. However, it's very likely that this ideological opposition, if indeed it really existed, is just a surface reflection of contradictions deep within the social essences of human beings, always present, but making themselves felt with unusual keenness in the twentieth century.

In all epochs, the dissension caused by human disparities, by the variousness of human aspirations, has had a highly significant influence on history's course, since of course these differences give rise to differences in people's values.

On the one hand, humankind could not have progressed without venting its characteristic irrepressible energy, propelling us onwards in a never-ending quest for new achievements, new goods, and for dominance over others. ... In this connection it's important to note that such striving gives rise to a corresponding conception of freedom, namely as freedom from all restrictions on one's initiative and the advancement of one's own "ME," the self which emerged long before the era of bourgeois revolutions. This side of human character manifests itself especially saliently in Western culture.

But there is another tendency no less deeply embedded in man's character, namely that whereby he *is* capable of limiting himself, and resting content, indeed, with rather little, provided that little is guaranteed, lies the within framework of tradition, doesn't require him to overexert himself, and moreover that those around him are subject to the same rules, which last proviso is especially important. In every age, millions of people have been satisfied with life lived under such conditions. This tendency also gives rise to a corresponding concept of freedom, but one entirely different from the foregoing. It also has its version of equality: not, as in a *laissez faire* bourgeois society, in the sense of material wealth, but of universally humble living conditions, the same for all. Throughout history, the appeal of such an attitude to the majority of people served as fertile ground for the growth of various communistic ideas and of most religious doctrines.

It is appropriate to interpret what is happening at present in the world as resulting from the ever-present collision of these two tendencies. To repeat: "Communism" has not gone away, alas, since it really never existed, but in one form or another still figures at one end of the spectrum of human aspirations. At certain times and places the "communistic," or, more precisely, "socialistic," tendency predominated among society's prevailing norms, and at others less so, for instance in protestant countries. But both options, among the many intermediate ones, have always and everywhere been there for the choosing. Failure to take this bipolarity of societal development into account, entails an incorrect evaluation of what is occurring in the present, resulting in wrong decisions affecting the lives of multitudes.

At each stage in the development of a society, its established "rules of the game," embody a compromise[4] between "communalism" or "collegiality"—I can't think of more precise word for the second of the two extreme tendencies—and the overflowing energy of individual personalities. Some of these rules may fit well with the traditions and cultural and mental peculiarities of the society in question, and others less so, calling forth or quieting social tensions, promoting social progress and security or hindering them.

What is happening in our country today? To what extent are we able to facilitate the process of reaching a compromise embracing degrees of individualism and communalism maximally concordant with our traditions, needs, and potential?

I'm deeply convinced that at the present stage of unfolding of the historical process, given the way societies are presently organized world-wide, development will proceed overall in the direction of a socially oriented, liberal economy. In the various other writings where I have advanced this view, I have justified it in terms of the particular stage of scientific and technological progress and level of production we have arrived at, along with the emergence of a planet-wide "universal market," rendering impossible in the twentieth century a transition to any form of social organization hindering liberalization and a socially oriented economy.

However, although this may perhaps explain the collapse of the socialist system imposed on the Soviet Union and eastern Europe, it's hardly adequate as the basis of an interpretation of what's occurring at the present time world-wide, so even less as a means of indicating future developmental trends. All of the formerly socialist countries of Europe are embarked on programs of liberalization[5] and the establishment of free enterprise systems. However, the associated reforms, enacted from above, have not yet yielded the expected results. And this the case not only in Russia, but also in those countries which began the liberalization process earlier than us, and where the "protestant ethic," to use Max Weber's phrase, was once part of their tradition. Poland and Hungary and other countries have experienced a decline of industrial output, and the accompanying increase in unemployment has led to a reduction in the standard of living and social unrest.

The explanation for this is to be found, of course, in various objective circumstances, such as the breakdown of relations between countries, a lack of economic expertise and appropriate economic infrastructure, and the ending of traditional economic cooperation, among others. I would say, however, that highly significant in this respect is the circumstance that over the lives of two or three generations, the populace of the countries in question became habituated to living under conditions of guaranteed stability, guaranteed health-care, and so on. This arrangement had

---

[4] Such as so-called "communitarianism," perhaps.

[5] However, since 2018, some twenty-five years after this was written, several of those very countries have been demonstrating populist anti-liberal tendencies.

proved attractive to a large proportion of the labour force, and all the more so for their being touted as members of a "privileged class."

Nevertheless, in all of the formerly socialist countries liberalization has taken root and there will be no going back. This statement doesn't tell us much, however: how long will the transitional process take, and what sort of liberal economies will in the end emerge? At present, answers to such questions are lacking.

What *is* clear, however, is that these emergent economies will not be copies of Western variants, although to say what they *will* look like in Hungary, Ukraine, etc. and, most of all, in Russia, is very difficult. Setbacks followed *perestroika* in every country of the former Soviet bloc, hindrances indicating that a free enterprise system on the Western model might not be achievable in the foreseeable future. In its refined American or West German form it failed to materialize. Thus it becomes important to understand why, even in countries with a traditionally Western outlook like Poland or Hungary, Western values failed to take hold.

I am certain that—to repeat somewhat—the answer is to be found in the abovementioned "socialist values" holding sway in those countries for decades, especially the condition of guaranteed and undemanding employment, established as a very attractive perquisite for the mass of citizenry. Having become embedded in people's consciousness, these values inevitably gave rise to various forms of resistance to the process of intensification of market mechanisms already begun. Nonetheless, we *are* moving in the direction of a liberal economic system. Although it will resemble Western models, it won't be identical with them, but will differ in certain respects, just as happened in Japan and other rapidly developing nations around the Pacific able to forgo the intial accumulative stage of the European variant. It is a far from simple matter to predict what form of liberal economy will replace that of Russia's socialist system: for that purpose it would first be necessary to carry out a precise sociological investigation as the initial step of a prognosticatory analysis of the possible directions of development.

However, at the present time, unfortunately, we do not have an adequate scientific infrastructure uniting sociological and economic research, so we're lacking the wherewithal for such an initiative. And that's why we should be extremely cautious about making categorical assertions.

A Liberal Economy and the Intelligentsia's Responsibility

More than once Russia has helped Europe evade dangers and remain intact. This occurred, for instance, when, ground down by the Mongol incursion, Russia served as a bulwark against Genghis Khan's horde, and, similarly, during the wars of Napoleon and Hitler. The year 1917 was one of terrible warning not only to Europe but to the whole world. Ultimately, however, that warning had a salutary effect on civilization as a whole.

There isn't, and in fact cannot be, one and only one interpretation of the October Revolution. Marxists see it as a culmination of a class struggle, with a victorious proletariat establishing a new order of life over a vast territory. This view is associated with the doctrine of an implacable ideological opposition between those peoples who have made the "socialist choice" and others keeping to the capitalist path. According to this view, an uncompromising antagonism arises, forcing the assignment of priority to confronting dangers from without, concentrating power at the centre, imposing a single prescribed worldview, and so on. All other interests have then to take second place. Ideological opposition and the pressure of external danger were essential to the "workers' and peasants'" government as justifications for totalitarian rule and the "path to slavery" of which Hayek wrote so brilliantly.

But there is another interpretation of twentieth-century history.

The first period of capitalism, the "Klondike era" or that of "savage capitalism," can be thought of as the apotheosis of free and unlimited initiative, embodying the *laissez faire* principle proclaimed by the French Revolution. The many evils of that initial era of capitalism have been described by Dickens, Balzac, and other great writers of the time. A systematic economic analysis of the phenomenon was carried out by Marx and his followers. In that epoch of unregulated market relations, an extreme, hypertrophied, unrestricted outpouring of energy by driven individuals was made manifest—as an extreme hot-house efflorescence issuing from the anti-social part of the established biosocial system of laws, one might say. Many people saw the ugliness of the era and looked for alternatives, of which Marx's proposal represented but one.

There were other prophets. One of these was Eduard Bernstein,[6] perhaps Marx's best pupil, yet vilified not only by Lenin but even by the renegade Karl Kautsky.[7] Observing the rising irrationalism characteristic of the emerging twentieth century order, Bernstein argued against opposing it by violent revolutionary means, being convinced that a gradual growth from an internal socialistic germ would ineluctably transform that order. I would be inclined to put it rather as follows: in a free enterprise society, the elements of a more socialistic orientation might be expected gradually to emerge out of the very experience of life itself in the spheres of economics and of social relations. But that's not all. Roosevelt is supposed to have once said that no one yet knows what a free enterprise society looks like precisely.[8]

In addition to Bernstein and Roosevelt, there were others, such as Keynes and other intellectuals, converging on the idea that in the course of further societal development, "humanizing factors" would begin to take effect jointly with an overarching use of Reason. This trend was supposed to be given impetus by the enhanced productivity resulting from the continual growth of technological complexity and an accompanying demand for more and more highly qualified personnel. Human society should thus progress, so the theory goes, in parallel with

---

[6] See also Chap. 12.

[7] Karl Johann Kautsky (1854–1938) was a Czech-Austrian philosopher, journalist, and Marxist theoretician. He was an outspoken critic of the Bolshevik revolution. See also Chap. 6.

[8] The exact quote has proved elusive. Perhaps this had to do with the deemphasizing of free enterprise during Franklin Roosevelt's "New deals" of the years 1933–1938, aimed at ending the Great Depression.

technological development, although there are other factors at play. And the role of a society's most important institution, namely its government, should also grow in sync with the growth of civilization's resources. The government should continually be alert to infringements of the people's freedom, though not in the sense of early protestant capitalism, but rather in that of Thomas Aquinas, namely freedom in the form of liberation from evil.

Looked at this way, the Russian Revolution takes on a wholly different complexion: instead of opting for a compromise between the two extremes, the Bolshevik revolutionaries declared, and then realized by violent means, an extreme kind of societal order, that of the anthill, one might say, diametrically opposed to that of the Klondike Era. Such an order could only arise at that time by virtue of being antithetical to the awfulness of the prevailing savage capitalism. Ultimately, however, Nature intervened, as it were, to demonstrate the impracticability of such an extreme form of social organization as a resolution of the eternal conflict between extremes. We were horrified at what happened in our country, and no one would dare risk repeating the experiment. But the world's societies have not rejected the positive lesson to be drawn from our experience: they now understand that a nation's economy cannot, above all in times of historical difficulty, do without government input.[9] Russia has yet once more shielded Europe from potential error by serving as a testing ground for an experiment of inestimable value to civilization. It's unlikely that the present European liberal political economies could have been established without the lesson afforded by the USSR.

And now once more we are in the midst of a revolution, and once more Russia is playing its eternal role as a "laboratory for political and social experimentation." Although an incurable opportunist, I hold the deep conviction that revolutions have never brought, and cannot bring, anybody happiness, and that any kind of social reformation must take into account the risk of spontaneous combustion into a real revolution, which can only be retrograde for the society in which it occurs. In Russia in 1986, there still remained the possibility of reform from above, of a gradual liberalization of the economy, and a de-ideologizing of the nation. And a little later the meetings at Novo-Ogarevo[10] provided another chance, and I thought I saw a light at the end of the tunnel. All such opportunities fizzled, however, as a result of the operetta of the putsch[11] organized by a bunch of political incompetents, and the subsequent fall of our great government. The period of *perestroika* came to an end, and a revolution began, with unpredictable outcome. There ensued one of the most dreadful periods of Russia's history: the dividing up of the country's wealth,

---

[9] In the West this lesson was learned also, at least by the postwar generation, from the experience of the Great Depression of 1929–1938.

[10] A meeting at Gorbachov's presidential residence outside Moscow in 1991, at which representatives of nine Soviet republics tried to reach agreement on the preservation of the Union in some form. In the end, several of the representatives refused to sign the resulting treaty.

[11] The August 1991 attempt by several hardline members of the Communist party of the USSR to wrest control of the country from President Gorbachov. It failed after only two days, mainly due to a campaign of civil resistance in Moscow led by Russian President Boris Yel′tsin.

when morality, patriotism, and Russia's security all flew out the window, and with a savage baring of fangs the system of biosocial laws began dictating conditions. I imagine much the same occurs in every "time of troubles."

However, in such fateful times, when life's paradigms are in flux, or, to use a scientific term, history's trajectory is undergoing a bifurcation, even apparently insignificant circumstances can lead to a radical change in the course of history. Trotsky was correct when he wrote that if he and Lenin had not been present in Petrograd in the Summer of 1917, the October Revolution would not have taken place; the war would have ended in January 1918 and our history would have gone off in a different direction. Churchill expressed this thought a little differently: "The Russian dreadnought has sunk at the entrance to the harbour."[12] At such times the responsibility of the intelligentsia is especially great.

And so at the present time, the intelligentsia, enlarged, perhaps, to include all those engaged in intellectual pursuits, are of ever-increasing importance for the future life of our society. If only the members of that social stratum were capable of uniting around a common understanding of our present situation in order to formulate a general plan for a desirable future, including an appropriate list of priorities, they might be able to exert a positive influence on the fate of our country and propagate a set of views as to appropriate national goals, so-called. Although it's tragic when a people's social life is distorted by ideology, nevertheless without some conception of national goals lending a definite perspective to the future, the people and their society stand defenceless, thereby putting the nation itself at risk and rendering the preservation of its culture impossible. An appropriate worldview—one must agree with Rousseau here—cannot be imposed: it must ripen in the bosom of the people. It may be necessary, however, to speed up this process and steer it away from savage extremes, and that is the responsibility of the nation's intellect.

In that regard, not only is enlightenment needed, but also the setting of an example to all. When certain persons claiming to belong to the intelligentsia and to represent the people's conscience, find themselves in positions of a certain degree of power, and then proceed to grab for themselves a big slice of the social pie while omitting to explain clearly the ends their advocated revamping of society is supposed to serve, when they are only interested in securing their own social position, then do we have the makings of real tragedy! Likewise when the mass of intellectuals makes convulsive attempts to shield themselves from hunger in order to survive in the midst of increasing poverty and chaos brought on by a power vacuum.

Despite such misgivings, it remains true that only a nation's intelligentsia is capable of formulating a plan for a "desirable future."

That plan should not be utopian, however, but take into account the current reality, since, as I have argued before, utopian illusions are more perilous even than any idea-less chaos. Every kind of utopia must give rise to one or another variant of totalitarianism; utopian thinking denies liberalism in principle since it prohibits alternatives. In order not to fall once again into the utopian trap, it's necessary to

---

[12] Also quoted in Chap. 2.

review all practical developmental scenarios and the effort needed for realizing them, especially of those uninvolved in the process of general redivision of wealth—or spoils, if that is the more appropriate term. The latter are always in the vast majority, and only through their pressure on those few directly involved in divvying up the economic pie can some influence be brought to bear on the fate of the nation and its people.

Possible Future Scenarios

What's occurring now[13] in Russia does not fit into the framework of any known economic theory—not that of Keynes, nor Friedman, nor Marx. None of these theories provides a recipe for extricating ourselves from the present crisis, for the simple reason that they are purely economic in nature. Our country is at present undergoing a rapid transformation from one socio-economic and political system to another. Such situations are always unique. Furthermore, under the conditions prevailing in such a precipitate restructuring of society, it is impossible in principle to separate out the purely economic issues, and hence exceedingly difficult to predict in any detail how events will unfold. In our country a new type of proprietor is emerging: real owners of real private capital. And new social groups and classes are crystallizing out. A nascent process of capital accumulation is under way. And all this is proceeding very rapidly at every level of the social hierarchy, in parallel with an overall restructuring. Nothing comparable has ever happened in Western Europe or America, so looking to them for guidance is dangerous. We must think for ourselves!

While similar processes of redistribution of social goods have in the past always involved violence and, as a rule, an essential lack of fairness, what is now occurring in our country is all the same quite different in kind from anything known to history. It is distinguished, in the first place, by the scale and celerity with which theft and the concomitant accumulation of individual fortunes in the billions have sprouted out of a vacuum. And in the process of dividing up the fantastic wealth of our vast country, those closest to the former centres of power got the first bite. This explains the bitterness of the struggle for power at present, when classes are no more, though certain cliques have formed. Thus at present purely economic considerations are no more than a backdrop to the action. They will assume greater weight later on, when the mafia and agents working on behalf of foreign interests have been taken out of the game. To say today, therefore, anything about what sort of society will emerge as the end result this nation-wide reconstruction is next to impossible, especially since the mass of people subject to the upheaval have yet to have their say.

But maybe they'll keep silent. Most people are tremendously weary from all the demagoguery and despair they've endured, and the intelligentsia, in fact all reflective

[13] Just prior to 1993.

people, are more and more giving up on politics. At the moment, the ratings of politicians are trending downwards: no one trusts anyone! Under such conditions chance plays a leading role, in addition to occasional lightweight populist slogans. One recalls how recently, although a majority voted for preservation of the Soviet Union, a few well-placed individuals nonetheless decided everything according to their own lights!

I believe that the short-term future, i.e. the next few years, will unfold in one or the other of two ways, which I call that of "weak power" and that of "strong power"; these terms are relative, of course, but everyone will understand me if I say that those at present running the country are wielding only weak power, since there is no respect for the law, there's no justice for victims of crime, and so on.

In view of the cheapening of our natural resources—a particular consequence of the fantastically inequitable valuation of the ruble—it has become unprofitable to sell anything in our country. Anyone with money refrains from investing it in manufacturing, and buys up every natural resource on the market—oil, coal, metals—to sell abroad. They then deposit the foreign currency thus gained in Western banks or else use it to buy rubles at the current extortionate rate, and again buy for peanuts everything they can, and so on, at each turn taking several hundred percent in profits. In pursuit of such profits people are prepared to indulge in all sorts of nastiness; I think it was Marx who made this remark. In any case we now see its truth demonstrated to us directly.

Our country is undergoing a chain reaction causing its natural resources to flood over the border. Yet stopping the flow seems to be difficult, even impossible, without resorting to extreme measures of some sort. It's a consequence of the process of privatization, under cover of which our natural resources are being looted and a bourgeois class of compradors[14] is growing apace.

All this makes clear that the scenario of a "weak government" is what we are facing at present. This is similar to the well known "Argentinian variant," where a country rich in natural resources becomes entangled in a web of poverty, exploitation of its people, and ownership of land by a few, for the benefit of whom its natural resources are exported abroad. Under such conditions bourgeois compradors grow improbably rich on profits from a siphoning off of the nation's resources. Such a country's capital ends up in one or more of the "secure" countries; we are already seeing Russian *nouveau riches* buying villas and yachts and enrolling their children in the most prestigious universities in those countries, while our own country, as represented by its people and culture, continues on a downward course of degradation. Our corrupt civil servants and our comprador bourgeois have more and more a sense of absolute impunity. And our country's present course is precisely gauged to serve the interests of the section of society made up of such people.

Especially dangerous is the continuing ruination of the country's internal market. Selling anything whatsoever inside Russia is becoming less and less profitable, and the mass of people lack the money to buy anything. There are at present, one might

[14] Agents working on behalf of foreign interests.

say, two almost non-overlapping tiers in the arena of financial dealings: the upper tier is occupied by those involved in the export market and their underlings. These buy imported beer at several hundred rubles a can, dine in fancy restaurants, buy up works of art, and travel abroad. Those in the lower tier have only poor quality goods available to them, scantily supplied workers' eateries, and kindergartens unable to make ends meet. And the twain meet less and less, somewhat as it used to be when the powers-that-be were sequestered from the rest of us by green fences.[15] Only now it is all more out in the open, the ideological fig leaves having disappeared, and the glaring difference in standard of living between the two tiers is fast becoming grotesque.

This situation is highly dangerous also for the rest of the world since Russia is not Argentina: the power of its substantial nuclear arsenal is not likely to be reduced over the next several years, while social tensions will inevitably continue to rise if the present trend continues. At the moment the mass of the people is passive out of sheer fatigue. But this is the calm before the storm. Villas on the *Côte d'Azur* for a few, while for the vast majority poverty such as has not been known since the Civil War:[16] the yawning disparity cannot last for long. Such an invidious gulf between a wealthy few and the impoverished masses must lead to social disruption; even now society seems to be teetering on a knife edge. And the lack of a true leader will make matters worse. But by "leader" I don't mean national pseudo-leaders like Gamsakhurdia[17] or Dudayev.[18] There is at present simply nobody in Russia with high enough overall national approval! No one to hold society back from a descent into the maelstrom!

Starting from our present scenario, and assuming nothing of significance intervenes, there is but one logical development: further weakening of the government leading to an increase in the centrifugal tendencies of society, until there occurs, perhaps, a complete collapse of the government and a reversion from an "Argentinian" to a "Yugoslavian" variant, with its infinitely more terrible mayhem.

For our country to avoid such a fate, a strong ruling hand will have to appear from somewhere, one having the people's trust.[19] Under present conditions it would be impossible for such a person to establish himself in power at a single stroke—by

[15] Around the residences of the Soviet ruling *élite*. See also e.g. Chap. 4.

[16] The war between the Reds and Whites 1917–1922.

[17] Zviad Gamsakhurdia (1939–1993) was a Georgian politician, dissident, scholar and writer. He became the first democratically elected president of Georgia of the post-Soviet era. He died while in office.

[18] Dzhokhar Dudayev (1944–1996) was a Soviet Air Force general and post-Soviet Chechen leader, the first president of the Chechen Republic of Ichkeria. His drive to create an independent Chechnya led Russian President Boris Yel′tsin to launch the first Chechen war. The second Chechen war (1999–2009) of Vladimir Putin eventually led to the establishment of Ramzan Kadyrov as president of a pro-Moscow Chechen government with a certain degree of autonomy.

[19] One might say that this is in fact what occurred, more or less, with Vladimir Putin's appointment to the presidency in 2000, and his steady increase in popularity (aided in particular by the second Chechen war and government control of most of the media).

means of a army coup, say, as happened in Chile. Any such violent action would in our case lead to bloodshed, ensnare the people in a web of chaos and end up annihilating Russia as a nation. Do our current politicians understand this? At the moment they keep their own counsel.

Once such a person is in power, he will need to undergo a maturing process, which, if not cut short, will need to be carried on for some time in conditions close to anarchy. Thus the new ruling class will need to assimilate power to itself gradually, relying in the meantime on some form of societal support for its system of views, for example by promoting national goals attractive to the people. At the present time a transvaluation of values is taking place in our country, and since anyone wanting to establish themselves in power must have the support of the people, he or she must understand the new values and be able to state clearly what we should sacrifice and to what end.

In the absence of such "social power" no ruler can last. But even with it, such a ruler will have to be strong in the sense of having clear aims and the ability to put an end to the growth of corruption and the despoliation of the country's natural resources. Only then will the people rally round and begin to have belief in the future, and thence a desire to work. The ensuing halt in the fall of production will then lead to the establishment of an internal market, the essential requirement for any country's economy, and its government, to prosper.

But at the moment there is in evidence no kernel from which such rulers might sprout. Nor do we see any intellectual milieu capable of formulating appropriate national goals. The authority of the politicians now conducting affairs is in steady decline, and among these there is in any case no one capable of formulating unifying ideas. In 1991, when the Novo-Ogareva meetings were in progress and we began to take hope for the future, one could see how those politicians who understood the tragedy of the collapse and sought alternatives were enjoying a rise in popularity. For example, the country-wide popularity rating of Nazarbaev,[20] whom many saw as a likely prime minister or even president of a future Union—perhaps not Soviet, but certainly not socialist—rose steadily precisely because he made such unifying proposals. But then the absurd putsch took place, followed by the catastrophe of complete collapse, greeted so joyfully by the likes of Burbulis,[21] whose only aim was "to have nobody above us," as Gennady Eduardovich himself expressed it. The putsch triggered a transformation of the evolutionary reconstructive process into a sort of revolutionary bedlam with unpredictable outcome, catastrophic even for the instigators of the destruction of the government. Should we instead perhaps look

[20] Nursultan Nazarbaev (born 1940) has been president of Kazakhstan since the office was created in 1990, when Kazakhstan declared its independence from the Soviet Union. His regime has been called authoritarian, and several human rights organizations have accused him of abusing human rights.

[21] Gennady Eduardovich Burbulis (born 1945) is a Russian politician. A close associate of Boris Yeltsin, he was a highly influential political figure in Russia in the late 1980s and early 1990s, and one of the main architects of Russian political and economic reform.

for our future popular, intelligent, and energetic rulers among present republican or regional politicians?

Impediments to a Liberal Economy

But even if rulers should come to power who *do* have an understanding of which national goals are most appropriate and an ability to convey them clearly to the people, thereby legitimizing their authority to implement measures essential to our future, they would still need to be able to jump certain hurdles, which I shall now enumerate.

The first is the creation of an efficiently functioning state capitalist sector. In every liberal economy, American, British, or what you will, the non-capitalist sector produces a significant portion of the gross domestic product (GDP). But for the foreseeable future of our country, the state sector—state capitalism being still some way off—will be responsible for the lion's share of production. This is unavoidable, whatever the government in power. A massive state capitalist sector is certain to be established independently of the government's say in the matter; for this claim the example of Britain provides good justification, and is, moreover, highly instructive. Over the whole period of rule of that country's neoconservatives under Mrs. Thatcher, her government strove to reduce the state's contribution to the British economy, yet in the end managed to cut the state capitalist portion of the GDP only by a few percent, and now it has begun to grow once more! It's true that England differs in essential ways from Russia: England never attempted to leave the United Kingdom and declare its independence, while Russia, in its own peculiar fashion, has succeeded not only in "shoeing a flea," but also declaring independence of itself![22]

The formation of organizational structures for state-controlled industry must proceed as a regularized search for the most rational way to manage a liberal economy—if you like, as a search for that optimal compromise mentioned at the beginning of the present chapter.[23] That is why our country's path to a liberal economy cannot bypass the stage of "developed state capitalism." The state sector of Soviet times must, however, be subjected to a cardinal restructuring. The system of industrial monopolies must be dismantled, and the state sector converted into a state capitalist sector organized corporatively into diversified industrial conglomerates capable of holding their own in the market. This last is of particular importance, and represents the basic difficulty of the conversion. The conversion process should be oriented towards a gradual corporatization able to attract investment in the extension

[22] This smacks of Putin's attitude, whereby the USSR is identified with Russia. Many of the inhabitants of other Soviet republics had different ideas—were in fact overjoyed at the prospect of shucking off the Soviet yoke.

[23] That is, between the two extreme tendencies of society.

and modernization of production by members of the public; this should be the sole reason for issuing shares at this initial stage. Thus the very first task of a strong government will be the development of state-controlled industry and its integration into the system of market relations, and *not* the destruction of the existing state sector through dismemberment out of excessive fealty to market mechanisms, which will in any case play their part in the eventual liberal economy.

The second hurdle is the re-organization of *agriculture* on a market model, that is, one of competition between various kinds of proprietors. A complete retreat from cooperative management of agriculture—collectivization turned on its head—would be deadly for us, since our country is unable to feed us all without the use of high quality, modern farm technology. We should ensure the establishment of collectively owned properties, government-run estates, individually owned farms, and peasants' small-holdings in proportions that would become clearer with time, or, more precisely, from the evolving behaviour of the market.[24] The task confronting a strong government will be the devising of a strategy of encouragement of such forms of agricultural enterprise; thus reconstituted, even at its present technological level our agricultural sector should be able to feed the nation. Incidentally, there is at present no national goal of greater priority than that of shielding the nation from the long-term hunger that will surely befall us if we don't import agricultural goods. The security and development of the town and the market is necessary in the first place in order to effectively address the problem of the countryside! And that problem cannot be solved without purposeful action on the part of the government!

The third hurdle is the creation of a new governmental, above all regional, power structure, taking into account objective tendencies of the provinces towards the consolidation and cooperation necessary to the formation of an integral economic organism, aimed at producing the wherewithal of our nation's civilization, as well as establishing local self-sufficiency. This is a task of stupendous difficulty, and those at present in power would be incapable of accomplishing it, even had they any such intention.

In looking for answers to all of the above-enumerated constructive questions, it would be essential to appeal to society at large, to arrange wide-ranging public discussions, and use the resources of a de-politicized press cognizant of the urgent need to formulate national goals, i.e., a genuinely patriotic press.

Over and above all else, we need to understand what exactly contemporary Russia *is* as a nation. Not merely in the sense of a geographical entity, nor as just a body-part ripped from a single organism,[25] but also in terms of a national politics relative to its own government, and of its cultural, as opposed to merely

[24] Something along the lines of the author's recommendations has in fact transpired: whereas in the Soviet period inadequate harvests were the rule, following a transition period, Russia has become a major exporter of grain.

[25] Here again the author is espousing the view, held also by Putin, that the USSR was a "single organism," which view was most emphatically *not* shared by the majority of inhabitants of several of the component Soviet republics.

geographical, dimensions, which, taken altogether, it would be natural to call "Russia."

Russia's Geopolitical Situation and Consequences Thereof

No matter what the particular scenario from which future events unfold, nor what unforeseen situations should arise, Russia's future development will inevitably proceed in the direction of consolidation of a socially oriented, liberal economy, this being the clear general trend of human history. But what cost the transition will incur, whether it will take place relatively quickly or be drawn out over several decades, and which of our current values will be preserved—such questions are another matter altogether. And it is equally difficult to predict what form it will take, that liberal economy, what compromise will be reached between the two polar-opposite human tendencies I spoke of at the beginning of this chapter. Much will depend on how well the nation's brain—its intelligentsia and those in charge—understand the distinctive features of our contemporary situation, and also on choices made by ecological, economic, and political groupings world-wide, which one hopes would be realistic and not based on illusions. To what ends will the efforts of the government and the people be directed? Will they be joint efforts?

It think it very important that any government coming to power in Russia appreciate the particular characteristics of our contemporary geopolitical situation and search expressly for a means of exploiting it. This is a chance we cannot afford to miss: if, God forbid, we *should* miss it, history will not easily forgive us, where by "us," I mean not only the government and the parliament, but also the intelligentsia.

For there is a new game beginning in town, i.e., on the planet, one without precedent. New trends are under way. Although as yet barely acknowledged, they have begun their "work."

First of all, instead of the former *two* main centres of *military* power, we now have *three* centres of *economic* power: the USA, the European Union, and the Pacific Rim region. As far as the third of these is concerned, I wish to stress that I'm not referring just to the technological revolution wrought by Japan and the "economic tigers," i.e., Taiwan, South Korea, Singapore and Hong Kong, but am including also a rapidly developing China, with its current 12% yearly increase in GDP. Relations among these three centres of economic might will be fraught, multifaceted, and diverse. However although conflicts and confrontations are inevitable, cooperation amongst them must nonetheless increase: such is the imperative of the times, dictated by the state of the world's economy and ecology, and the pressing need to ensure the preservation of human civilization.

Relative to this triangle, Russia occupies a pivotal position as the linchpin holding the three together in a single whole. The cheapest and shortest communication and travel routes among the three centres lie through the Russian vastnesses. But Russia's northern geographic situation, very advantageous today though is, is not the main thing: Russia's natural resources represent one of the most significant means to world-wide prosperity. And, finally, there is the fact of Russia's still being the world's second nuclear power and possessing a highly developed aerospace industry subject to demand world-wide. But all these advantages can become sources of prosperity for Russia only if her people are able to see her potential and shrug off that mood of hopelessness induced by the recent years of political intrigue. If they would only revive in themselves that erstwhile respect for their own creativity!

Yes, but in order to realize its great potential, Russia needs honest, energetic, educated and intelligent rulers. By itself that still won't be enough, however: the intelligentsia needs to step up and take responsibility, instead of taking up the destructive stance it has taken over the last seven years, collaborating not only in the disintegration of the single-mindedness of the majority, but also in the collapse of industry and the government. Maintaining such a position is bound to lead to disaster for the country and for the intelligentsia themselves. But do they understand this? Do they understand that today we need ideas, a new map of the world, an appreciation of Russia's potential—in sum, an understanding of what the word RUSSIA means today! Not the Russia of the Romanov tsars and not Soviet Russia, but a country that has borne the burden and glory of a thousand-year history.

I sense that the mood of "liberation euphoria" is beginning to wear off, that the intelligentsia is coming to its senses a little and once more taking up creative work. More and more people are recognizing the necessity of converting the territory of the Russian Federation into an integrated, powerful economic and political organism, which is the only defence against the economic and political genocide that certain "third-world" countries have lapsed into through not having learnt the lesson.

We need vigour sufficient to the formation of a new world-view, and a public elucidation of the sense of what has just occurred. And, finally, we need to draw up constructive plans for a series of concrete projects or programs aimed at cementing the Russian Federation, at converting it into a well-integrated organism—in addition, of course, to finding a way of realizing such plans even if we are, as at present, unable to form a government capable of implementing them. At the present time, we can call on private capital and a burgeoning, solidifying middle class. Our country must also become attractive to foreign investment. The time has come for a direct appeal to society.

One possible such project is what I call the "Northern Band" project, a multifaceted one, certain parts of which I have described elsewhere. My previous articles on this topic attracted a certain amount of attention in America and Europe, but, alas, not here at home. So what exactly is the essence of this program?

When the Islamic wars closed the route to the Middle East via the Mediterranean Sea, the safest route from Europe to Byzantium became that by which the Vikings travelled to Greece,[26] thereby elevating Kievan Rus′[27] to the rank of a European state to be reckoned with. Russian princes were then able to use the new geopolitical situation of ancient Rus′ to their advantage. Today something analogous is taking

[26] From the north via the Black Sea.

[27] The original Russian princedom founded by Vikings from Scandinavia in the ninth century.

place: the shortest, quickest and cheapest route from the Pacific Ocean to the Atlantic lies through Russia—a circumstance whose worth would be hard to overestimate. Opening such a route from Europe to the Pacific—from England to China!—would facilitate not only the transit of raw materials but also semi-manufactured goods, the preparation of which has the potential to translate into hundreds of thousands of jobs and billions in foreign currency revenue. For instance, many of St. Petersburg's armaments factories might be converted into termini for the assembly of high tech components from Japan, Korea, and Taiwan,.... And a northern sea route would cut by half the present shipping time from Europe to Japan and China, to say nothing about the possibilities for air transport. A fibre optic cable should be laid as soon as feasible between St. Petersburg and Tokyo since if it is laid through the Middle East we stand to miss out on billions. Planet Earth cries out to be connected up from ocean to ocean, and no satellite connection would be able to compete for cheapness and efficiency with a cable link through Russia, a fibre optic cable capable of handling the demand for the flow of information.

Similarly large-scale are the problems associated with the export abroad of goods from Siberia, the north-eastern region of European Russia, and the Urals. By way of a solution I would propose the creation of a version of the Persian Gulf in the ice-free estuary of the Indiga River.[28] Such a project was actively entertained in the 1920s, but back then the Vorkutinsky Magistral[29] hadn't been built. At its closest approach this railway line's distance from the Indiga estuary is only two or three hundred kilometres. In those days, of course, the rich oil and natural gas reserves of our North were unknown and the Urals had not yet been industrialized. Nothing was there but the forests of Pechora.[30] It is now urgent that the project be revived.

But in touting the "Northern Band" project I don't envisage merely its use in solving purely economic problems. The wider region around the North Pole is one of the Earth's ecologically most vulnerable regions, and the state of its health affects the ecology of the whole of the northern hemisphere. Responsibility for that region falls squarely on the USA, Canada and Russia, the countries of the "band."

Then there is, finally, my take on the *security* of a future integrated world, which met with significant interest when aired in the West. Although this is a large, albeit specialized, topic, I should say something about it here since it represents a theme organic to the "Northern Band" project. Of central concern here is the creation of a universal information network using satellites. Without this it makes no sense to talk of the ecological, political, and so all the more, military, security of the planet. Note that such a system would play a positive role, a stabilizing one, only if it were indeed a collective, that is, universal, enterprise. *National* information networks, on the

[28] A river in the north of European Russia of length about 193 km, flowing north into the Barents Sea.

[29] A 1500 km railway built in the 1920s and 1930s using the labour of prisoners from the Gulag, linking the town Konosha in the far northern Arkhangelsk Province of Russia with Vorkuta, a town above the Artic Circle and Europe's most easterly town.

[30] The name of a region and town in the Republic of Komi, part of the Russian Federation. It was formerly an Autonomous District of the USSR. Vorkuta is one of its towns.

other hand, lend themselves to enhancing the efficacy of aggressive and defensive arms, thereby provoking military destabilization.

In order to realize such a spatial information network for use world-wide, it will be essential for the Russians and Americans to pool their rocket technologies. And then, naturally, such a system should not remain exclusively the property of the two rocket superpowers, but must be run by an open consortium with representation from any government wishing to participate, and have its actions subject to scrutiny by the United Nations.

Even if only at the planning stage, the "Northern Band" project represents an enormous amount of national capital, and realizing the project may breathe new life into many regions stretching from the Pacific Ocean to the Baltic Sea, while at the same time strengthening the sinews binding our nation's body.

On Formulating National Goals

Above I spoke of only one possible project for developing Russia's potential, namely that concerning her natural and intellectual resources and the developmental effects of changing her relative geopolitical situation. There are a great many other promising ideas for development, no less far-reaching. Public debate on such ideas is by itself already of great importance: a country in the process of discussing large-scale projects for its own development, a nation capable of looking ahead, cannot be regarded as the mere ruins of a communist phalanstery, since this indicates that it is beginning to develop a rising sense of confidence in its own powers. Self-confidence and self-respect would represent a big step forward, given the present pervasive hopelessness.

The formulation and analysis of the perspectives of a large-scale developmental program, i.e., of a set of projects, seem to me to be the most pressing duty of the present government. But it must be kept in mind that work on realizing such a program should be nothing like the sort of activity formerly engaged in by Gosplan[31] in connection with so-called "grand building projects of communism." The aim of an analysis of possible projects should be that of determining their consequences once realized and their conformity with explicit national goals, in the form primarily of a many-sided assessment of the projects' perspectives—precisely an evaluation of the real prospects for their realization and nothing more! Such assessments would then provide a basis for determining governmental priorities and appropriate regional and fiscal policies, and afford an indicator of scenarios of most efficient capital investment, and a *sui generis* scientific estimate of the risk to private capital. . ..

[31] Abbreviation of "State Planning Committee," the agency responsible for central economic planning in the Soviet Union.

This last is especially important. Such large-scale enterprises as the "Northern Band" project are sure to attract the attention of foreign trustees of investment funds and individual investors. To ensure this, however, we shall need not only fresh ideas, but a profound, all-round study of the situation, moreover at the government level. Clear recommendations as to internal government policies will also be necessary. A wide-ranging public debate on such developmental programs will be essential in order for the public to gain an appreciation of their prospects and also to get a sense of their own power to influence decision-making at the top, thereby fostering a belief in the future. This may indeed represent the most important aspect of the implementation of such programs.

Such activity will require energetic leaders and a competent parliament. And then a feedback mechanism will be set going: work on such programs will tend to have a "civilizing" effect on the powers-that-be, in turn tending to consolidate their position. The trend of thinking of those in power will necessarily have to be re-oriented from pure politicizing to concrete practical goals. And furthermore such projects will have a definite integrative effect on the nation, one of consolidation into a single economic and political organism—especially if suggestions initiated by people from the far corners of our vast country are taken into account. But this is just one stage in the determination of national goals. For the most part they will arise of their own accord out of the real aspirations of the people and their understanding of the world. And by debating publicly the various projects and their prospects, thereby calling on the interest and vigour of the people, the intelligentsia will shorten the time needed to realize them many times over.

Discussion of ways to avoid famine by means of a project whose specific stages are laid out so that its prospects are clear represents just one component of an over-arching process of defining a system of national goals, transforming our country's inhabitants into citizens and uniting them into a nation. And it is equally essential that in the developing twenty-first-century world Russia be established as a cultural entity with its own view of itself.

Russia in the Twenty-First-Century World

Today we look to the West. And not without reason, since Western nations first showed the way to the liberalization of economies, to the need for integration, and, finally, to a socially oriented economy adapted to the way of life of the mass of the people. Many Western economists, ecologists, and others working on problems to do with the future of civilization have come up with notions of certain universal formulae, or "universals" of twenty-first century life. However I believe many of their ideas to be both wrong and dangerous.

Certain definite universals of human existence undoubtedly exist. Since human beings interact with Nature as a single biological species, it could hardly be otherwise. Universals independent of race, locale, and other such contingencies arose in people's consciousness as manifestations of the logic of the universal evolutionism[32] which led to the appearance of rational man on Earth. One of these grand universals is, for example, the commandment "Thou shalt not kill!" common to all peoples. Universals are even now in the process of emerging out of the current practice of human life. Aren't world-wide technological development and the establishment of elements of planning in liberal economies both instances of such universals?!

But then there is a law of divergence which is also a universal. This law codifies the important property of the evolution of human society according to which human social characteristics are continually diverging. For instance, throughout the evolutionary process the variety of forms of human cohabitation and the peculiarities of spiritual life has been continually multiplying. There are thus limits to universality and that is why mere blind aping is likely to be harmful to a society.

And that is why for us the West represents just an experiment to be noted, but not necessarily to be emulated—and likewise the East. For Russia especially, since its territory links the two Western regions of economic and cultural power, the two most important centres of the future information society.

I've been speaking of the twenty-first century. But do we really know what the phrase "the twenty-first century world" means? We have yet to unpick this phrase, to understand how we view the planetary society of the margin of the second millennium. It will not be those whose lives are today organized around mere production who will be in the vanguard of history, but those whose mentality is based on the universals of the civilization of the twenty-first century and its requirements.

So those are the positions to be taken in thinking of our national interests, difficulties, and possibilities and in trying to determine what Russia *is* exactly.

Without illusions, but with the severity and pitilessness of true patriots—only thus will the intelligentsia be able to acquire a real understanding of the situation, to fully grasp the actual circumstances in which we find ourselves. Only thus shall we be capable of understanding how, in the course of an unlikely process of intermixing of peoples, a sweeping ethnogenesis, taking place throughout the territory of the Soviet Union, a huge proportion of the people of our nation became lumpens, out of the blubber of which our "communal scum" arose, as well as everything else standing in the way of our assuming an honorable place in twenty-first century society. Only unsparing honesty can help us develop new standards fitting for the Russian reality of the end of the century, so different from that of its beginning.

* * *

I call my position one of "limited pessimism," which I justify on the one hand by the tremendous potential I see in my country and its people, while on the other nursing grave doubts as to our present competence to exploit this potential. Meeting with people who have devoted their whole lives to politics, I see that the tenor of their thoughts is such as to impede peaceful debate of the country's future—the sort

[32] See Chap. 9 for the author's definition of this phrase.

of discussion that would exclude individual and moreover immediate gain. The way of thinking of the present wave of politicians is, perhaps, one of the most terrifying legacies of the communist era.

Nevertheless we must think hard about Russia's place in the twenty-first-century world, and do everything we can to make sure it is a worthy one.

Chapter 14
The Year 1993

Tomorrow Hasn't Yet Dawned

The year 1993 has arrived, and I am finishing up my book, which I began in a mood of lightness and even euphoria, writing for myself just as, in my youth, I wrote verse. The present book was written on those days when I had no urgent work to do or else found it difficult to do what work I had, when I felt like relaxing in my own company, alone with my past and my thoughts. And also when I was sickened by what was going on in the present; at such times I would leaf through the pages of my life, astounded that it had all gone as it did and not otherwise, not only as far as I myself was concerned but also my country.

I am by nature a "lark," always waking very early. But as a youngster I used to be lazy about rising and liked to lie idly abed, if possible, thinking of something pleasant. But now such lazing in bed doesn't work for me: it only needs my brain to switch off from something needing doing for thoughts to arise that are usually very sad since I no longer look towards the future. And then my mood turns sour, sometimes for the whole day, hindering me from working. So nowadays when I wake up early, which happens nearly every day, I rise hurriedly and sit down at my computer, which has become a substitute for Lake Ladoga, my former meditation in that remembered place having been replaced by a retreat into the past. I don't think about the future; the time is not right, and I don't wish to get upset. Nor do I contemplate the fate of what I write.

But one day when I was in the frame of mind that in a certain children's story book is evoked by the words "There's neither crowing nor pecking for me since I have chicken flu," I used my printer to print what I had written. My very first reader was my wife, whom I asked to check my spelling and catch the many other mistakes. She it was who first told me that perhaps my reflections might be of interest to a wider readership than just my nearest and dearest, and that they should be published. So I gave what I'd written to those few of my acquaintance who could be trusted to give an honest opinion, and they echoed Tonya's words. I quickly read through my

N. N. Moiseev, *How Far It Is to Tomorrow...*,
https://doi.org/10.1007/978-3-030-96651-5_14

typescript and understood that I had unwittingly produced a document containing information of a high degree of authenticity. I saw that it contained, in essence, fragments of the history in Russia of the intelligentsia, natural science, and even philosophy, of the last decades. Furthermore, that this history was propounded in the context of a specific life, that is, as part of that same three-quarters of a century that I had already lived through and had changed the character of a vast populace out of all recognition. I thus present my life to those wishing for an insider's view of that history, as it were, with all its tragedy and enchantment.

After re-reading the typescript more attentively, I decided not to change or add anything, fearing to spoil the pure, spontaneous spirit of the original. Should I perhaps have kept it to the first thirteen chapters? But thirteen is an unlucky number, and moreover the last, thirteenth, chapter had been written a year ago, whereas now I was thinking of possible readers. That is why I sat down to write this concluding chapter, which proved more difficult to write than all the others: I wrote with difficulty and without good cheer since now the future had to be considered. It was as if I wanted to say something encouraging to my readers, especially the young ones.

Academic Advisor

Victor Hugo wrote a novel with the same title as these concluding reflections. As a child I loved Hugo's novels, especially his *Ninety-three*.[1] I read it over several times, each time feeling compassion for the victims of that fearful year of the French Revolution, which, even though still a little child, I understood as the story of a catastrophe more than a tale of heroism. As a result I have ever since heard something ominous in the sound of the phrase *"quatrevingt treize."*

Two hundred years ago the Terror was unleashed in France, the Vendée rose against the Jacobins, and frightful savagery engulfed the country. People were bundled into barges and the barges sunk in the Loire. And this was all perpetrated in the name of foolish, senseless slogans. People's psyche was changed; from a formerly kind, peace-loving person, a neanderthal[2] emerged, and members of the intelligentsia, called "philosophers" at the time, turned into executioners. Something of the sort occurs, maybe, during all revolutions. After all, our country also went through a *"quatrevingt treize"* phase, when people were transformed into monsters for whom life, everyone's life, was worthless. And especially horrible was the fact that the perpetrators' own lives were held at nothing!

[1] Hugo's last novel, published in 1874, following the invasion of France by Germany and the upheaval of the Paris Commune. The novel concerns certain counter-revolutionary revolts of 1793, at the height of the French Revolution.

[2] The context for the use of this word here may jar, since the neanderthals are not known to have been especially blood-thirsty. There is, moreover, a theory to the effect that *Homo neanderthalensis* was driven to extinction by the more blood-thirsty *Homo sapiens*.

And that explains why, on the New Year's Eve just past, I proposed a toast whose meaning was perhaps not readily grasped by my guests: "May God grant that the coming year 1993 be for us nothing like the French year 1793!"

I would like to believe that our country contains a sufficient number of wise people for another catastrophe to be circumvented. It's unlikely that the people of *any* country would be capable of withstanding *two* such revolutions, two "ninety-threes!"

But that's only what I'd *like* to believe. A sober analysis tells us something else: at no time since the Time of Troubles[3] has our people been so humiliated, so cut off from the future, so sunk in grief, as now. The main evil besetting us is hopelessness in the face of a lack of any ideas as to how we might become whole again. Everyone is busy trying to scramble out of the hole we're in by him- or herself. What I find most distressing is the absence of anyone of the stature of Minin or Pozharsky,[4] someone in whom we could believe, who was worthy of our trust—that's what's most terrifying today. And the people have not yet had their say. What form will that take?

In March 1986 I retired.[5] I resigned from my duties in the Academy of Sciences, and declined an appointment as chair of my department. Since my material situation had been made secure by the new law concerning Academic counsellors, I had hoped not to need to find work on the side to supplement my income, and thus be able to devote myself wholly to the peaceful pursuit of science or, more specifically, to answering those questions I'd long been interested in but had not had the time for. There was much I wanted to write about. I made sure to preserve all my scientific contacts. I was completely clear about my plans: I would continue my investigations into the theory of self-organization and, recalling the great covenant binding us to the view that it is no longer possible for a theory to meet modern standards unless painted on a mathematical canvas, attempt to express the general methodological foundations of universal evolutionism in terms of a system of mathematical models. By such means I hoped to be able to come to grips with the main problem preoccupying me: how to determine the right approach to the study of the stability of the biosphere as a single whole, how to lend precision to this endeavour, and thence evaluate humankind's competence to secure its future—together with that of the rest of the biosphere, of course! I wanted to sketch the outlines of an appropriate scientific program aimed at achieving these goals.

[3] See Chap. 13.

[4] In 1612, Kuzma Minin, son of a Russian salt manufacturer, and Dmitry Pozharsky, a member of the Russian nobility, led a successful assault on Moscow to liberate the city from the invading Poles under King Sigizmund II. In the following year Mikhail Romanov, a distant relative of Ivan the Terrible, was made tsar, thus ending Russia's "Time of Troubles" and inaugurating a dynasty which would last till 1917.

[5] See Chap. 10.

So my wife and I set ourselves up in a little house some 60 km from Moscow, where I had thought to spend a good part of my time alone with my computer—and with my wife, naturally!

But reality intervened with its inescapable modifications of man's schemes. *Perestroika* began, greeted joyfully by most of the populace and especially by the intelligentsia, who, like almost everyone else, failed to grasp the real meaning of what was happening. Each citizen had to determine what position to take up *vis-à-vis* the new political environment. I myself wrote a long letter to M. S. Gorbachov containing three assertions: First, that it was essential that the liberalization of the economy pass through a stage of "developed capitalism," during which our branch monopolies would be replaced by government-capitalized corporations capable of competing in the international market; Second, that all possible forms of private property be legalized, subject to regulation by a "land court"—a crucially important instrument of any citizens' society—with the mandate of preventing any kind of degradation of any lands whatsoever, land being, of course, humankind's most valuable resource; and Third, that the means be found for preserving and exploiting the greatest asset and main achievement of our socialist regime: the realization of universal literacy among the populace, and the intellectual capital we consequently have at our disposal. In fact, this last provides the key to solving both our economic and social problems. That alone can help us consolidate our higher technological sector and so retain our Union's membership in the club of industrially developed powers.

The implementation of such recommendations would require a new level of government to be set up, since without general government-run programs, not to mention the will and energy of the whole nation, nothing of what I suggest could be achieved! For similar reasons, a strategy must be determined for carrying out the necessary changes to our society, changes which have been long ripening and without which our great nation will become a stagnant backwater in the world.

Such opinions were very far both from those held by people who would later call themselves democrats, and likewise those of the group of party activists who launched *perestroika*.

Although I had handed in an envelope containing the above recommendations to a person at the reception area of the office of General Secretary Gorbachov, when I had a chance to talk to him a couple of years later, he said he hadn't received it. I have no reason not to believe him: as he himself once said to me: "A bureaucracy is a bureaucracy!"

I made no attempt to keep quiet about my ideas. On the contrary, I worked hard at airing them by means of articles and lectures. As time went on my views came to diverge more and more from the officially adopted course, from what was considered fashionable, and from what the "elders of *perestroika*" said and wrote, which was absolutely against the retention of state power. I was happy not to have been numbered among them.

The Physico-Technological Institute, where I had been a professor for over 30 years, proposed me as a member of the Supreme Soviet of the USSR, and my candidacy was supported by the Moscow Institute of Forestry and several other organizations of the Mytishchi[6] district electorate. I expounded my ideas and my views on *perestroika* to a large meeting, and . . . withdrew my nomination as deputy. It had become quite clear to me that, for reasons of health and age, but primarily because of the way I thought, I was unable to enter politics.

I don't have the qualities essential to a politician. I tend to believe what people say, am unable to find my way through the finespun cobweb of personal interests, invent gambits to suit one party and neutralize another, etc. In a word, I am unable to do anything a politician needs to do to achieve his or her aims. I am likewise unable to belong to the faithful of any party: I can agree with its aims, but no more, can share the views of one or another group of people, but absolutely not every group. My wife supported me in all this; in fact, she was more categorical in that respect than I myself.

This explains why I did not give up my Party ticket when the mass exodus from the CPSU[7] was taking place. I had joined the Party at a difficult time, when on active service at the front, had joined along with those who were defending the country from Fascism.[8] Obtaining my Party ticket felt for me more like a confirmation of the importance of Russia, rather than the Party, to my life. I have never been a hard and fast Party adherent, have always had my own view of things and avoided every kind of political or Party activity. And so it is that now I consider it impossible to reject anything from my past in order to toe one or another political or party line. What has been, has been. So let my Party ticket, inscribed with the letters VKP(b),[9] and now lying in the pocket of the shirt given to me by Lieutenant-Colonel Fisun when we were together in the Sinyavinsky Marsh,[10] remain in the drawer of my writing table.

At the time of the first meeting of freely elected Councils,[11] my wife and I were at the sanatorium "Desna."[12] It was the last time we'd have the opportunity of buying vouchers for a four-week stay at a sanatorium under the observation of doctors. Now the sanatoria are reserved for salespeople working at commercial kiosks or perhaps miners.

---

[6] A town in Moscow Province.

[7] Communist Party of the Soviet Union.

[8] Nazism, that is.

[9] Anagram for the All-Union Communist Party (of Bolsheviks).

[10] A marsh not far from St. Petersburg (then Leningrad), where in September and October 1942, German troops cut off and surrounded Red Army troops forming part of a Soviet operation on the Volkhovsk front, aimed at relieving German pressure on Leningrad. It is estimated that in the course of attempts to break out of the encirclement, the Red Army suffered over 100,000 casualties, of which over 40,000 were killed, taken prisoner, or unaccounted for.

[11] The "Congress of People's Deputies"?

[12] Named after the Desna River, a major left tributary of the Dnieper.

However that may be, at that time we spent a lot of time walking and even more watching television. To view the first meeting of Councils without cuts or unanimous votes in favour was so new and unprecedented that one could hardly believe it was really taking place right there and then in the Kremlin. I watched goggle-eyed. It was extraordinarily interesting, yet at the same time I felt sad for those who had gone into politics in the hope of doing something useful for their country.

I saw how powerless the good-hearted and courageous idealist Andrey Sakharov was when he expatiated to a jeering crowd of scoundrels on what was in his heart and what he had been meditating on throughout the long years of his ostracism, humiliation and exile, and which coincided with what the vast majority of our compatriots had been thinking. I saw an angry and repellent Landsbergis,[13] considered a member of the intelligentsia on the grounds solely of his musical expertise, as well as others without any understanding of the present and no ideas for the future, who had been elected for reasons beyond me and who mounted the dais merely from a desire for self-advertisement. I was dismayed by the fact that I saw no kernel of an idea emerging to justify these goings-on in the Kremlin. Was it possible that the people who had trumpeted the words "socialist elections" failed to appreciate what lay behind this phrase? I was soothed only by the fact of my decision: Thank God that I wasn't in that auditorium! How easily I might have been there! What would I have thought then? And what part would I have played? Like Sakharov, one more victim of crucifixion? I wouldn't have been able to endure it. But how much worse, how heart-rending, to merely be among the ranks of the silent majority.

It seemed to me that I had a good idea of parts, at least, of a program of directed social development, of gradual liberalization, that would allow us to avoid an eruption of nationalism, the disintegration of the Union, and another revolution, which was what I feared most. To repeat: I have always been an irredeemable opportunist and have always feared above all else, even when still young, the elemental forces let loose by revolution. But is it possible at this stage to convince those gathered in that Kremlin auditorium that *perestroika* is real and should be allowed to continue rather than risk a wrong turn towards general woe and bloodshed? The crowd of know-nothings and understand-nothings called "the people's chosen representatives" will otherwise bring more violence and misery on us.

[13] Vytautas Landsbergis (b. 1932), musical expert and politician. Entered politics in 1988 as one of the founders of a movement for independence of Lithuania from the USSR, becoming Chairman of the Supreme Council of Lithuania in 1990. Lithuania, along with several other Soviet republics, proclaimed independence of the USSR in 1991.

The Putsch Operetta

At the time of the August putsch,[14] I was in Pereslavl-Zalessky,[15] having gone there with my old friend Georgy Nikolaevich Korsakov. He was a French citizen, having been born in Paris. That was in 1921, so we were almost the same age.

I first met Georgy Nikolaevich in the late 1960s in Bordeaux at the conference "Cybernetics and Life," organized by the International Life Institute. We were introduced by the president of the institute, Professor Maurice Marois,[16] a kind and independent-minded person such as is not often encountered among Frenchmen. Korsakov, who was then the director of the patents department of the computer firm Honeywell Bull, was ready to be of assistance whenever I visited Paris on computer business.

It was at a later time, however, when he was in Moscow on an errand connected with his job, that we became good friends. We were sitting having dinner in my apartment, when he said something that utterly changed my attitude to him. What he said was something like: "The circumstance that I, a grandson of Admiral Korsakov,[17] after whom a town on Sakhalin[18] is named, am forced to live in France and not here where I belong, I forgive you. That our Kursk (or perhaps it was Oryol;[19] I don't recall exactly) property, renowned for its exemplary and more than merely profitable husbandry, has been transformed into the devil knows what, incapable of feeding even those working it—that I also forgive you. But the fact that at the hotel Metropole (or perhaps it was the National; I don't remember what hotel he was staying at then) they began addressing me politely only after I had shown them my French passport—that I can never forgive you." I likewise cannot forgive the Soviet leaders for this. The destruction of the sense of personal worth of individual Russians, in fact of all citizens of Russia,[20] is one or the filthiest crimes

---

[14] The 1991 *coup d'état* by hard-line members of the Communist Party of the Soviet Union (CPSU) opposed to Gorbachov's reforms, aimed at wresting control of the USSR from him. They were opposed in Moscow by a short but effective campaign led by Boris Yel'tsin, and the *coup* collapsed after only three days. It is thought that this event contributed to the demise of the CPSU and the dissolution of the Soviet Union.

[15] A town in Yaroslavl Province, 140 km northeast of Moscow.

[16] Maurice Marois (1922–2004). Founded *L'Institut de la Vie* in 1960. From 1969 head of the Histology Department of the *Faculté de Médicine Saint Antoine (Université Paris VI)*.

[17] Nikolai Alexandrovich Rimsky-Korsakov (1852–1907) was a vice-admiral in the Russian Imperial Fleet. He was an uncle of the composer Nikolai Andreevich Rimsky-Korsakov, and was married to a niece of the composer Pyotr Ilyich Tchaikovsky.

[18] Russia's largest island, lying north of Japan.

[19] Oryol and Kursk are two cities (and districts) on a line running south from Moscow.

[20] That is, of the USSR. After centuries of accretion and conquest of territory, Russia became identified with its empire in the minds of many of its citizens.

of Bolshevism[21] and the source of innumerable calamities. So that is why Korsakov and I became friends: concerning the most burning questions, namely one's attitude to Russia, what it means to be a self-consciously Russian individual, and the need to return to him or her a feeling of self-respect, we were absolutely like-minded.

But to return to the putsch, which found us in Pereslavl, or, more precisely, in its environs, in a little Finnish dwelling used to house visitors to the Institute of Programming Systems of the Academy of Sciences, and in which Korsakov and my wife and I happened to be relaxing at the time.

Early one morning, on the television fragments of *Swan lake* were being shown instead of the usual news, and we understood that something serious was happening in Moscow. However, no one, not even those in charge at the institute, knew what exactly. But when, a couple of hours later, we saw Mr. Yanaev[22] with trembling hands like those of an alcoholic suffering *delirium tremens*, we understood at once what had happened: there had been a putsch organized by moral, physical, and intellectual impotents, an attempted *coup* by drunkards and scumbags. We, my wife and Korsakov and I, went for a walk, all feeling that the situation could not end well. Although certain the putsch would fail, we feared that the consequences might take the form of a great tragedy for our state, however unpredictable in detail. On this all three of us were agreed, and also that tragedy was inevitable. But what did indeed eventuate in Belovezhsky Forest[23] took us by surprise.

One could hardly imagine such a thing occurring even in a nightmare. A state that had been more than a thousand years in the making was undone in a single evening. This was not merely irresponsibility but a stab in Russia's back, and perhaps even that of world civilization. And moreover I couldn't see any reason for it: after all, a referendum had just been held in which the people had voted for the USSR to be preserved, followed by the meetings at Novo-Ogarevo,[24] considered by many to be a light at the end of the tunnel, as they say nowadays. And suddenly this! And the ridiculous proclamation of Russia's sovereignty. This seemed to me to be as illogical as, say, proclaiming England's independence of Britain. One day the appropriate documents will be made public, and we'll know what hidden springs of action were at work.

But what if there are no secrets, nothing but the play of men's passions and ambitions? I was led to the idea that this might be the case by the following episode.

---

[21] There are those who would counter that the lack of a "sense of personal worth" among the Russian masses dates from before the revolution, as evidenced in some of Chekhov's stories and other pre-revolutionary Russian literature.

[22] Gennady Ivanovich Yanaev (1937–2010), Soviet politician. In 1990 elected Vice-President of the Soviet Union with Gorbachov's backing. Led the *coup d'état* of August 1991, which failed after only three days. Pardoned by the State Duma in 1994.

[23] A large natural park 340 km southwest of Minsk, famous for its primeval forest, where, in December 1991, at a government dacha, an agreement was signed by the leaders of Russia, Belarus and Ukraine abolishing the USSR and creating in its place a Commonwealth of Independent States.

[24] See Chap. 13.

In March 1992, I was at a mini-meeting with G. E. Burbulis.[25] I don't recall why, but the conversation turned to the significance of the Belovezhsky tragedy. Gennady Eduardovich began saying something to the effect that the collapse of the Soviet Union was a good thing. But suddenly he broke off, raised his hands to the ceiling and said: "Surely you understand that now there's no one standing over us!" So perhaps that was the real reason; after all, Burbulis was then the third-ranked person in the government.

From that time on, I was more and more drawn into an active socio-political life; in fact the times were such that it was impossible to remain on the sidelines. Thus I wrote articles addressed to the public even though I set little store by their usefulness, and kept postponing and postponing my scientific plans to a more propitious time, should it ever come. And in any case in such times as these who can get worked up about the stability of the biosphere!

One More Try

In the Autumn of 1991, in November, I think it was, a general meeting of the Academy of Sciences of the USSR was held, the last such. The main items on the agenda were questions of the Academy's fate and how it now stood with respect to Russia. There began long and, as always, tedious discussions. I got up and gave a short speech, whose gist was, first, that the latter question was of no importance since the answer had been decided for us: it was clear that the Academy should henceforth be called the *Russian* Academy, as in olden days. The main problem, I said, is to put the intellectual potential of the Academy to work on behalf of Russia, just as this had been done during the war, when various Academic committees had played a not unimportant wartime role in the enhancement of our country's defence capability and the mobilization of her resources.

Today, however, Russia is no longer a single organism, I continued, but a large piece of territory torn away from the whole. And, although this is a great tragedy, we now have to take it as a given. To reject the *status quo* is to invite bloodshed and deepen the tragedy. We have to get over the loss of Sevastopol′,[26] that bleeding wound, and the other humiliations to which those now in power have subjected us, and, gritting our teeth, begin working to secure a good future. We must learn how to live in what is for us a new country, but still *ours*, for which we and only we are responsible.

And our first task is to transform the land that we now call Russia into a genuine modern state, into a single economic organism with all the appropriate infra- and superstructure. This has clearly not been grasped by the people who have come to power and are attempting to run the country, and may even be beyond their

[25] See Chap. 13.

[26] Since that city now belonged to an independent Ukraine?

comprehension. This task must be confronted, since only then will the extent of our particular responsibility, namely that lying on the shoulders of the scientific intelligentsia, become clear. It is essential that we find a way to exploit the intellect, knowledge and energy of members of the Academy, and also of the young people, especially those in science, who stand behind each of us. I am especially worried, I said, by the possibility that the nation will be engulfed by a wave of mediocrity, that in the consciousness of those young folk to whom, by sheerest chance, the keys to our future have been handed, there reigns a conviction of their own self-sufficiency, the first indicator of mediocrity and foreboding misfortune. How do we explain to them that their rise to the pinnacle of power is in no way attributable to their own particular talents, and that to fulfill the responsibility that has fallen to them, it is essential to have solid support.

I also said that the use of the banal term "crisis" is inappropriate. It is better to acknowledge that our country is going through another "time of troubles," and discuss how to re-organize our present life and formulate a far-reaching strategy for emerging from the present chaotic time. And that is why it is crucial to set up a council dedicated to the science of a future Russia.

My speech did not, however, make a great impression. Only Marchuk,[27] the last president of the Academy of the Soviet Union—it was he who chaired that memorable last general meeting—exclaimed: "Taking the initiative has consequences, Nikita Nikolaevich! You must write a letter to the government."

So I wrote a letter addressed to B. N. Yel′tsin, and had several members of the Academy sign it. I then laid it on Marchuk's table, where it remained undisturbed for a week. I retrieved it and took it to the chancellery of the Russian President.

The fate of that letter is revealing.

About six months later, Yel′tsin unexpectedly issued a decree authorizing the organization, under government auspices, of a consultative social council to analyze critical situations and vet projects the government was planning. More than a few qualified specialists were appointed to this council, which gave certain grounds for hoping that there was a chance some good would come of it. I was confirmed as its president. It was my hope that this complement of council members would, in the course of our new "time of troubles," become an instrument of strategic analysis that would prove indispensable. To allow glib dilettantes, deluding themselves into believing they are ready for anything, into the ranks of the council would be dangerous and I tried to keep such people out. I believe that the failure of *perestroika* was in large part due to a lack of understanding of the perspectives open to the nation, and of a clear appreciation of what was attainable rather than merely desirable! But even what was desirable was never made clear. In my imagination, I always return to the following instructive example of the sort of organization we needed.

[27] Gury Ivanovich Marchuk (1925–2013) was a prominent Soviet and Russian scientist in the fields of computational mathematics and atmospheric physics. He was president of the USSR Academy of Sciences from 1986 to 1991.

A number of memoirs tell the story of how, following the Japanese attack on Pearl Harbour, an informal group of anxious scientists came together in Washington. They organized meetings at which they discussed problems arising in connection with America's new war footing, and thought about how the various solutions proposed might be realized. Sometimes Roosevelt, Hopkins and other government figures took part in these meetings. Roosevelt usually just listened, occasionally asking questions, or requesting that one or other hypothetical situation be laid out in detail for him. Under the aegis of this group of experts various investigations were carried out, of the sort that would today fall under the heading of operations research or systems analysis, namely detailed scientific analyses of the problems that might arise from one or another decision. Following Roosevelt's death, Truman formalized this activity, and the celebrated Rand Corporation was born, an entity that would later play an important role in the Cold War. It is still operative: presidents and governments pass away, but the Rand persists, furnishing the US Federal Government, and in recent decades also certain large corporations, with objective information of the greatest importance, and working independently of contemporary politics.

It was my hope that I would be able to produce a new version, under the different conditions obtaining with us, of the American experiment, thereby creating an analytical instrument that would help our country find a way out of its new time of troubles. I spent the whole of the year 1992 on a multitude of efforts to realize this idea. What I failed to take into account, however, was that Yel′tsin was not Roosevelt and Burbulis was not Harry Hopkins.

There was absolutely zero reaction to the great many letters I wrote to Yel′tsin, Gaidar,[28] and Burbulis. I was given two hearings before the Supreme Council and addressed several meetings of the Institute of Philosophy, the House of Scientists, and sometimes in my own apartment, since the council I presided over had neither premises nor staff, nor, for that matter, any financial means whatsoever. But not a single one of the rulers over this sublunary realm, or even of their advisors, deigned to show any interest in what I had to say. A. B. Yablokov,[29] ecological advisor to the President, said to me one day: "You're setting up new councils when we already have a whole institute of them." We nevertheless managed to put out two brochures, titled "Strategies for Survival," and "Problems of Regional Management." (I would like to thank the rector of the UREA,[30] Professor Bim-Bad, and the director of the Institute for Regional Problems, Professor Ailamazyan, for finding the means to defray the associated typographical expenses. Without their aid it would have been impossible to print these pamphlets for distribution at the time of the hearings

[28] Egor Timurovich Gaidar (1956–2009), Russian economist, politician and governmental liberal reformer. One of the main leaders and ideologues of the reforms of the Russian economy carried out in the early 1990s.

[29] Alexey Vladimirovich Yablokov (1933–2017), Soviet and Russian specialist in zoology and general ecology, and political activist. From 2005 leader of the "green" faction of the political party *Yabloko* (Apple).

[30] The University of the Russian Educational Academy.

before the Supreme Council.) Incidentally, these hearings also failed to have any observable impact; the deputies were preoccupied with their political struggles and in any case seemed to have scant understanding of anything outside politics.

I grieved at the astonishing lack of interest of those running the country in providing any kind of support that might be seen as apolitical, such as support of science or scientists working independently of the government or any party affiliation, and without any departmental or political bias.

But it was Gaidar who astonished me most of all. I used to see him at seminars in the Institute of Systems Research in CEMI.[31] As an old professor, I always paid close attention to the up-and-coming young. I liked him. Later we met in the editorial offices of the newspaper *The Communist*, where he was editor of the economics section. He was then working under Latsis,[32] a person incomparably better educated and more capable than he. Alongside him Gaidar became almost invisible. But all the same I was glad when he was made premier.[33] After Silaev,[34] whom I had known for many years as the Minister of Aircraft Industry and someone who could stoop to hiring Simonov[35] as general designer in the Sukhoy Design Bureau, this was a big step in the direction of the "intellectualization" of the government. Along with many others, I was glad that we now had a more or less well-educated government.

There were, however, one or two personages in the new cabinet of ministers about whom I had my doubts, and all the more so since one of these had grown up right before my eyes—a cocksure lad whose mama was always saying how clever he was! But apart from such minor misgivings, I was overall happy with the formation of such a youthful government. I wrote a couple of letters to Gaidar, enclosing some of my articles and copies of my proposals. There was no reaction. I discovered that a former secretary in my old department was now working as Gaidar's secretary, and I asked her if she would pass my request for an audience on to him. But that *démarche* was also without result.

I then set about determining in detail what exactly "Gaidar's team" represented and who was lending it support. Rather than go into everything I unearthed, I shall confine myself to a few of my conclusions on that score.

First, in place of general ignorance, we now have semi-knowledge–but I don't know which is worse. Those who ran our country before were clever and cunning

---

[31] Central Economico-Mathematical Institute.

[32] Otto Rudolfovich Latsis (1934–2005), journalist, member of the Central Committee of the Communist Party under Gorbachov, advisor to Yel′tsin.

[33] Of the Russian Federation.

[34] Ivan Stepanovich Silaev (born 1930) is a former Soviet, now Russian, politician. He was Minister of Aircraft Industry from 1981 to 1985, and Premier of the Soviet Union from September 6 to December 26, 1991, just before the official dissolution of the USSR on December 31, 1991.

[35] Mikhail Petrovich Simonov (1929–2011) was a Soviet aircraft designer, famed for designing the Sukhoy Su-27 fighter bomber in the late 1970s. He joined the Sukhoy Design Bureau in 1970, and was Deputy Minister of Aircraft Industry 1979–1983. But here the author seems to be taking a negative view of him.

peasants, but they were aware of how ignorant they were, so now and then invited genuine specialists to advise them. They partly listened to what these had to say, and partly twiddled their moustaches. Thus although the decisions they then made were largely based on established rules of the game, they were nonetheless prepared to listen to advice and occasionally acted on it. But now the people who have come to power believe, on dubious grounds, that they are well-educated, as a result of which they exhibit the familiar "syndrome of self-sufficiency." Instead of independent advisors, they have aides, whom they recruit from the field of their acquaintance, and who also lack a real education. Whence the wave of ill-educated, self-opinionated, mediocrities characteristic of "semi-science," now breaking over our country.

But here I am not recommending that the government be run only by scientists or highly educated people. Absolutely not! I was very impressed when I once read Napoleon's statement to Berthier,[36] his brilliant chief-of-staff: "You'll never make a war leader," which he followed with the explanation that the qualities of a field-commander and a chief-of-staff must of necessity be very different. A good scientist is usually a poor leader since a crucial quality of the former is to be always doubting. His task is not to overcome doubt but to exploit it in the interests of obtaining new knowledge. A person who has to make quick decisions, on the other hand, must be able to conquer doubt, have the courage to court risk; for him or her the most important thing is to "go on the attack." That, anyway, is my take on what Napoleon said to Berthier. But I would like to add something on my own account: a genuine statesman should be intelligent enough to rely on his general staff, each member of which should be better educated that the statesman himself.

Thus when I at last understood that this is not how things stood with us, when I saw that instead of a home-grown version of the Rand Corporation, we have only a limited set of consultative agencies for whom the Western technology of paper-shuffling is the height of wisdom, I decided to quit the game. But that turned out not to be on the cards!

The Presidential Council

Out of the blue I was informed that I had been chosen to join the ranks of the Presidential Council. No one had said anything to me beforehand about the possibility of my becoming a member, and to this day I don't know why I was accorded the honour. But—why pretend otherwise?—I was pleased to be chosen. And furthermore the idea of working in the Council prompted a certain sense of optimism and, most of all, a hope of realizing the idea of using the nation's scientific potential to transform Russia into a integral, economically vigorous, world power. What Russia needs is an organization, called, perhaps, "Science in the Service of

[36] Louis-Alexandre Berthier (1753–1815), 1st Prince of Wagram, was a French marshal and for a time Napoleon's chief-of-staff.

Russia," similar to the Rand Corporation, like it depending first and foremost on the work of temporary collectives of our best minds, assembled individually from all over the country. Such an organization *will* be created, if not by me then by someone else, since without it, given our present circumstances, the nation simply cannot become a power of the first rank. And working on a high professional level, if possible without any politicking, such a body should be capable of examining the various developmental alternatives open to us, warding off utopian ideas, and making objective judgments concerning the place of our country in the world society of the twenty-first century.

Having few illusions, I expected to encounter significant impediments. In particular, there would most probably be members of the council not sharing my political views. In the first place, I don't consider myself a democrat in the primitive modern sense according to which "a cook can run the government." It seems to me that in a country like ours the principle "one person, one vote" may easily lead to absurd situations tending to confirm Cicero's view that democracy always degenerates into chaos. I would favour representative government only in the case where the democratic principle operates at the lowest level where people know one another. I think there should be a determinate *élite*—not exactly in Plato's sense, but in the sense that the government should be in the hands of professionals capable of exploiting the intellectual resources of the nation.

Secondly, as follows from the "scheme of evolutionism" I have been preoccupied with for the past twenty years, the guiding role of the intellect, and, through it, of the institutions of the citizenry and, primarily, the government, should grow in parallel with the growth of productivity. This is especially important now, when we are on the threshold of a global ecological crisis. It follows that the body guiding the government's actions should have a determining role to play in directing the development of industry, education, science, and the country's infrastructure generally. But this guidance should definitely not be effected by means of a hierarchy of command. In fact the necessity of exploiting people's talent and initiative most effectively precludes, at least in most situations, any kind of command structure of management. Such a structure is in any case unnecessary since the market mechanism is sufficiently flexible and finely tuned for it to be capable, without too much interference, of satisfying practically all of society's social, political, and ecological needs.

I saw my task as that of cooperating in a process of transformation of Russia's inhabitants into citizens of a single, multinational state, capable of demonstrating a sense of responsibility for their country and a willingness to participate in the framing of national objectives—in particular, of course, in the creation of an institution conditionally named "Science in the Service of Russia's Future."

For now, however, this remains just a dream. At the meetings I attended, only political gambits were discussed, and I had nothing significant to say in that regard. I was unable, for the time being at least, to get a discussion started of possible plans for me to actively work on implementing.

All I had to show for my endeavours was a scientific seminar I organized on the theme of "Russia in the 21st Century World." This attracted a set of interesting people, who, setting aside their political likes and dislikes, attempted to make sense of what was happening to the country and distinguish utopian goals from attainable ones. But here there were again difficulties. It's one thing to meet once a month and have clever conversations, and a quite different thing to ensure that any good ideas arrived at are communicated to the intelligentsia at large—and not just the humanistic intelligentsia but also the technological one, which today represents a great potential force for good. At present the members of the latter are not very active, but they have corporative interests that, I would say, coincide with national ones, and moreover generally speaking have a sense of community. It is precisely to these people that thought of the future is germane, and the members of my seminar were in effect addressing just them.

At the present time it is nor easy to organize the publication of an anthology like *Milestones*,[37] but nevertheless enough good people may be found to do something worthwhile in that direction. Naturally, this is all small-scale, but one should keep in mind La Fontaine's fable of the frog who fell into a jug of milk!

* * *

During the two or three months of the editorial preparation of my manuscript for printing, an incident of great moment occurred, yet another catastrophe. I don't think that throughout history there has ever before been a situation where a president fired on his own parliament.[38] People whom I knew personally were killed. A lad of a romantic cast of mind, just graduated from Phystech,[39] was killed by an OMON[40] trooper near the protesters' campfire. A second-year female student was killed by a sniper's bullet fired from the rooftop of a building as she walked home. And a highly talented director of scientific-documentary films from a St. Petersburg studio was killed in the course of the risky business of recording some of the action on film.

The horror of it was that people were now shooting at one another who had recently been like-minded, who had stood shoulder-to-shoulder just two years earlier. And, just like the catastrophe of the dissolution of our great state, the events of early October 1993 can be traced to personal ambition and a struggle for power. No one had anything to say about Russia's fate, about the implications for Russia's future of the attack on our "White House." There is apparently no end in sight to unrest in Rus′.

---

[37] A collection of essays by Russian philosophers of the early twentieth century about the Russian intelligentsia and its role in Russia's history. Published in Moscow in 1909, it was reprinted four times. In 1990 another 50,000 copies were issued.

[38] On October 4, 1993, following orders by the President of the Russian Federation Boris Yel′tsin, the army shelled the building housing the Russian Supreme Council (or Parliament), and took it by storm, arresting the leaders of the resistance. This ended an executive-legislative stand-off between Yel′tsin and the parliament, dating from December 1992.

[39] Faculty of Physics and Technology, Moscow State University.

[40] Acronym for Detachment of Special Service Police.

Like many others, I believe in a president having considerable power. There was a time when we might have had one such, when the people needed to hear words of hope, words they could believe. They waited to hear such words, but when the time was ripe they were not pronounced.

Now we live in a different country. . ..

Moscow, 1989–October 10, 1993

Epilogue (added in 1999)
On the Threshold of a New Century

Chapter 15
Henceforth Life Will Have to be Lived Differently

Starting the Conversation

More than five years have passed since the first edition of the present book appeared. Having begun writing it for those closest to me, and for myself as a form of relaxation and refuge from everyday pressures, at some stage I understood that something rather more substantial than I had planned was emerging, a distinctive document. It had turned out to be not just the story of my life, with its joys and sorrows and professional activities, but also a depiction of the highly non-trivial fortunes of that portion of the Russian intelligentsia to which the baton had been passed on chiefly by their parents, the representatives of the former, pre-revolutionary, technological intelligentsia. In that sense my life has been rather typical for a sizeable section of Russian society, and may therefore serve to illuminate a facet of the life of our country that has not received proper media coverage. For that reason, perhaps, my reflections may interest a wider circle of readers than I had in mind initially.

* * *

Every member of the social stratum to which I regard myself as belonging was possessed of exceptional resilience. I would not hesitate to venture that that quality originated in a true devotion to country, handed down over generations, and imbibed with our mothers' milk. So many troubles beset my generation that it's a wonder any of us are here to tell the tale. Summary arrests and liquidations of whole families, the stress of war, which we had to bear while enduring a long-lasting relegation to the status of social pariahs, outlasting which was very far from being a simple matter. And there was also, of course, impoverishment. . .. But most of us managed to survive, battling our way through the debris of the twentieth century, and not just survive, but maintain our belief in Russia, or, more precisely, the Soviet Union, which preserved our Motherland under a new name. I have a great many accounts stored up of the fates of my fellows, all confirming the above description of that particular social stratum of Russians.

N. N. Moiseev, *How Far It Is to Tomorrow...*,
https://doi.org/10.1007/978-3-030-96651-5_15

Here, for instance, is an abbreviated biography of just one "typical" person of the Russian intelligentsia of my generation.

Georgy Evgenievich James-Levi was my peer at university, and of the same age, 1917 vintage. A gifted mathematician, he ultimately specialized in geometry. In 1942, while leading a company into action near Leningrad, he was wounded and taken prisoner, but by the Finns, thank God, rather than the Germans. In 1943 he was released, and soon found himself once more at the front, this time in the neighbourhood of Königsberg, not so far from where I myself was at that time. There he was again wounded, only this time they had to amputate a leg, rendering him unfit for further active service. In the hospital he met Katya, a signaller, who had also had a leg removed, having stepped on a mine. They subsequently lived a long life together, Katya dying just last year. There are three children, five grandchildren, and seven great grandchildren—Yura always liked prime numbers. Although he didn't become an academician, he successfully defended his Candidate's dissertation, since when he has worked as dozent in the Mathematics Department of the Lipetsky Polytechnical Institute, becoming one of the most respected of its teaching staff. He still teaches, when health permits, in the ninth decade of his life! Not long ago he sent me a paper on projective geometry, in connection with which he expressed hopes for further research.

He belongs to the set of people descended from families of the old Russian technological intelligentsia: his father was a railway engineer and one of his grandfathers a doctor. His immediate family's fate was tragic, although, unlike a great many, his parents didn't perish in a Stalinist prison: in the Summer of 1941 they were on holiday in Bryansk Province,[1] when they suddenly found themselves in territory occupied by the invading Germans. The name "Levi" affixed to the surname James signified to them that an ancestor on some remote branch of the family tree was an English Jew. For the "civilized" Germans this sufficed to brand them as expendable, and the parents and their very young grandson perished when the hut they were staying in, on the outskirts of a village near Bryansk, was set on fire!

How many such biographies I know! Despite everything, the people of my generation retained their desire to live, work, and be concerned for their country. They maintained their belief in their Motherland—that's what I want most of all to emphasize.

I'm often surprised by the fact that although the people of my *milieu* and generation, and even those a little older, have endured so much injustice and grief at the hands of the Soviet regime, they show much more tolerance of our communist past than younger folk, those of the next generation, who, having quit the Komsomol[2] and gone on to make careers for themselves, have never experienced what it's like to be social outcasts.

---

[1] Bryansk is a city about 380 km southwest of Moscow, the administrative centre of Bryansk Province.

[2] The Communist Union of Youth, defunct following the collapse of the USSR.

That devotion to one's country seems to be characteristic of the Russian intelligentsia. In the late 1950s and early 1960s, I was working in France with a group of Russian engineers who had lost everything when forced to emigrate in the 1920s.[3] Thoughts of Russia were constantly on their minds, and they forgave the evil Bolsheviks much for having at least preserved the integrity of their country. I imagine that if they were here today, they would share my attitude to those power-hungry types who are at present sacrificing the nation to their own narrow interests, turning it into one of the world's backwaters.

I believe that for our sense of Russia's inherent worth, for our sense of Russia as a supreme value, we have to thank our parents, who, rearing us, were able to inculcate in us an attitude of respect and love for our people's spiritual traditions. It's precisely because of this that we are able to more objectively evaluate what has occurred and is now taking place.

And moreover that is why I am entirely justified when I speak of my peers as constituting a generation of patriots and toilers, like those of our fathers and grandfathers. For this reason my account of that generation is important for those to whom one day we in turn shall hand on the baton.

I firmly believe that day will come.

And I pray to God that those who receive the baton from us will have easier lives than we, even as I doubt it.

The Fate of the First Edition of My Book and Plans for Its Enlargement

The fortunes of the first edition of my book *How Far It Is to Tomorrow...* were somewhat dismal. From the information I received, I inferred that it had not found a readership. It wasn't distributed for sale since the cover bore the word "Complimentary." Nor was it to be found in the libraries I searched. I donated one of my own copies to the *Leninka*.[4] This would seem to indicate that the publisher hadn't done even the necessary minimum as regards distribution of the book! I was given author's offprints, and those of my acquaintances to whom I passed on a copy found it interesting, and likewise those of *their* acquaintance to whom they subsequently loaned their copies.

In our present time of troubles, I can only guess at the wider fortunes the book enjoyed, but haven't the energy to pursue the matter.

So, convinced that the book hadn't found its readership, with the help of my friends in Phystech[5] I offered it *gratis* on the Internet—in Russian, naturally! And all at once miracles started happening: the book acquired its own peculiar readership,

[3] This period in the author's life is described in more detail in Chap. 6.

[4] The Lenin State Library of Russia, then the largest library in the USSR.

[5] Faculty of Physics and Technology of Moscow State University.

and mainly outside Russia. I began receiving letters from Europe, America, and Australia. I have by me a list of misprints sent to me by an Australian. A colleague of mine at Pushchino-on-Oka[6] was sent a copy printed off the Internet in California, and there were other such occurrences.

How many Russians there are scattered about the world! And my book turned out to needed by some of them; they actually read it, which is the best possible reward for an author's labour. By the grace of the publisher—the publication costs were borne by the Soros Foundation[7]—there were no copies available in Russia!

In view of all this I decided to prepare a new edition. There was an appropriate occasion for this: in 1997 I turned 80, and to mark this the INUEPS[8] decided to re-issue some of my writings in a small printing, including, primarily, these *Free Reflections*.

But more than a few years had passed since I completed work on the first edition. I concluded that book with the words "But now we live in a different country," and that is certainly true. We have not only become distanced from the romanticism of the first years of *perestroika*, but have lived through the most difficult years of our lives. It was easier on the frontlines during the war!—*and* during the famished immediate postwar years. We have all got to learn how to live differently! In preparing a new edition, therefore, it was impossible for me to limit myself to just correcting misprints.

And all the more so since now there is much to be said about the years that have passed since the cavalry attack, following that series of idiotic mistakes going by the name *perestroika*. We can now view those years from a more neutral standpoint, as it were, that of the coming new century, consider them as belonging to the past, and qualify them with past participles notwithstanding having lived through them. Much of what happened has now become far easier to understand. So that is why I decided to write this "epilogue" to my *Free Reflections*, entitling it "On the threshold of a new century."

However, I decided not to change anything in the first part of my *Free Reflections*, completed in 1993. As I have already said, that first part is an distinctive document best left in its original form, and this supplementary material is a natural outgrowth of it, that is, of the book I called *How Far It Is to Tomorrow*.... The additional material consists not so much of a description of the next few years of my own life, as of my further reflections about what has happened and the future: a collection of mature commentaries and ideas concerning what is at present taking place in our country and in the world, and thoughts about what fate has in store for Russia.

---

[6] A pre-revolutionary estate outside the town of Pushchino, in Moscow Province, now in use by the Russian Academy of Sciences.

[7] George Soros (born 1930) is a Hungarian-American-Jewish investor and philanthropist. He has donated many billions through his philanthropic agency Open Society Foundations, in support of progressive and liberal political and other causes.

[8] The International Non-governmental University of Ecology and Political Science, founded in Moscow in 1992.

This last I regard as most important: the opportunity remains open for Russia and its people to make a choice. And not taking advantage of that opportunity is out of the question!

It is not up to the government, nor any other power structure, to do the choosing, but to the nation. If it finds a readership, perhaps my book will in some way help in making that CHOICE.

One More Attempt to Look Back

It has become fashionable lately in our country, in "decent society," at least, to talk of the market mechanism as a panacea for warding off economic and other disorders that may threaten our country. This is blinkered thinking and may give rise to insuperable obstacles on the path of Russia's future development. Yet I feel our cumulative experience may at last have taught us something. We are, at least slowly, on the way to reaching an appreciation of the narrowness of our focus and its consequences.

Every life experience begs to be made sense of. Looking back from our present standpoint in the last years of the twentieth century, we see many things differently, and evaluate the past otherwise than five or so years ago. Reasoning about the past leads to fresh views of the present, and although it is unlikely for these not to be mistaken to some extent—"history never taught anyone anything," as, I think, Hegel said—, they nevertheless make possible a more realistic evaluation of the successive "reforms" being implemented and their implications.

Yet it can also happen, albeit much more rarely, that certain judgments formed and pondered at some time in the past, turn out after all to be confirmed in the present. Such retrospective confirmation then acquires practical significance, providing new arguments for evaluations of present, and, especially, possible future, states of society and its developmental tendencies and possibilities.

In an earlier chapter[9] I mentioned returning home one evening in the company of the late Academician Glushkov from a routine meeting of the MIC[10] or some subcommittee thereof. Crossing Red Square in the late evening, or rather nighttime, we talked of many things. The main object of our conversation, however, was the problem of the "convergence" of a rigidly controlled socialist economy to a liberal market economy. This topic was then on the minds of many, and much debated among our *milieu*, and not just among kitchen dissidents, but in auditoria and at scientific seminars. These were serious scientific discussions of a fully constructive character. We talked of using programming methods in connection with mechanisms of market type, and related things.

[9] See Chap. 10.

[10] Military-Industrial Complex.

We even wrote notes to the policy makers and published articles in scientific journals on the problems of "convergence," although at that time the term "convergence" was considered distasteful and didn't normally figure in our lexicon. Although all this activity yielded no tangible result, our discussions nevertheless laid the groundwork for "convergence," groundwork which the democrats or, more accurately, "deconstructionists," as we called them, refrained from using, attempting instead to grow new vegetables in old clay.

In those years the prevailing ideas of how "convergence" was to be achieved were rather primitive. Mostly people thought and spoke about the possibility of combining in equal measure socialist—mainly in terms of planning—and market economic principles, and about how such "convergence" might be achieved more or less abruptly through decisions of the policy makers. I think this was approximately the view held by A. D. Sakharov. I discussed the topic with him on more than one occasion, but he didn't accept my objections.

I believed then and continue to believe that such a quick transition cannot work in principle. It is not only that such a change is of necessity very complex technically; it also requires complicated mental adjustments on the part of the people, especially after 70 years of Soviet rule, which has left us more than merely unprepared for taking on board any of the organizational elements of capitalism. In this regard, our experience of *perestroika* was qualitatively different from the restructuring of the years of the NEP,[11] when traditions of individual enterprise were still alive. On the other hand, I have always been convinced that the future structure of our economy should combine planning, to be firmly imposed and implemented, with market elements. Only by means of such a symbiosis can we arrive at an economy that is both efficient and capable of overcoming the crises periodically besetting both market and socialist economies. Even then, however, the path to such an economy will be long and thorny. It will be necessary to solve a multitude of preliminary problems of an organizational or methodological sort.

But this symbiosis should be quite other than that imagined by Sakharov and the many others who have thought seriously about the "convergence" problem.

I set out my system of views in the book *Socialism and Information Technology*, published by *Politizdat*.[12] That was the last book I wrote under "socialism." Naturally, my views have evolved to some extent over the nine or ten years since then, but that book represents a significant stage in their formation, and all of the following discussion has developed out of what was expounded in it.

In fact, I had begun thinking along these lines much earlier, prompted chiefly by my encounter on a personal level with problems arising in connection with

[11] The New Economic Policy launched by Lenin in 1921, allowing small-scale free enterprise as a temporary expedient aimed at reviving the moribund economy. Recovery rapidly ensued. The NEP was ended by Stalin in 1928.

[12] Portmanteau word for the Central Publishing House of Political and Party-Historical Literature of the former USSR.

scientific and technological development and its introduction into society, problems with which my colleagues and I were confronted directly.

The pursuit of scientific and technological progress is fundamental to humanity's existence; I would call it a *natural process*, a component of the overall development of the biosphere. The natural inclination of the human element to create a "second natural environment" is entirely analogous to the wolf's natural propensity to hunt. The development over the past several hundred years of, first, technology and then science, giving rise in turn to more sophisticated technology, continually changes the structure of society, imposing on it an ever-changing organizational form.

The development of science and technology is a two-faced Janus: a source of societal progress, but also of potential danger, even doom. However, there is no stopping the process of technological improvement of the means of production and of the practical appurtenances of life. Even if it *were* possible, halting the process would in any case be mortally dangerous! Humankind has always striven and continues striving to create a "second nature." The central problem for the future of civilization is that of controlling the impulse, i.e., using it to secure the homeostasis of humankind as a species while avoiding the possible side effect of its degradation. And this will be possible only if all of humanity is privy to the facts—yet one more argument for "convergence."

Among the group of specialists who constituted my intellectual *oecumene*, this truth had been grasped long ago, or at least the truth that scientific and technological progress had become the most important factor in the development of an economy. This was in any case clear from published statistics. Already by the 1960s the economic growth of the developed nations, including the Soviet Union, was more than half due to technological development. And we, those working on problems of application of computer technology and other new technologies, understood that for a country to fall behind in the tempo of technological restructuring meant something akin to death! I mentioned this earlier[13] when describing the development of computer technology in the former USSR. To be able to shoe a flea is only the beginning. The main thing is to mass produce those shoes and ensure that after being shoed a flea can still hop!

Furthermore, the transition to new or, as they now say, higher, technologies, using computers with miniaturized components, etc., requires an appropriate new societal structure, makes new demands of people, and of their culture and educational system. I shall now list a few of the characteristics of a social and civilizational restructuring essential to a successfully functioning society in possession of higher technologies. The reader will see to what extent they are in conflict with one another and hence how difficult to realize.

First is the highest level of technological discipline. This doesn't mean merely precision; the following general principle must be added to the list of universal human values: "The best quality a person can have is an ability to carry out assigned work at the highest level of competence."

---

[13] See Chap. 7.

The second is a principle of Western liberalism: in every sphere of activity continually be on the lookout for new ideas, etc.

The third is the Russian peasant's age-old principle: sacrifice everything for the sake of the production and manifold increase of seed material. Only now "seed material" stands for knowledge, education, craftmanship, and culture.

I would add to these three the cult of attention to detail. Pedantism is not a common trait of the Russian character, but without it one cannot achieve the necessary purity of product or competitiveness. In connection with cleanliness, for instance, I remember my grandmother saying that she estimated the cultural level of a family not by the rich trappings of their apartment but by how clean and tidy their toilet was.

I believe that in the absence of an adherence to such principles, the possibility of creating a "second natural environment," without which the future of our race is unimaginable, must be illusory. But combining and adhering to those principles will require the directed activity of the government and citizenry, a specially formulated program and the efforts of the whole nation. I would even say that such a program, formulated in the context of "convergence" and the transformation of the planet into a single organism, should be the chief task of government.

Our *entrée* to the Post-industrial World

That human development is entering a new stage is an objective fact: humankind is developing swiftly, and our planet is being transformed into a single organism, that is, into a system with certain definite general objectives and possible programs for achieving them. This transformation of planet-wide society, seen in the effects of technological advances on that society, poses a multitude of difficult questions to the people of every nation. In the West this effect had been observed as early as the 1960s, and was endowed with a whole series of scientific implications once given the name "post-industrialism."

The problem of "convergence," the process whereby *our* country is to be included as an equal partner in world society, is the most basic of all those arising in connection with Russia's transition to post-industrial status. This transition is bound to take place; the only question is what the condition of our nation might be when it is near completion.

In the late 1960s the phrase "post-industrial society" had not penetrated to Russia. We were ignorant of the fact that in the West, primarily in Western Europe, the characteristics of a post-industrial society were subject to intense debate. Yet we also were preoccupied at that time with the very same issues. We were concerned, in particular, with the path the Soviet Union would have to take in order to enter a completely new phase of human societal development, although whether our rulers would have heeded us back then is another question! It was felt that that new society would necessarily be very unlike the Soviet socialist or the American capitalist versions. At that time this theme was very widely discussed. And I wish once more to emphasize that this wasn't just dissident gossip in communal kitchens, but serious science-based activity, although, it's true, we didn't rush to advertise it.

There were a great many concerns prompting these conversations. First among these, perhaps, was the fate of our computer technology, that is, of the work we were immediately involved in. In those days the subject of Kosygin's reforms[14] was also prominent in our conversations as representing a first step in the right direction for our country. These called forth a certain optimism and the hope that that those "on high" were beginning to understand a thing or two!

But that optimisn soon faded: just as now, so then, no one "on high" wished to face reality soberly, to see what was actually going on in the world around them, and register the fresh avenues of possibility opening up. There were many who understood very well that in the epoch then dawning, the Soviet Union simply had to become integrated into the world's economic system; to end up outside that system would mean our country's being side-lined to the periphery of world development. But in order to avoid this outcome, to come to the party, as it were, a qualitative transformation of the whole of our socio-economic system would be necessary, especially of its system of economic management. Of course, this could not be achieved overnight; extraordinary intellectual and organizational efforts would be needed.

Unarguably pioneering work in this direction had been done by Yu. P. Ivanilov,[15] late professor in the mathematics department of the MIPT,[16] and originally a student at Rostov University, where I supervised his Master's thesis. He then became my graduate student, and I was involved to a certain appropriate extent in both his candidate's and doctoral dissertations. Later on he became a deputy in the Supreme Council of the RSFSR[17] and then a member of the Russian parliament. Not long after the shelling of parliament in 1993, he died of complications brought on by a cruel and incurable disease. Over the last years of his life I closely followed the development of his ideas, which turned out to be extremely useful to me; I learnt a great deal from my former student.

In addition to his other good work on complex technical systems, Yu. P. Ivanilov was the first to come up with the concept of a programmable method of agricultural management, or, as they now say, a method of targeted planning and development of complex technological systems. He gave his first talk on this topic in 1964 or 1965 in my seminar at the Computer Centre of the Academy of Sciences. Later

[14] Alexey Nikolaevich Kosygin (1904–1980), member of the ruling clique of the USSR during the Brezhnev era. The 8th five-year plan (1966–1970) incorporated Kosygin's economic reforms, including greater freedom of initiative in (government-run) enterprises and less prescription of levels of production from above, which led to improvements in several economic indices.

[15] Yury Pavlovich Ivanilov (1931–1995), Soviet and Russian mathematician and pedagogue, best know for his work in mathematical economics. Succeeded N. N. Moiseev as professor and dean of the Faculty of Control Theory and Applied Mathematics of the MIPT. People's deputy to the Supreme Council of the Russian Federation 1990–1993.

[16] Moscow Institute of Physics and Technology.

[17] Russian Soviet Federal Socialist Republic.

the academicians V. M. Glushkov and G. S. Pospelov independently, or nearly independently, developed similar ideas. They wrote important monographs on the subject, for which, alas, there is today no longer any demand. These books laid the foundations of the theory of developmental planning and management. I would say, however, that as far as the underlying ideas are concerned, Ivanilov went further than the latter two respected academicians: he was interested in competitive, in essence market, mechanisms for realizing government programs, one of the most important paths to "convergence."

In those years my former student Ivanilov influenced me directly. It was in the course of conversations with him that I understood that the key to solving many problems of management is the realization that every complex system, be it governmental of corporative, is inherently hierarchical, and furthermore that, although the interests or goals of those standing on its different rungs never coincide, in most situations neither do they conflict with one another.

Such questions were also of interest to the late Professor Yury Borisovich Germeier,[18] at that time head of the Department of Operations Research at MSU. I believe that our published joint work on this theme represented a first step in the direction of creating a new control theory for hierarchically organized systems. Subsequently, our work was considerably extended by members of Yury Borisovich's school, culminating in the excellent monograph *The Theory of Non-antagonistic Games*, appearing not long before his death.

But everything achieved in that field had thus far been purely theoretical, or even merely philosophical. The burning question remained: what steps, executed calmly, without causing cataclysms or revolutions, needed to be taken to come to grips with that new, barely understood world of "convergence"? This question had also been much bandied about; I went into it with both Ivanilov and Germeier, and was thereby greatly helped in establishing my position on that very difficult question of political science.

I came to the conclusion that, as the first step along the Russian or, more precisely, Soviet, path into the post-industrial epoch, we should liquidate our branch monopolies, replacing them with government-run, but in practice independent, corporations, capable of competing in the market both externally and internally, that is, among themselves—in other words, that we should organize something akin to the syndicates of the NEP era. Against the background of such corporations, it should then be possible for private enterprise to develop, a crucial step since without it the transition to market mechanisms and, most significantly, the market mentality, would be impossible! It was important that the old framework, kept operative by means of government investment and capable of formulating and then implementing government-designed programs of development, should be preserved no matter what form of *perestroika* was envisaged. And not just preserved, but improved using all the management science that up till then had been accumulated in our country. Yes, in *our* country.

[18] See e.g. Chap. 7.

During the latter period of *perestroika*, in the late 1980s, I sent a memorandum about all this to M. S. Gorbachov, and met with him twice. Naturally, our talks were brief, but in one of them I said word-for-word the following: "While you still have an emperor's power in your hands, end the monopoly enjoyed by our branch industries: it is ruining any chance of our being able to engage in technological competition with the West." Alas, no one paid any attention to that thought of mine, even though given voice a number of times and expounded in detail in two books I wrote back in the days of socialism: *Socialism and Computer Science* and *Paths to Becoming*.

I am convinced that those ideas are still relevant today.

The Skewing of Social Life

I was very cautious in my appraisal of Gorbachov's actions in connection with his *perestroika*, and of those surrounding him in the halls of power, and even more of those members of the intelligentsia called the "foremen of *perestroika*."

I have known Gorbachov for a long time, in fact since the mid-1970s, when he was secretary of the Stavropol′ Regional Committee of the CPSU.[19] His work in that post earned him my respect: not only did he know everything about the region and its many-sided economy, but in many other ways demonstrated qualities superior to those of his colleagues holding similar positions in the Party hierarchy—and in my time I have had experience of quite a number of regional secretaries. Then when he became the Central Committee's Secretary for Agriculture, and I again fell into his orbit, his judgements seemed to me professionally weighed and completely reasonable.

But as head manager of the process for restructuring our vast nation? I was saddened to discover that in this case the hat didn't fit!

But neither did it fit those who came after him. Like all those of my circle of acquaintance, I understood the necessity, or rather inevitability, of profound changes to our country. They had to come sooner or later. But I quickly understood that Gorbachov and his colleagues lacked a clear idea of the sort of society they would like to see emerge at the finishing-line of *perestroika*. What exactly did he want to restructure and why?! He seemed unable to make this clear to the people, without whose help transformations on that scale could never be crowned with success.

I especially didn't trust the "foremen of *perestroika*," so-called. These were for the most part dissidents—destroyers, I would say. The majority of them were motivated by personal interests. I was looking for constructive arguments, so shunned the lot of them. In this, as before, I acted, frankly speaking, like a "born opportunist."

---

[19] Communist Party of the Soviet Union.

Nevertheless, like most people engaged in intellectual labour, from time to time the hope would surface in me that together with Gorbachov we would all somehow rise in the world, and that our science would be useful to the nation in this respect. For this reason, despite the exodus of many Computer Centre faculty to positions in commerce or abroad, I continued for several more years with my scientific work (now having to do with computer models of processes in the biosphere), at the same time participating in various publicistic enterprises, including some initiated by those same "foremen," although here I independently sought alternative ways to reform the structure of our society.

But somewhere around the turn of the 1980s into the 1990s a significant insight came to me: What they were calling *perestroika*, especially as prolonged under Yel′tsin, and the means being used to prosecute it, made me think of a slippery staircase, leading not up, but down, reminiscent of the stairs in Sverdlovsk which the last emperor of Russia descended.[20] And now our new president—now of Russia—has begun a new descent. Many political parties have sprung up, likewise without clear goals, as if in imitation of social cliques. The mass media have been intimidated by threats to restore communist management methods. But the communists themselves are more cunning: understanding that they have very little support and are unable to influence the course of events, they are refusing to admit losing an already lost cause. In Zyuganov's[21] place, I would probably do the same: better to retain the leadership of the opposition than lose everything and be forced to quit the political scene forever, although, of course, such behaviour is mere unworthy politicking.

At the same time, we now see that the political struggle has gradually ceased being a struggle between different parties, or, even more so, ideologies, and become more like a conflict between clans, with questions of ideology, the future of Russia as a state, our national culture, or even just the survival of the nation, all consigned to the periphery of the interests and activities of those in power.

This is all taking place right in front of our eyes and has clearly registered with the great mass of the people. In the 1990s Russia entered the epoch of "the war between families," in the words of the Sicilian mafia.

But surely things cannot continue thus.

What is to be the fate of our "deformers" and their reforms? Where are they taking us?

---

[20] On the night of 16–17 July 1918, Tsar Nicholas II and his family, together with those few who had chosen to accompany them as far as their final place of exile in Ekaterinburg (later renamed Sverdlovsk), were taken to the basement of their appointed dwelling and shot and bayoneted to death by a squad of Bolshevik guards, ostensibly to avoid having them fall into the hands of approaching White forces.

[21] Gennady Andreevich Zyuganov (born 1944) has been the leader of the Communist Party of the Russian Federation since 1993.

I Try Playing the Oracle

In order to live a worthwhile life, one must somehow imagine one's future. Everyone does this to some degree.

But to go public with one's prediction of the future, and, even more, of the outlines of the future course of development of our society in these times of rapid change, would be a thankless enterprise of dubious worth. But on the other hand to just follow the narrow path above the chasm without attempting at least to see what's a metre ahead is much worse since mortally dangerous!

So that's why I have made the attempt. But in order not to be taken for a mere fantasist or summarizer of others' writings, I first carried out a rather detailed survey of what had already been written about our possible future by those regarding themselves as futurologists or political scientists, and found a series of ideas corresponding to some extent with my imaginings, although my prognosis, determined as it was more by my professional attainments, was identical with none of theirs. I therefore took the risk of expounding my own views as to the future, and in the midsummer of 1996 wrote it all down.

An abridged version appeared in the newspaper *Green World*[22] under the title "Russia's Agony" and subtitle "Does Russia Have a Future?" The full text was published by the Moscow publishers *April* on the recommendation of Professor V. T. Loginov and at the expense of the Gorbachov Foundation[23]—one more favour from Mikhail Sergeevich. It appeared as a slim volume of six or seven leaves with the title *Does Russia Have a Future?*.

Although the edition was small, my booklet resonated quite well in certain quarters and sold out quickly, even though Loginov had no contacts of any sort with distributors, so that it was available only within the Gorbachov Foundation. A certain number of copies were acquired even by the Federal Council.[24] The response I received was substantial.

At about that time, by way of preparation for the next international conference continuing the work of that held at Rio in 1992, the organization Green Cross International (GCI),[25] in consort with our Lands Council, began work on an Earth Charter. Speaking as president of GCI, Gorbachov made the proposal, very sensible, I would say, that each national sub-organization of the Green Cross issue a

[22] A Russian newspaper concerned with ecological problems and programs, appearing first in 1990.

[23] A non-profit organization founded by Gorbachov in 1991, dedicated to researching the *perestroika* period and current issues of Russian history and politics, and financed by Gorbachov himself and through donations.

[24] The upper house of the Russian parliament.

[25] A global independent non-profit and non-governmental organization working to address the interconnected global challenges of security, eradication of poverty, and environmental degradation through a combination of advocacy and local projects. It was founded by M. S. Gorbachov in 1993, in order to continue the program initiated at the 1992 Earth Summit held in Rio de Janiero.

description of its view of the contemporary planetary situation. Since I was president of the Russian Green Cross, this proposal was of direct relevance to me.

In fact the text published under the title *Does Russia Have a Future?* was precisely what was needed by the politicians who were to frame the Russian version. (Unfortunately, work on the Earth Charter was carried out mainly by politicians.) However, that text far exceeded the agreed restrictions as to length, and in any case no politician would be likely to read the whole of such a work. It thus fell to me not only to rewrite my text, but to shorten it to about a tenth of its original length, and it was in that form that, after being translated into English, the Russian view of the planetary situation was presented to the GCI. Incidentally, I don't believe my text was much consulted by the politicians. Politicians everywhere remain politicians, even when they are attempting to write an Earth Charter: they are little interested in an objective systemic analysis of problems since they have their own agendas. But to be frank, I wasn't thinking of them when working on the text.

On the Use of Jubilees and Old Men's Pleasures

With the passing years life's customary pleasures fade: travel, holidays, and especially the joys of sport. Such is life's logic. Hence the importance of striving in old age to preserve the little that keeps up the interest in living so well expressed by the innocent line of the children's song:

> . . .tomorrow another day will dawn!

More than forty years ago I wrote the following lines on the occasion of the birthday of my friend Igor′ Pavlovich Lyubomirov, since departed life, alas:

> Already forty! How long
> You've been around, old friend!
> You likely recall how things were
> Back in our times, how she. . ..
> And that one—the other girl;
> But enough of stirring up old deservings.
> We can't have our wives deriding them,
> So for clarity's sake let's be silent. . ..

Yes, back then forty seemed a respectable age, and it never occurred to me that the path was still ascending and not near the peak! In fact, in 1956 the peak was still far off. *Very* far off!

* * *

I really began to feel my years, perhaps, only when just over seventy, when certain ailments began circumscribing my activity. Then the defining stimulus for living was provided by the feeling that I was still needed—of being needed on account of what I was doing, namely thinking and writing. In other words, the stimulus depended on the existence of people interested in what I had to say to them.

When I was younger, other stimuli acted, though none relating to the pursuit of a career. Those have played an insignificant role in my life. More than once I've declined offers of distinguished and important posts. And it was no accident that I was the first active member of the Academy to invoke the new law and in 1986 retire from all administrative positions.

For me the main thing, perhaps, has always been "interest": I wanted to understand what hadn't yet been understood, look at old problems from a new angle. . .. And also—why pretend otherwise?—there was the sporting side: wanting to be "first over the line" as my grandfather would have said, although I would describe it more accurately as a striving for self-affirmation.

With the passing years the latter kind of striving has gone completely; I stopped needing to "affirm myself." On the descending branch of life's trajectory, boosting one's self-esteem is no longer a priority! But on the other hand, the question of the extent to which people need what I do gradually became the most prominent one. After all, it's very disagreeable to feel, in one's advanced years, having lived one's life, that one is spending the little time remaining doing things of no concern to anyone! I reckon it my great good fortune that, as it has turned out, my reflections, transferred to paper, find readers. Who cares about physical incapacity if the mind's preserved!

I began my tale with a description of a memorable night on the shore of Lake Ladoga. When alone with Nature, I have never felt lonely. It was precisely on such occasions, when I happened to be completely alone with myself, free from all distraction, that fresh desires and thoughts arose.

And now, in the mid-1990s, after several medical operations, I spend as much time as possible in the countryside outside the city, where Lake Ladoga has been replaced by rural peacefulness and . . . by my computer, my silent interlocutor and preserver of my "I."

*   *   *

On August 23, 1997, I turned 80. My wife and I hadn't invited anyone to celebrate the occasion, but were pretty sure there would be people coming to visit us in Abramtsevo despite its considerable distance from Moscow: my children and grandchildren, and perhaps some of my old friends.

So I got behind the wheel of my old Zhiguli and we drove to the market in Khot′kovo, as well-provisioned as any Moscow market. In case a crowd should descend on us, my wife bought a variety of foodstuffs. And that's indeed what happened: over the next two days, Saturday and Sunday, there was a procession of 104 people through our *dacha*. At that time my wife's niece was staying with us, and thanks to her we were able to cope with the crowd of well-wishers: I think they all had plenty to eat and drink.

To be honest, this "invasion" was a welcome birthday present for me, and all the more so in that it had been unexpected. Perhaps the attention of one's close relatives and the appreciation of outsiders give old men the greatest pleasure.

Then in September there were two more noteworthy events.

On September 24 an official celebration was held in the House of Scientists. A great many people gathered, and many nice things were said. And the following week, in the *Leninka*, now renamed the Rumyantsev Russian State Library, at the initiative of its management aided and abetted by the Computer Centre and the INUEPS,[26] an exhibition of my *oeuvre* was mounted. Those involved did a great job of organizing some hundreds of my publications on shelves. But I myself had to go into hospital just then, and the opening had to wait till March 1998. On that occasion more than a hundred people gathered and once again many kind words were spoken.

[26] The International Non-governmental University for Ecology and Political Science; see earlier in this chapter.

Chapter 16
Through the 1990s

A brief accounting of my life,
as seen by its "performer"

The Shock Therapy Nightmare

Gaidar's economic reforms[1] pushed many people working in the scientific and engineering spheres, including me, in fact a great many members of the intelligentsia generally, to the edge of a financial and moral catastrophe. It even came down to the question of physical survival: how to get enough to eat. And in my case this was taking place in my eighth decade, when I had expected my life to be finally settled and a fresh start out of the question!

Up till then, what with my academic salary and old age pension, I had considered myself pretty secure, financially speaking; at least I had no need to think about moonlighting in order to earn a crust. Even more, my wife and I had enjoyed the privilege of staying at a sanatorium each Winter. For me personally, most crucial had been the apparently unthreatened stability of life, thanks to which I was able to pursue the scientific questions I myself considered important. And, what is of greatest import in connection with feeling comfortable in one's own skin, I had been convinced that the scientific pursuits in question were not merely of interest to myself, but also relevant to my country's needs. The requirement to feel what I do to be of use has always played a primary role in my worldview.

But suddenly, as it were in an instant, both sources of comfort dried up. The money I continued to receive each month from the Academy, my pension, and my wife's pension were together sufficient for at most a week, and even then only if we

[1] Implemented in 1992 in the face of an impending famine and a bankrupt economy. In particular, Gaidar's voucher program for privatizing state assets (the distribution of shares in Russia's industries on a per capita basis) was blamed for much of Russia's wealth ending up in the hands of a small set of influential businessmen, the infamous "Russian oligarchs." But many, such as Boris Nemtsov, Deputy Prime Minister under Yel′tsin, were of the opinion that Gaidar's "shock therapy" was necessary in order to avoid civil war.

N. N. Moiseev, *How Far It Is to Tomorrow…*,
https://doi.org/10.1007/978-3-030-96651-5_16

were very modest in our food requirements. Our savings evaporated. I felt strong enough to take up some kind of paying work, but what could I do? Not what my people, my country, needed, but something whereby I might earn enough to pay for the basics of life! However, although this had become a matter of urgency, at the time I was unable to imagine anything I might be able to do for which anyone would pay me.

We were faced with having to get through the nightmare of the Summer and Autumn of 1992, and I at last understood what it means to "live on the smell of an oily rag." Nothing I was qualified to do had any prospects of generating an income, and I felt justified in my earlier suspicions of the actions of the "foremen of *perestroika*" and the Gaidarists who succeeded them. To repeat: the question of physical survival became pressing at a time in my life when I had thought all practical details of living a calm and ordered life in my last years had been satisfactorily arranged. I was overtaken by panic. There was no demand for any kind of scientific work: the Gaidarists were of the opinion that foreign science sufficed. Publicistic work and lecturing were still remunerated more or less, but for me to obtain such work someone would have to order it, and in the prevailing "market conditions" obtaining such work was no easy matter! In any case I couldn't stoop to approaching appropriate sources of such employment like a beggar, or rather didn't know how to go about it.

But my wife—thank God!—kept her head: from her wartime experience of living in the countryside, she knew what to do. The backyard of our *dacha* was quickly transformed into a source of sustenance, perhaps even the main one. Then she started running chickens, another such source. And then war veterans were officially permitted to travel *gratis* on public transport in the city and on suburban electric trains.

By such means, we little by little got up on our feet again.

Which by the Summer of 1992 enabled me to return to thoughts of writing a BOOK, thoughts continually on my mind over the 50 years since that June morning on the shore of Lake Ladoga, when I first conceived the idea. Now, at last, I had the time, and the conviction arose in me that such a book would turn out to be needed—if not now, then soon—and I began hoping that one day I would see it published. Though not immediately, of course!

But the main thing was that I now had SOMETHING TO DO, and, as I said, I managed to convince myself that it was a *needed* occupation. I wrote the bulk of the book over the Summer of 1992, taking for its title the line from the extraordinary poem of Georgy Ivanov: *How far it is to tomorrow. . . .*

M. S. Gorbachov and His Court

One Autumn day of 1992, I received a phone call from Gorbachov's office requesting my presence at the premises of the Gorbachov Foundation. When I arrived at the Foundation, which was accommodated in a building on Lenin Avenue,

Mikhail Sergeevich greeted me very cordially; after all we had known each other nearly 20 years! And there ensued an unhurried, substantive, and, to me, interesting conversation.

The main topic of our talk was the Green Cross International (GCI),[2] an organization whose founding had been agreed upon by heads of state at the celebrated United Nations Conference on the Environment and Development held in Rio de Janeiro in the Summer of 1992. Within the Gorbachov Foundation a working group had been formed with the mandate of producing a science-based blueprint of the new international organization. It had been proposed that Gorbachov be chosen president, and this was duly arranged a few months later, towards the end of 1992. The working group was supposed to help the president prepare a speech on the organization of the GCI, to be given at an international conference scheduled to take place in Kyoto at the end of 1992. Gorbachov wanted to enlist me as a member of this working group.

I agreed with alacrity. Not only was the proposed work a good fit with my scientific interests, but the promised payment represented a most opportune supplement to our drastically reduced income.

I spent the Autumn of 1992 writing a rather substantial essay on the specifics of contemporary mutual relations between Nature and human society, concluding with a list of fundamental problems the new international organization should address. Unfortunately, the original text of my essay has been lost; through negligence I omitted to transfer it to the new computer I acquired at that time. It was precisely in that essay, intended as a draft of Gorbachov's address, that I formulated in more or less finished form the system of views fundamental to the whole of my later methodological work on the relations between Nature and Society.

The final editing of the penultimate draft of Gorbachov's address was done by A. S. Tsipko.[3] This was difficult since it meant accommodating, to both speaker and anticipated audience, the ideas worked out by the various professionals comprising the working group. I was pleased to see that one of the main conclusions I had drawn had found, albeit in somewhat simplified form, a place in Gorbachov's speech; I had recommended that GCI should not so much concern itself with separate partial programs—since these could be left to the many other ecological organizations such as Greenpeace—as with programs for a new twenty-first-century civilization and a strategy for society's transition to the epoch of the noosphere.

For my work on Gorbachov's address I was paid 1000 dollars, which greatly eased the financial problems facing us in the Winter of 1992–1993.

However, M. S. Gorbachov did something else for me of much greater significance than merely relieving, through his proposal, our financial crisis that Winter, or at least postponing it for a while. I remain very beholden to him, although not

---

[2] See also Chap. 15.

[3] Alexander Sergeevich Tsipko (born 1941), Soviet and Russian social philosopher and political scientist. At present (2019) head of the Institute for International Economical and Political Research of the Russian Academy of Sciences.

only to him. He himself doubtless remains unaware of how important to me were his concern and what he did for me in 1993. But allow me to relate the events in the order in which they occurred.

I had had several several urological operations and, subsequent to these, two contradictory diagnoses, about which, apart from the doctors, only my wife knew. In the Spring of 1993 another operation was needed, and she insisted it be done abroad, even though they cut flesh just as well in Russia. But it turned out to be crucial that a qualified arbiter resolve the conflict over the diagnosis, the ideal such person being the preeminent French oncological urologist Professor Le Duc, working at the Hôpital Saint-Louis in Paris. However, the possibility of traveling to Paris to be operated on in a French hospital was in practice beyond my wildest dreams. An application to obtain financial support from the Presidium of the Academy of Sciences was dismissed out of hand; they wouldn't even deign to discuss the idea with me. My situation seemed hopeless. But it wasn't, for there are kind people in the world. As always, help came entirely unexpectedly from a quarter whence I'd have least anticipated it.

One day I was attending a seminar in the Gorbachov Foundation and happened on Mikhail Sergeevich himself. He asked me how I was doing, and I told him of my medical tribulations. His reaction was immediate: "I'll phone Mitterand right away." And a few hours later he informed me that the President of France had arranged for my travel and operation at the Hôpital Saint-Louis be paid for by the French government.

Aftermath of the Bombardment

On New Year's Eve 1993, I raised my wine glass for the first toast, an obscure one to the effect that our country not have to endure a year ninety-three. I had just remembered Hugo's novel, and had felt that the the best thing I could wish for in our new time of troubles was that nothing like the year 1793 in France should figure in *our* history!

But as things turned out, I proved to be a poor oracle.

History has its own logic, and, by following its logic, perhaps, one may have been able to predict far in advance of October 1993 how events would unfold. Or perhaps we see the logic of that unfolding only in retrospect. I had indeed foreseen something of what would happen, but had hoped bloodshed might be avoided.

I now think that the shelling of our parliament[4] or some other similarly violent development was inevitable. The situation had become tense, and the sense of impending tragedy had been growing sharper with each passing day. The people in the process of destroying Russia, or, more precisely, the great Soviet Union, were unable, out of overweening pride, to tolerate the existence of a responsible

[4] See Chap. 14.

and powerful parliament—could not countenance, more generally, sharing power and wealth with anyone else. Whatever our Supreme Council[5] may have been like, and I'm not inclined to say anything praiseworthy about either it or its leader, it was nonetheless *the* organ of government representing the nation, and it was appropriate to relate to it as to the supreme power. Yel′tsin and his crew should have shown respect: for themselves, their people, and the constitution, after all! And not heap shame on the concept of democracy for which earlier the people had stood firm at the walls of the White House.[6] It was not by chance that everyone of my acquaintance whom I respect and to whom I talked in the course of those tragic days, perceived what had happened as a catastrophe. After October 1993 we really had begun living in a new country.

In fact the "victors" turned out to be the vanquished—vanquished by the nation's contempt, and never again able to win its trust.

The shelling of parliament was a national catastrophe which will echo down the years and even, perhaps, the generations. The horror of the event consists in large part in its predictability. History's logic is such that the Supreme Council's 1993 finale represented a fully "natural" outcome of the Belovezhsky tragedy[7] and the insinuation of a self-contradictory totalitarianism into our lives.

One now hears the following popular aphorism: the Bolshevik-Communists have yielded to the Bolshevik-Democrats. Amongst the latter we find the repellent "defender of human rights" Kovalev,[8] who is, however, far from being the worst of them.

But now, on the verge of 1999, if we compare the two types of Bolsheviks we see clearly that the former were more intelligent, more modest, and . . . more honest, although, of course, their honesty also was less than merely relative.

* * *

The stairs go down: this feeling, prompted by the shelling of the Supreme Council, lay like a curse on my life, like a massive weight oppressing me with a sense of hopelessness. And I was far from alone in feeling this.

I ended the first edition of this book with the words: "Now we live in a different country." The bombardment of the Supreme Council really did mark off the past from the present, and even more from the future. This was yet one more national tragedy: that's how I together with many of those closest to me understood what had taken place in the Autumn of 1993.

Before the shelling, we, i.e., those who reacted to what was occurring as did I, had all been hoping that Russia was not yet entirely done for, that we would be able to once more get up on our feet, and, most importantly, the time would come when we'd once more have real work to do, work needed by the people and by the state.

---

[5] Russian parliament.

[6] Supreme Council building.

[7] See Chap. 14.

[8] Sergey Adamovich Kovalev (born 1930), a Russian human rights activist and politician, and former Soviet dissident and political prisoner.

And, although from 1994 the situation in our country became relatively stable, or at least inflation decreased to a reasonable level and the exchange rate for the ruble became more or less stable, this stability was not much of a consolation since the rate of fall in production was likewise stable, as were the rates of fall in the standard of living and of the rise in the feeling of despair. It was becoming more and more evident that the people who had come to power were under-educated, since lacking systematic analytical skills, not to mention appropriate management qualifications, and were, in addition, handicapped by the complex of "self-sufficiency." But the worst of it was that these people had sought power for the sake of power alone, and had absolutely no ideas for bettering their country. It was becoming obvious that in the Autumn of 1993 we had taken a further step downwards! But not only economically speaking: the descent had been started long before by the ruling *élite*, when in Ekaterinburg they ordered the last emperor of Russia downstairs.[9]

On the eve of the new year 1993, I proposed a toast expressing a wish that the coming year be nothing like the French year 1793, but, alas, my wish didn't come true. There was, of course, nothing as awful as a barge carrying a crowd of arrestees being sunk in the Loire, but all the same blood was spilled, and not somewhere offstage, but in my country's capital city.

Every year, step by step, we have been descending, but since 1993 the rate of descent has increased. In 1994 there was another convulsion: the terrible and unjustified Chechen war.[10] Although one can appreciate the reasons for the rising of the Vendée,[11] understanding the Chechen war is impossible! Can it be that no one will be punished for this? And for how many decades will the memory of this crime hang over us? And the shame of defeat? Not in distant Afghanistan, but here at home. The effects of this senseless slaughter will determine the fate of a whole generation, at least.

The Last Elections

By this I mean the elections of 1996. I hope, and it's a miracle that I *can* still hope, there will be elections again in the year 2000, but who knows!? The shelling of parliament and the Chechen war are ominous signs of our present political situation.

[9] To be executed, along with his family, in the basement of the house they had been confined to.

[10] The "first" Chechen war was waged by Yel′tsin's government with the break-away republic of Chechnya from 1994 to 1996. The Russian government declared a ceasefire in 1996 following military setbacks. A second war was launched in 1999, prosecuted ruthlessly once V. V. Putin became president of Russia in 2000, resulting some years later in defeat for the Islamist Chechen separatists.

[11] The war lasting from March to December 1793, between a royalist army raised in the Vendée region of France and the army of the new French republic, eventually victorious, resulting in the death of tens of thousands.

But I still want to believe that Reason will eventually prevail and we'll abide by our constitution, which, however imperfect, is nonetheless ours.

But to return to the 1996 elections: In early Autumn of 1993, my membership in the Presidential Council was terminated. This was done in the best Bolshevik manner: when my pass into the Kremlin expired they simply refused to extend it, saying, without any further explanation, that my name simply wasn't on the list of those eligible. I was chiefly offended by the fact that I had considered Filatov,[12] the mouthpiece of the Council, to be a member of the intelligentsia, or at least someone who knew how to behave decently. From that time on I avoided further contact with him.

But every cloud has a silver lining: I soon understood that once again I'd been lucky, that my exclusion from the Presidential Council had in fact been a boon to me. That was my thought when the Supreme Council was bombarded: had I remained in the President's Council, I would have had to follow the precedent of the late Gefter[13] and submit my resignation, which would have been hard for me since I don't like to get into the bad books of the powerful of this world too much! But all the same I was very curious to know precisely how I'd fallen into royal disfavour, and I think I now know the reason.

For the whole of the year and a half that I was a member of the Council, in every one of its meetings I kept my peace—was "as quiet as an iced-in fish" as they say in Odessa. The only time I opened my mouth was to answer a question put to me by B. N. Yel′tsin. At that moment the Council was absorbed, if I remember rightly, by the rather strange question as to what sort of actions might improve the president's public image. For some obscure reason Boris Nikolaevich turned to me and asked me for my opinion, and I told him exactly what I thought.

I am utterly convinced that it is essential that all of us, i.e., the Russian people, have an appropriate perspective on events, and at the present moment that is not the case. Although our present situation is difficult, it is not as hopeless as many think, and it's very important that our leader confirm this opinion as fact, that he express it to the people as often as possible, as well, of course, as justifying it. The people must believe that they have muscle, that they can achieve much more, that there are many successes yet to come. We're not Americans, so can't confine ourselves to family affairs; it is important for us to feel there's someone out there we can depend on!

---

[12] Sergey Alexandrovich Filatov (born 1936), Soviet engineer, and, since 1990, Russian politician.

[13] Mikhail Yakovlevich Gefter (1918–1995), Soviet and Russian historian, philosopher and publicist. A human rights dissident in the USSR from 1970.

That day I was sitting at the great oval table between two mayors: Sobchak[14] to my left and Popov[15] to my right. When I came to the end of my two and a half minute speech, Sobchak said nothing but raised his hands heavenwards as if to say "What are you talking about!?" On my other hand, Gavriil Kharitonovich exclaimed rather loudly and judgmentally: "Well, Nikita Nikolaevich, as always you have your own peculiar view of things," to which I riposted, *sotto voce*: "So have I been brought here to be other than I am?"

As I now think, remembering the looks on the faces of Boris Nikolaevich and some of his advisors sitting to his left, that two and a half minute speech was the cause of my falling out of royal favour. But who cares about being in or out of favour! How transitory and unimportant such things are against the background of what is happening in our country, against the background of the swelling grief of our people!

Following the shelling of our parliament, and even more so after the Chechen war was launched, I was overtaken by a feeling of despair and helplessness. I could envisage no agency capable of affecting the course of events even slightly. And I would say that that sense of apathy and helplessness was also filtering irreversibly into the souls of my friends, indeed into the souls of everyone I happened to talk to. By comparison with most, I myself was even relatively more optimistic—enough, at least, for me to be able to reflect on and write about our situation.

I think that this apathy became widespread not only within the relatively narrow stratum made up of the academic intelligentsia with whom I was in contact, but began to infect a much wider section of the population. I inferred this from the way the election of the president proceeded, and from the general lack of interest in discussing political questions, among other indicators. How different it all was from the period of *perestroika*, especially in its initial stages. The era of hope was at an end, and for many years people would once again be preoccupied with their personal troubles. Today the main problem for most people is that of mere survival.

And then came the year 1996. I had hoped the presidential elections would rouse up the masses to greater political participation, but no such thing occurred. The people had apparently grasped that they were not being given a real choice. It was clear that it would make no difference which candidate won, since no real change in essentials would result. Things would remain just as before.

I made assessments from the sidelines: it was entertaining to see how the incumbent party puffed itself up, and how hard Zyuganov[16] tried *not* to win. One thing above all else was clear to me: so far no real leader has emerged in Russia,

---

[14] Anatoly Alexandrovich Sobchak (1937–2000), Russian politician, co-author of the Constitution of the Russian Federation. In 1991 he became the first democratically elected mayor of St. Petersburg. He was earlier a mentor and teacher of both V. V. Putin and D. A. Medvedev at Leningrad State University. He died suddenly in 2000 in suspicious circumstances.

[15] Gavriil Kharitonovich Popov (born 1936), is a Russian politician and economist. In 1990 he became the first democratically elected mayor of Moscow, resigning in 1992.

[16] Leader of the Russian Communist Party; see Chap. 15.

and the WORD has not been enunciated that would galvanize the people to embrace fresh sacrifices.

In the absence of these we won't make it.

* * *

I didn't vote in the first round, but in the election of representatives to the *Duma*,[17] I voted for Yavlinsky.[18] Although he is, I think, a Gaidarist, he is more intelligent than Egor Gaidar and a more decent person. His most recent actions in the *Duma* are worthy of the greatest respect, but he has a long way to go yet to recover from the ignominy of his "500 days" program.[19] But in himself Yavlinsky is neither here nor there; it's his party *Yabloko* that's of interest, occupying as it does an advantageous centrist political niche and making an impression on a wide swathe of the public, including the technological intelligentsia. I believe that, given the right circumstances, *Yabloko* might very well attract a significant portion of the electorate.

In Russia today there is a large section of the populace who see no future for themselves, largely consisting of the scientific and engineering and technological intelligentsia but including the class of skilled workers, that is, precisely the people to whom the nation owes the establishment of its industrial and educational power. These already understand that they have interests in common, and might well serve as a significant political base for *Yabloko* if only the latter's political program and slogans were adapted to their interests. This is not difficult to do since the interests in question are in line with what *Yabloko* stands for, and if reflected in its political platform would render it not only more attractive but more scientifically based.

According to my calculations, the number of such people not yet identifying themselves with any particular political movement lies between 10 and 12 million. But it should also be easy to attract members of the teaching profession and medical workers. For this to happen, however, the party's platform must be carefully reformulated in conformity with its general aims. Is *Yabloko* capable of managing this in good time?

However, in the second round of the elections my wife and I decided to participate. The considerations determining our preferences were, however, not at all political but rather purely emotional. It was nauseating to be exposed to all the agitprop on behalf of B. N. Yel′tsin, whose rating had in the initial stages been one of the lowest. One can imagine how many millions of greenbacks and other banknotes had been paid to the media to ensure their loyalty, although in any case the result had been determined beforehand.

Well, we voted for Zyuganov.

---

[17] The Russian equivalent of a House of Representatives.

[18] Grigory Alexeevich Yavlinsky (born 1952), a Russian economist and politician, is still, in 2019, the leader of the Russian Social Democratic Party *Yabloko* ("Apple"). He ran for the Russian presidency in 1996, against Boris Yel′tsin, and in 2000, against Vladimir Putin.

[19] A plan, formulated in 1990, for a rapid transition from the Soviet Union's centrally planned economy to a free market economy, to take an estimated two years.

This was, of course, quixotic in the extreme. For, in the first place, Gennady Andreevich is, like his fading party, in no way remarkable. That party was unable to put forward any kind of well-founded, attractive platform for the further post-industrial development of our country. Zyuganov's rating remained reasonably high for a considerable time, but then, in light of his television interview with that most unpleasant of newscasters Kiselyov,[20] it became clear that he himself didn't actually want to win. It was following that interview, in fact, that we decided to go to the polling booth and vote for him.

My Scientific and Other Activities

I retired on the eve of *perestroika*, relinquishing thereby all administrative cares both in the Computer Centre and my MIPT department. I then had the leisure to arrange a rather comfortable life for myself. I no longer had to be at work every morning by 9 am, and, having a computer, printer and even photocopier at home, had no need of a secretary. But my seminar at the Computer Centre continued to meet once a week, and I had some very interesting ideas for further research. Not only was I freed from all official duties, but also from worry over having to get a crust. Is it possible to imagine a better life for a scientific husband?!

But gradually, as *perestroika* progressed, this life began to "shrivel." People became more preoccupied with getting a crust. Seminars were held more and more rarely and fewer and fewer people attended—understandably, since one can't investigate even the most interesting of scientific problems on an empty stomach. Thus it was that I was finally left almost one on one with my computer.

But that wasn't so bad, after all. I still had two areas of research that I could pursue under the regime of "a scientific institute consisting of a single person." One of these had to do with methodological problems of research into the mutual interactions of Nature and Society, which I had been thinking about since the end of the 1960s, and the other was ecological education, whose importance was growing by leaps and bounds.

Earlier I had had two books published by the Academic Publisher: *Man, the Environment, and Society* and *Developmental Algorithms*. These were brought out in quite large printings and received favourable reviews. I also had a few articles in the journal *Problems of Philosophy* devoted to universal evolutionism.[21] I was even invited to join the editorial board of this academic journal; it seemed the philosophers were ready to accept me as one of them.

---

[20] Dmitry Konstantinovich Kiselyov (born 1954), Russian journalist. Appointed by President Vladimir Putin in 2013 to head the new official Russian government-owned international news agency *Russia Today*. Noted for his hate-filled anti-Western tirades on Russian television.

[21] See e.g. Chap. 10.

In the Spring semester of 1992, Professor V. S. Petrosian[22] proposed that I give a course at Moscow State University on problems of universal evolutionism, and in the Autumn semester of the same year a similar proposal was made to me by G. A. Yagodin[23] of the International University. Using these lectures as a basis, I wrote the moderately sized book *The Ascent to Reason*, which on publication quickly sold out.

This and the success of the two lecture courses heartened me: despite all the Gaidarist shenanigans there were still people in Russia prepared to spend money on books of a methodological character. So I decided to bring order to my thoughts about methodological problems of science and the mutual interactions of Nature and Society.

By 1995 I had completed a large monograph in which I attempted to present a coherent exposition of my worldview. The book was titled *Contemporary Rationalism*. The Russian Foundation for Research in the Humanities financed the printing of 1000 copies, and the International Non-governmental University for Ecology and Political Science (INUEPS)[24] another thousand. Thus *Contemporary Rationalism* was published with an initial printing of 2000 copies, which is not to be sniffed at given the times we're living through. I am very grateful to the late Academician Nikita Tolstoy[25] and his deputy Professor Semyonov,[26] as well as Professor Stepanov,[27] rector of the INUEPS, for arranging the financing of these printings.

The book sold out, and in 1996, at the initiative of the Institute of Philosophy and the journal *Problems of Philosophy*, a round table discussion of its contents was held in the Polytechnical Museum. And then in 1998 the publishing house AGRAF reissued the book under the new title *Path to the Obvious*, in an edition of 3500 copies. Thus my work in the field of methodology to a certain extent filled the breach that had opened in my life when I forfeited the possibility of carrying out computational investigations using the computing facilities of the Academy's Computer Centre, since the work in question amounted to an examination of the same problems of mutual interaction of Nature and Society, but now from a different perspective. The fact that the above-mentioned publications turned out to be interesting to a fairly wide circle of people served as a stimulus to continue my work.

[22] Valery Samsonovich Petrosian (born 1942), Armenian academic chemist. Deputy chair of the Supreme Ecological Council of Russia 1990–1996.

[23] Gennady Alexeevich Yagodin (1927–2015), Soviet and Russian government and social activist. Rector of the International University in Moscow 1991–2001.

[24] See also Chap. 15.

[25] Possibly Nikita Alexeevich Tolstoy (1917–1994), Soviet and Russian linguist of slavonic languages, and full member of the Soviet Academy of Sciences from 1987.

[26] Possibly Yury Ivanovich Semyonov (born 1929), Soviet and Russian historian, philosopher, and ethnologist specializing in primeval human society.

[27] Stanislav Alexandrovich Stepanov; see below.

At the same time I was working on a different, though related, subject, namely, as noted above, ecological education. This was a logical offshoot of my work on modelling the mutual interactions of society and the biosphere, since the latter gave one an appreciation the vast potential of contemporary natural science for solving problems of human ecology, i.e., those problems whose solution is essential for rendering the life of humankind in its home, the biosphere, secure. While science cannot and will probably never be able to say how society *should* be structured, how it should conduct itself in order to be sure of preserving a homeostasis with Nature, on the other hand she—science—is already today able to tell us what *shouldn't* be done! Scientific knowledge qualifies scientists to speak up about what's necessary, though without making any claims as to sufficiency!

In other words, contemporary natural science is capable of indicating, even if not with absolutely *total* accuracy, the forbidden line that people should not step over under any circumstances!

But where is the guarantee that people will heed the scientists' advice and observe the taboos they prescribe, even given the importance and persuasiveness of their judgments? There are a great many examples demonstrating the lack of any such guarantee. What could be more noble and convincing than the Sermon on the Mount? Yet that failed to prevent the annihilation of the Albigensians,[28] the burnings at the stake of the inquisition, among the many other crimes committed in the name of Jesus Christ!

Thus injunctions made just once are never enough. Humankind must be persuaded of the absolute necessity of observing the taboos in question, as covenants enjoined on the people by those in possession of the distilled experience of humankind that is scientific knowledge. And the essential knowledge and ability to nurture must be possessed in the first instance by those who frame the laws, design programs of economic development, serve in the organs of government or in the courts. ... To put it briefly, those with a humanistic education must also be expert in ecology!

Of course, restructuring education this way may not by itself be sufficient for the securing of humankind's future—such conditions may indeed not exist—, but it is most indubitably necessary.

That is how I lit on the idea of an ecological-humanistic education. But it turned out that I was not the only one to whom the idea had occurred.

During the *perestroika* period, that is, somewhere in the second half of the 1980s, at some small-scale meeting or other I entered into a discussion of problems of ecological education with the well known biologist Professor Nikolai Fedorovich Reimers,[29] now deceased. We bewailed the fact that humanities subjects are,

[28] The Albigensian Crusade was a 20-year military campaign launched by Pope Innocent III in 1209 to eliminate the Christian heresy of Catharism in Languedoc, southern France. Albi was one of the towns of the region whose inhabitants were liquidated. It is estimated that altogether around 20,000 people were killed over the course of the crusade.

[29] Soviet zoologist and ecologist, and one of the Soviet Union's most prominent conservationists. Lived from 1931 to 1993.

generally speaking, so very far removed from the questions of humanity's survival that are so troubling to us, the natural scientists.

We agreed that, taken as a whole, ecological questions are primarily those of culture, rights, morals, the behaviour of people, etc.—in other words, that all ecological problems come down in the last instance to human ones. So we began to think about what an appropriate syllabus for a humanities institute dedicated to ecological problems should be. Under present conditions, however, our ideas seemed to us far too utopian, especially during *perestroika* with its atmosphere of an approaching crisis in all our lives, signs of which were becoming more and more evident. We were both certain that, given these turbulent times, it would be impossible to organize any such institution! Not even in principle! But we were wrong, thank God.

It turned out that similar ideas had been brewing simultaneously in the depths of the Ministry of Higher Education, so that when we were invited to a meeting arranged by Stanislav Alexandrovich Stepanov in premises on Lyusinovskaya Street and asked if we would participate in the founding of a humanistic ecological university, we were more than ready to accede.

Although realizing the projected institution turned out to be far from easy, and I remained sceptical for a considerable time, I was finally won over by S. A. Stepanov's energy and organizational ability. The day came at last when the project had progressed to the point where it had become time to call a meeting to assign the various university "portfolios." Stepanov was unanimously chosen as rector, Reimers as dean of the ecological faculty, and I as the university's president, a post involving duties then very unclear. Eventually I became Chair of the scientific council and something like scientific advisor. I was given the responsibility of formulating the university's ideology. So the process was under way, as our last Gensec[30] used to say.

In September 1992 the new university, called the International Non-governmental University for Ecology and Political Science, welcomed its first student, in 1996 the first baccalaureates were graduated, and in 1997 the first five-year specialists. Despite the vagueness of my presidential duties, the university soon became my main source of occupation.

At around that time, in early 1993, to be more precise, the Russian Green Cross[31] was founded, and I was chosen its president. And then at the end of that year UNEPcom, our national committee in support of UNEP,[32] was founded and I became its president also. Of all the work I was saddled with from 1994, I remained chiefly preoccupied with ecological education and associated methodological problems. I especially valued contacts with teachers, those remarkable people who, in

---

[30] Short for General Secretary of the Communist Party of the Soviet Union. Gorbachov was the last such.

[31] See Chap. 15.

[32] The United Nations Environmental Program, an agency of the UN coordinating its environmental activities.

our abhorrent times, without spiritual coherence or enlightened goals, continue to remain themselves and impart to children the genuine knowledge without which no individual has a right to be called human.

This activity necessitated giving public speeches, participating in international conferences, etc. In the next, concluding, chapter, I reproduce excerpts of my published material relating to ecological education.

Thus in 1994 I once again had serious BUSINESS to attend to!

* * *

The logic of such work is such that one set of questions suggests neighbouring ones, and one becomes willy-nilly involved in fresh reflections, such as those to do with the fate of my Motherland, the peculiarities of our nation's trajectory over the past ten years, and various future perspectives.

Although many of my talks and public lectures have been published, much unpublished material remains, nevertheless, on my computer's hard drive, and I consider it expedient to include some of that material in the present book.

Chapter 17
The Nineteen-Nineties

The View from the Brink of a New Century, and Questions Arising

The 1990s are coming to a close. What lies ahead? What fate awaits our nation? Will we be able to crawl out of the slough in which history has enmired us? These are the questions agitating us all but to which few are prepared to vouchsafe answers.

In my worst nightmare I could not have imagined what has happened to my country, to my people. And today, sitting in my little *dacha* in Abramtsevo and looking out at the rain falling aslant, obliterating the nearby hills and smearing the outline of the forest, a memory comes to me involuntarily of another evening, one of late Autumn 1942.

In an old abandoned hut somewhere not far from Shlisselburg,[1] around a barrel in which a fire of dry fir logs roared, a few men were sitting listening to a disquieting news summary from the Information Bureau. Outside wet snow was falling; it was cold and gloomy. We had just come from the aerodrome and were warming ourselves around the fiery hot barrel. The day had been a good one: all airplanes had returned from their sorties. Yet no one felt especially upbeat: we were listening to distressing news from the other end of the vast frontline.

The enemy was racing to the Volga. There were a few remaining areas where ours were holding out, isolated pockets of resistance, solitary houses where our people were fighting to the death! A little longer and they would be forced back into the cold Autumn waters of the Volga, and then what?

In the hut a grim silence reigned. But in thought we were all there in those Stalingrad pockets of hopeless, as it seemed then, resistance. What will happen now? Well, we'll continue fighting, but the question remains: what comes next?!

[1] A town at the head of the river Neva; see Chap. 1.

N. N. Moiseev, *How Far It Is to Tomorrow...*,
https://doi.org/10.1007/978-3-030-96651-5_17

A senior lieutenant, unable to tolerate the oppressive silence further, stood up and went to the darkening window. "No, that cannot happen!" He spoke for all of us.

If only I were able today to pronounce those words with the same conviction! Perhaps I would then be content.

So what exactly happened to our government, how could it happen that after our tremendous Victory, after becoming the world's second greatest power, we suddenly, in a matter of hours, as it were, lost everything: our might, international prestige, and our more or less secure material lives, our country transformed into the back of beyond?

But what seems to me most terrible is not the collapse of our economy. Nor does our chief calamity lie in the weakness of our present government and the politicking of our ruling *élite* with their inability to understand the real needs of our country. No, the most terrible thing for me is the general apathy of our nation, the lack of genuine will and individual self-respect. Most terrifying is the feeling of hopelessness that has overcome the many millions of our people, the feeling of having to just give up. So how did all that come to pass, and why did the collapse occur with such speed?

I have thought a great deal about this question.

First of all, the collapse did *not* occur overnight. It was the culmination of a lengthy process that had begun back in Khrushchov's time, and perhaps even earlier. I call this process a "systemic crisis." In what did it consist?

Systemic Crises

A system's crisis or a systemic crisis: such expressions are used fairly commonly in connection with large systems of various kinds. They might be large financial systems, economic or social systems, etc. One hears talk also of systems of government and their crises.

However, the meaning embraced by the concept is never closely examined, and lends itself to use in a wide variety of tonalities. For that reason, I shall begin by attempting to make clear the sense I shall give to the concept of a "systemic crisis."

Thus consider some large system—a corporation, for instance. It will have a determinate set of goals: stability, a good position in the market, decent profits; it may also have different types of goals, such as, for example, political ones. I'll call such a system System A.

In order that such a System A achieve its goals or at least follow a trajectory leading to their fulfilment, a certain subsystem B is necessary, its system of management.

Every system of management is predominantly made up of people.[2] Each worker in the management apparatus has definite official goals, the totality of his or her duties, whose fulfilment is necessary, though not always sufficient, for the overall

[2] Note that at the present time (2019) some functions of management are being automated.

system to attain its goals. But every person, no matter who he is or what he does, also has his own personal goals, which may or may not be clearly formulated. They indubitably exist, however, and influence his behaviour, in particular his work activity, putatively directed towards the attainment of the goals of the System A.

It's important to recognize that, as just adumbrated, in the System B there arise goals peculiar to it, in particular that of maintaining its own stability, of preserving itself. Hence the actual activity of the management system, System B, is, contrary to the accepted view, always a compromise between the interests of System A, the "official" aims of System B, and the aims specific to the latter system. The quality of the work being done by System B, which determines how well System A functions, depends on the structure of this compromise.

This is all objective fact. Never, or almost never, do the interests of the System B completely coincide with those of the System A. We have to resign ourselves to this fact, at least for the time being.

The structure of a compromise of the above sort, that is, the mutual relations between the interests of the two systems, is never constant. For a variety of reasons it changes over time. And it may even happen that aims specific to the System B will displace its putative "official" aims (or duties). In other words, the management apparatus may under certain conditions begin working exclusively on its own behalf! It may even begin to completely neglect the aims of System A, its original *raison d'être*.

Such a situation is what I shall understand by the phrase "systemic crisis." In such circumstances it is necessary to reorganize the management system so as to eliminate the factors preventing it from carrying out its official functions.

Reasons for the occurrence of systemic crises have not been thoroughly investigated. Like the theory of organizations as a whole, they lie on the periphery of political science, if, indeed, they are considered at all. I believe that the first person to look into the problem was A. A. Bogdanov,[3] founder of the theory of organizations. As early as 1911 he had come to the idea of the necessity of a regular reconfiguration of the structure of an organization, as described in his book *Tectology: The General Science of Organizations*,[4] published in 1922. Restated in my terminology, his recommendation comes down to the requirement that the management apparatus of an organization, its System B, should periodically be restructured or renewed.

There are a variety of reasons for systemic crises: hypertrophied self-interest or desire to maintain the status quo, or just plain incompetence, for instance. Another possible reason may be an increasing lack of understanding of the interests or aims of System A or the means to secure these, in a changing external environment.

[3] Alexander Alexandrovich Bogdanov (1873–1928), Russian and later Soviet physician, philosopher, writer of science fiction, political scientist, and revolutionary. He opposed the Bolshevik government in its early years.

[4] *Essays in tectology: The General Science of Organization*, translated by George Gorelik. Seaside, Ca: Intersystems Publications, 1980.

Systemic crises are rather common phenomena. Overcoming them usually requires a partial or complete restructuring of the System B: changing the conditions under which it operates, or even completely replacing it.

In order for such a restructuring to take place it is necessary to have a certain other system operating at a "higher level." This may be, for example, the set of shareholders of a company, or a country's government. To take one instance: In the 1950s, following the death of Henry Ford Sr., the Ford Motor Company began to suffer scandalous losses. The US government of the time took over management of the concern, appointing a new CEO and new operations managers. And when these had restored the viability of the company, rendering it once more competitive, the government withdrew its protection.

Systemic crises may also occur at the level of government. In such cases there are again a variety of possible causes. Most often a crisis of this type occurs when the System B, in this case the executive branch, be it a president or an administration, has either lost sight of the aims of System A or doesn't understand them, as a result of which it is unable to work out an appropriate strategy for attaining current goals or setting up a functioning management structure. Note in this regard that while the long-term goals of the System A may be generally well enough understood, especially under conditions of war or in other critical situations, the chain of intermediate goals necessary to their attainment may require talent, wisdom, disinterestedness,. . ..

I believe that our country was long subject to a deep systemic crisis of the type described above. It had many causes, some of which have not been elucidated even to this day. And there existed no higher level system capable of diverting the stream of events, of introducing appropriate correctives into the structure of the System B so as to compel it to operate differently. An understanding of such circumstances is today a matter of life and death to our present government.

In fact it's clear that the systemic crisis continues to this day, despite the change of rulers and the fact of our living in a different country!

Below I shall apply the above in detail to the USSR and to our present situation in an attempt to support my point of view and draw certain conclusions for political science.

First we return for a time to the Soviet Union.

Indicators of Instability the USSR on the Threshold of a Systemic Crisis

We begin with the fact that in the 1950s there was no crisis in our country.

Although one may criticize the system of the Soviet Union on various grounds, the fact remains that in the postwar years it was not subject to a systemic crisis.[5] Nor, by the way, was the officially declared form of socialism in crisis at that time, although debates on socialism were necessary as a kind of ideological camouflage. One might call the Soviet Union's economic, social, and governmental system—its System A—a "system consisting of a single factory." The corresponding System B, so well known to all of us, was made up of the Party, together with the nomenclature, i.e. the Party, managerial, and administrative *élite*.

Here System A had a completely clearcut goal: to attain parity with the United States in nuclear warheads, missiles and other armaments! This was in fact essential to our future security. And it was this goal, and not communism, the phraseology of which served merely as a convenient backdrop, that the System B actually reached. And people worked hard to achieve this! They dared not do otherwise! The mechanism whereby System B worked to attain the chief goal of System A was very clearly worked out.

Thus although one might speak of the nomenclature's deficiencies or miscalculations, it cannot be said that the Soviet Union was at that time beset by a systemic crisis. The aims of System A were predominant in determining the work of the apparatus of management. And it cannot but be admitted that on the whole System B coped with its basic task rather well. Not only did we produce nuclear weapons and superb rockets, but were able, in an unprecedentedly short time, to establish industries which in terms of GDP brought us up to second place in the world. And even more, we showed ourselves capable of developing one of the world's best systems of education, a decent social security system, and a reasonably good standard of living compared with our present one.

Yet notwithstanding all that, the Soviet system was doomed.

There were flaws in the system which would inevitably cause it to be overtaken by a profound systemic crisis. As we shall see, for this to occur all that was needed was an external jolt.

* * *

Of course, a real compromise between Systems A and B did exist. It was maintained on the one hand by the stick of rigid discipline, even terror, and on the other hand by the carrot of certain rewards allotted to members of the management apparatus. Members of the latter automatically received not only decent material rewards, but also access to special medical services, transportation, and so on, and these blessings were lifelong, maintained even in retirement. This list of perquisites determined the spectrum of interests of System B, in particular its stability, and for a

[5] This statement may seem incongruous in view of the fact that in those same postwar years Stalin increased the rate of imprisonment in the Gulag, in particular of returning Red Army soldiers, and launched a vicious anti-Semitic campaign, ended on his death in March 1953. Evidently, by the words "not subject to a systemic crisis," the author is referring narrowly to the functioning of the Soviet military-industrial complex and the provision of social perquisites, and, as during the war, the complete alignment of most of the people, in particular System B, with Stalin and his government, i.e. with System A.

time facilitated the realization of the aims of System A. And in those postwar years the successful functioning of our System A was evident to all.

Over the course of several decades, I was closely involved with problems of the use of computer technology in various associated large-scale projects, mostly having to do with military defence, so I had a privileged, personal view, from inside the hothouse, of the extraordinary achievements which nevertheless foreshadowed crisis.

At the end of the 1950s, I happened to be a member of one of the first groups of Soviet specialists to visit various computer centres in western Europe. I returned home floating on air: we were the equals of our west European counterparts not just in mathematics, but also in computer technology. In several ways we were even ahead of them.

But things had changed by the early 1960s, when I spent almost the whole of one Spring semester in Paris, and was, in particular, in close contact with the NATO Centre in Fontainebleau,[6] then concerned with control problems of technological systems. I saw nothing much new there as far as my own narrow professional sphere was concerned, but observed something else of greater importance: the beginning of the computer revolution. Over the few years between my two trips to western Europe, the situation with respect to their computer technology had undergone a qualitative change.

Somewhere around the turn of the 1950s into the 1960s, electronic valves were replaced by transistors in computers, allowing the abandonment of those monster valves and a consequent compactification of computers. At the same time computers became much cheaper, although this was not yet our main consideration.

Computers using the old cumbersome valves were extremely unreliable. They were continually malfunctioning, and for complicated computations it was essential to have a staff of competent electronics engineers. As a result, for a considerable time computer technology was accessible only to large organizations able to afford qualified personnel. Moreover, access to computers as a rule required security clearance, both in the USSR and the West. With the advent of the transistor, the situation changed drastically. The computer as technological instrument became as common as mechanical calculators had once been, except that, of course, they were incomparably more versatile in terms of their technical possibilities.

Although a further 15 years would pass before the first personal computers made an appearance, many features of the computer revolution were already apparent. I mention just two such: first, computers began to impinge on various fields of activity, such as commerce, accountancy, control of technological processes, etc., and, second, the guiding impetus to the design and manufacture of computers became more and more that of producing universal machines rather than narrowly specialized ones.

---

[6] The Allied Forces Centre in Fontainebleau was moved to Holland following France's partial withdrawal from NATO in 1967. See also Chap. 6.

On my return to Moscow I made several attempts to communicate the impressions I had gathered during my lengthy field trip to France, and not only within the Computer Centre of the Academy of Sciences. I even wrote a short memorandum to the Science Department of the Central Committee of the CPSU.[7] There was, however, no observable reaction to my communiqués.

In this connection I also had a long conversation with Academician Sergey Alexandrovich Lebedev, our leading computer designer. At that time he and his institute were busy building some sort of specialized computer for control of rocket complexes and were also beginning work on building the universal computer BESM-6.[8] I don't now recall the details of our discussion, but I do remember his concluding sentence, a prophetic one, something to the effect that: "It's becoming ever more difficult for me to convince the authorities of the necessity of creating original computer technology useful for solving a wide range of problems, and I very much fear the BESM series will be terminated." And that, indeed, is what occurred a few years later.

Thus one cause of the systemic crisis was the new upswing in scientific and technological progress. The situation with our computer technology—its being allowed to lag—was just one indicator. The rush to utilize the new technology was evident everywhere in the West, in agriculture as well as industry, yet our system proved unable to meet the challenge! We shall see the reason a little later.

But there was a second reason, no less significant, for our technological falling behind: towards the end of the 1950s our country had achieved parity in nuclear missiles and other arms with its major adversary, the United States of America. Thus the main goal of our System A had been attained, and the stress of responsibility borne by our System B was consequently eased. The mechanism of the compromise between the two systems began to falter.

In order for our country to continue developing, it was essential to formulate new goals, and of course find the political WILL to realize them. But the nation's goals continued to be couched in the language of communist ideology, which was less and less able to serve as a stimulus to organizational and other necessary endeavours. The System B began to pursue its own narrowly selfish interests.

The Growing Crisis: Some Fresh Ideas

Thus it was that, as far back as the early 1960s, many of us, members of the scientific-technological and engineering intelligentsia, began to discern various symptoms of approaching trouble, and to seriously ponder our common fate.

[7] Communist Party of the Soviet Union.

[8] The *Bystrodeistvuyushchaya Elektronnaya Shchotnaya Mashina* was the first second-generation Soviet computer, using transistors. See also Chap. 10.

We had never been dissidents; I would instead call the set of people with whom I was then in contact "constructivists." We were not much exercised by problems associated with *glasnost′*,[9] and least of all by the right to travel abroad. Above all we were looking for ways to improve our system, its management subsystem, and its productive capability, ways to avoid degradation and defeat in the "Cold War." The latter possibility seemed to us especially dangerous, and haunted our perspective like a terrifying mirage, since we knew what that would mean for our country. But none of us imagined that the collapse would come so soon and so tragically.

However, our failure in that respect didn't make our criticism any the less severe. A difference from the criticism of earlier times was that it was no longer confined to kitchens but had become public. It was now voiced openly, at scientific seminars and other meetings. Once, as we were leaving a meeting concerned with the termination of the BESM series of computers, the closure of our own computer design operation, and the transition to the production of IBM computers, Viktor Mikhailovich Glushkov[10] said to me something like: "I don't know what they're jailing the dissidents for, but they should certainly jail us!" And indeed nothing nearly as harsh as the criticism given voice at that meeting in the Kremlin had ever been whispered in communal kitchens, although formerly a great many people had paid dearly for such whisperings with exile[11] and other forms of ostracism.

There was a wide variety of responses to the question as to how our nation's system might be improved, mostly concerning the need to restructure our System B and the choice of long-term goals for our System A. A noteworthy work on this theme was published by Academician Germogen Sergeevich Pospelov.[12] Having worked at developing programmable management methods, he was looking for ways to extend his ideas, so well suited to programmed planning, the design of complex systems and management of industrial enterprises, to management at the level of government. (Note that as far as his ideas went, they were very close to those of Aron and Marcuse.) There were other such initiatives, for instance that of Nikolai Prokofievich Fedorenko[13] at the Central Economico-Mathematical Institute of the Academy of Sciences, who oversaw the construction of the system SOFE[14] of levers for managing an economy.

[9] Meaning "openness to the public," this was the name of one of the main policies adopted by the political reform movement within the Communist party of the USSR during the 1980s and 1990s, associated with *perestroika* and M. S. Gorbachov.

[10] See Chap. 10.

[11] A euphemism for "imprisonment in the Gulag."

[12] Soviet scientist working in automated control theory and founder of the Soviet school of AI methods. Headed a department of the Computer Centre of the Academy of Sciences of the USSR. Lived from 1914 to 1998. See also Chap. 7.

[13] Soviet economist and chemist. Head of the Central Economico-Mathematical Institute of the Academy of Sciences of the USSR in Moscow from 1963 to 1985. Lived from 1917 to 2006.

[14] System of Optimal Functioning of an Economy.

However, the one who had the deepest understanding of the essence of the matter was the late Professor Yury Pavlovich Ivanilov[15] of fond memory. Late in life he was a member of the Russian Supreme Council, and his activity ceased together with that institution's in October 1993. He was the only one of my acquaintance able to call things by their name. He said to me once: "The real problem is that our System A has gradually drifted into a state where it is bereft of ideas for future development, and management without goals is not really management but only a means to preserve its staff's own security, which is entirely unrelated to the long-term objective goals of the System A!" My thoughts exactly.

I began sensing the growing crisis especially keenly following the failure of Kosygin's reforms,[16] on which so many of us had placed great hopes for a qualitative renewal of the whole of our system of management. The reforms collapsed with a thud, being in any case doomed for the same reason as that at the root of the Soviet Union's inability to meet the challenge of the new technological revolution. The systemic crisis was becoming evident even to the naked eye of non-specialists.

The lack of overall developmental goals and future prospects resulted in our "single factory system" having as its motivating interests those of the management of that factory—the System B, in my terminology. Such interests are always short-sighted.

By that time, the system of branch monopolies had become firmly entrenched. This had been an efficient form of organization when our "single factory system" had clearly prescribed goals. Under conditions of well-defined planning, competition would only have been harmful. In theory at least, all branches of the "people's economy" worked with precision and coordination as if they were workshops of a single factory. In such a system only one sort of competition made sense: that between workers as to who had best fulfilled the task handed down from above. And, as we have seen, for a good while that system worked without major breakdowns. But no sooner had the system's overall aims become blurred than the regime of branch monopolies was turned into a highly effective brake on any kind of development. Their monopolistic character served to stifle technological innovation, and even more so organizational improvement.

What, in effect, might be the most terrifying scenario for a management apparatus, i.e., for any System B? In the case where the dominant motive of that subsystem has become the maintenance of its own stability, the answer is easy. The subsystem's homeostasis is called into question whenever it is threatened with restructuring of some kind, be it technological or the replacement of key employees, since this will inevitably involve a reorganization of the management system and the appearance of people possessing necessary new skills. That is why, without vigorous pressure by those loyal to and in command of the interests of the System A, a "single factory system" must find itself poorly equipped to follow the twists and turns of a

[15] See Chap. 15.

[16] See Chap. 15.

scientific and technological revolution. For our former leaders to be able to exert such pressure, they would have had to possess a clear understanding of the goals of the System A, as well as the political will to realize them. Among our leaders at that time, however, there were none such.

And that is also why Kosygin's reforms failed: they aimed at providing enterprises with greater independence and restructuring the whole of the established management system of the economy. The System B proved unable to undertake this without significant pressure from above, and when this pressure was needed there was no one to apply it.

* * *

Many have fond memories of the Brezhnev era: full employment, a more or less acceptable standard of living, a secure tomorrow.... But this stability was actually only apparent. In fact the country was headed for disaster: the Brezhnev era was not just one of stagnation, but of a rapid process of degradation, and not only in terms of industrial output but also in moral terms and those of interpersonal relations. Here I'm not referring to the need for a technological restructuring: as far as technology is concerned we were already hopelessly behind. The country was being run according to the slogan: "You are commanded: Do not disturb!" The trouble with this was that the members of our nomenclature adopted this command as their own.

There were reasons for this. Since arms parity with the US had in fact been achieved, so that external threats might, apparently, be ignored, was there any further need to worry about the future? Furthermore, macroeconomic indicators were more or less satisfactory. The percentage increase of GDP had declined, but was nevertheless still positive! But for people working in science and technology that pseudo-secure picture was viewed in another light: we had clearly lost the "cold war," moreover on all fronts, and that surely bode a very bleak future. Given this circumstance, some of us began casting about for a qualitatively different model of development of our state. One widely considered such idea was that of "convergence."

Convergence[17] began to be discussed in the mid-1970s, and not only among us but also in the West. But at that time by the word "convergence" something rather primitive was intended, namely a structural organization of government combining the best of socialism and capitalism, an artificial sort of symbiosis. It was this version of convergence that was then being debated in detail by European theoreticians of postindustrialism—by Herbert Marcuse, for example. In the Soviet Union, it was discussed in a similar vein by A. D. Sakharov, among many others.

However, I consider such ideas pure eclecticism. You can't mix the immiscible. I have always inveighed against that primitive form of convergence. Twice I discussed this with Sakharov, both times feeling that he was far from having grasped the actuality of our situation. We found ourselves at cross purposes.

My own thoughts on convergence were in a different spirit altogether. For me the main consideration had to be what was before our eyes: it was obvious that

[17] See also Chap. 15.

the Soviet Union was dropping out of the planet-wide process of development, was slowly but surely moving out to the periphery of world society. I considered this fact as constituting the essence of the impending catastrophe. Given this, our main task, therefore, was to re-insert ourselves back into that process, called by some specialists "the world-wide distribution of labour and income." What I myself call "convergence" is the inclusion of the socialist world in that planet-wide system. It would seem that in fact we are now, at the end of the 1990s, living through such a process of convergence, although of a much harsher kind than what might have been back when we were a powerful nation with a functioning industry.

However, then the structure of our government, primarily our System B, was such as to make convergence infeasible. Thus, whether we liked it or not, in order for us to be included in the system of world-wide distribution of labour and income, we would have had to change our ideas concerning private property and fundamentally reorganize the way we managed our economy, while preserving as far as possible the organizational innovations and achievements in the social sphere accruing from our past experience.

But how were we supposed to achieve this, how overcome the natural resistance of our System B, how dismantle it and impose a new organization without bloodshed and the destruction of the existing government? These were troubling, even painful, questions.

And there was yet another question giving me no peace. How might it be possible, even in principle, to overcome a systemic crisis at the level of government? In this case there would be no higher-level system to take on the task of restructuring the System B. I could see only one way of "regularizing" such an impasse: the imposition of a genuinely democratic form of government that would guarantee a painless transfer of power from one ruling *élite* to another, from one system of goals to another. This may not have been sufficient, but I was convinced of its necessity.

* * *

The coming to power of M. S. Gorbachov was inspiring for many, including me, and I began rather intensively pondering the various possible paths to an appropriate apotheosis of *perestroika*. From my present vantage point, however, these reflections now seem pointless. Back then it may have been possible to imagine that they would be of use to somebody and that the end of the systemic crisis was imminent, but I was wrong.

The basis of my thinking then was the idea of a painless transition to a new organizational structure. Without any more revolutions! And without any more slogans like "...raze everything to the ground and the world shall rise on new foundations...." No more extremism of any kind, please! The restructuring process should be conducted with great care and everything thought through in detail.

I believed then that the following two particular, carefully planned, actions should be carried out. The first, which I would call "ideological," consisted in the following. The instigators of *perestroika* should make absolutely clear to the people what they envisaged at its end, what sort of political system the people might expect to find themselves saddled with, what form of economics, what relations with world society, and so on. In other words, before creating the new System B, the new apparatus tasked with realizing the aims of *perestroika*, and before delegating tasks to that apparatus, the goals of the System A should be clearly formulated. Without clarity as to the set of goals of the overall system, it would be impossible to halt the systemic crisis since otherwise those staffing the System B would, even under different working conditions, continue to pursue their own interests, or perhaps the personal interests of their rulers, which is much the same evil.

The second direction I would call the "constructive mode of behaviour." I held that there's no need to wait for some appropriate ideological basis to be formulated since there are completely obvious first steps to take, and they should be taken promptly—and all the more so inasmuch as we have past experience of them. The first, absolutely essential, step should be that of liquidating our branch monopolies and setting up a governmental corporation in the form of the syndicates of the NEP[18] era. These syndicates should operate under conditions of competition, since only then could there emerge stimuli to technological improvement and the securing of the necessary quality of consumer goods. This step had already been tested during the NEP era. In fact I might have proposed other organizational revisions on the model of our experience with the NEP, a remarkable piece of enterprise on the part of our then Soviet leaders. It was, and still is, my belief that, in rejecting, at the end of the 1920s, the principles implemented in the NEP, our country missed a unique chance of realizing a "third way," as they now say both here and abroad.

In addition to these two actions, the main and to me most obvious ones, I believed that we should undertake a well thought-out series of changes essential to the democratization of the nation and the emergence of a "higher authority," that is, to the creation of a system enshrining democratic institutions and capable of quenching the systemic crisis while still in embryonic form.

But although I had at various times given talks on our economic crisis along these lines at economics seminars, I failed to generate much interest or even comprehension. Meanwhile our systemic crisis worsened. I don't think M. S. Gorbachov grasped the complexity, the extreme intricacy, of our situation or its actual roots. This would explain his unsystematic and random actions, such as his anti-alcoholism campaign, which succeeded only in worsening the crisis. But on the other hand perhaps I was mistaken in identifying the aims of the one and only one president of the Soviet Union with those of the System A.

[18] The New Economic Policy (NEP), allowing a certain amount of small-scale enterprise in the Soviet Union, was introduced by Lenin in 1921 in the face of economic collapse. Supported enthusiastically by Soviet leaders such as Nikolai Bukharin, it was ended by Stalin in 1928. See also e.g. Chaps. 2, 3, and 6.

The Systemic Crisis of the 1990s

Then towards the end of 1991 the Soviet Union collapsed. One might have thought that with the resulting change of rulers, with the coming to power of new, energetic people, the crisis would now be overcome. But, as it has turned out, despite the tragic circumstance of the ruination of a thousand-year state,[19] and the new economic views of the new ruling clique, the systemic crisis continues to deepen, taking on an ever more ominous hue. But the essence of the crisis remains the same: the priority of System B's aims over those of System A, now represented by the people of Russia.

There are many reasons for this, the main one being the incapacity of the current rulers to formulate an appropriate set of aims for System A to pursue. In fact, they don't seem to even see a need for this.

The suspicion has long been dawning on me also that our new System B does not see its role as that of fulfilling the aims of System A, is not inclined to seek a stable compromise with that system. At the same time, many of our rulers are gradually coming round to the view that the incongruence of the actions of the management system with the objective aims of System A might very possibly end in overall destabilization. There is a growing understanding among them that if they wish to secure their own homeostasis for a greater of lesser interval of time, and, consequently, also that of System B, "stopping up the holes" is no longer an option even in principle.

It's impossible to secure the future without trying to look into it, that is, without having some idea of what sort of future might be desirable or of a STRATEGY for achieving it!

More and more of us are coming to a realization of this. There is talk of the need to work out a system of national ideas. But national ideas objectively exist; they arise spontaneously from the people, and cannot be conjured up out of thin air or, even worse, dreamt up by public servants. They must emerge as the result of wide public debate. Or else be achieved via a serendipitous find like that of the merchant Minin,[20] who was supported by the people, and whose third army proved able to accomplish what the armies of Godunov[21] and Shuisky[22] could not.

Didn't the slogan "The land to the peasants and the factories to the workers" play a role similar to that of a national goal in the Civil War? Are there not a great many comparable examples in history?—for example Roosevelt's ideas, supported by the American people, ideas whose implementation helped end the Great Depression of

[19] Ignoring the discontinuity of the October Revolution, apparently.

[20] Kuzma Minin was a respected merchant from Nizhny Novgorod who formed a Second Volunteer Army to take back Moscow from the Polish invaders in 1612, thus ending Russia's "time of troubles." See also Chap. 14.

[21] Boris Godunov was tsar of Russia 1598–1605, considered by many as having usurped the throne by murdering the only surviving son of Ivan the Terrible.

[22] Vasily Ivanovich Shuisky or Vasily IV (1606–1610), the last tsar of the Rurik dynasty.

1929–1933! And now the British prime minister Tony Blair is in the process of providing us with yet another example, as he assembles "the best minds of Great Britain to discuss the Third Way!"

But here in Russia there is absolutely no talk of the future, and not a word about our systemic crisis. We have been waiting a long time for someone to say clearly whither we're going and why, what the current reforms are in aid of, what awaits us in the future, and to propose a discussion of the various STRATEGIES for achieving that future.

The mentality of your average Russian is such that more than anyone else he or she needs to hear the right WORD! He needs to be transported. Only by means of a belief in such a WORD, will we, the Russian people, be able to endure the sacrifices yet awaiting us.

Appendix
Milestones-2000 Remarks on the Russian Intelligentsia on the Eve of a New Century

Preface

Before me lies the anthology *Milestones*, published 90 years ago, near the beginning of the present century.[1] The contributors were considered among the most intelligent people of the time: Berdyaev,[2] Bulgakov,[3] Struve,[4] *inter alia*. Such is their intellectual stature, that the thoughts and evaluations they recorded not only deserve attention and respect, but may be assumed to possess a high degree of objectivity. We can depend on them today in pondering tomorrow.

Over the century since they wrote, the world has changed unrecognizably. How everything has changed since then! The conditions of life and the very borders of countries differ greatly from what they were. Yet on rereading *Milestones*, I see that the text concerns the same kind of people as are now living in Russia; despite all the changes, we see before us much the same people or folk, with all their familiar deficiencies and good qualities. Except that at the beginning of the century the intelligentsia constituted but a tiny fraction of the population, since its members belonged, as a rule, to the class of "gentlemen"—from the point of view of the "simple folk," at least—, while now it is one of the most populous of social groups, of great weight and actively influential in the electorate. Thus if in the pre-revolutionary period the mentality of the democratic intelligentsia played a significant role in preparing the revolution, so even more in the present time may it prove crucial to determining the fate of our nation.

[1] The 20th.

[2] Russian political and Christian religious philosopher. See Chap. 6.

[3] Sergey Nikolaevich Bulgakov (1871–1944), Russian Orthodox Christian theologian, philosopher and economist.

[4] Pyotr Berngardovich Struve (1870–1944), Russian political economist, philosopher and editor. Initially a Marxist, he switched to liberalism, joining the White movement against the Bolsheviks. Lived in Paris from 1920.

N. N. Moiseev, *How Far It Is to Tomorrow*. . . ,
https://doi.org/10.1007/978-3-030-96651-5

In those pre-revolutionary years, Berdyaev distinguished a precisely defined subset of the intelligentsia he called the "discussion group intelligentsia," whose members lived for their discussion groups, were always opposed to whatever government was in power, and, while always discussing the "folk"—how they were oppressed and their possible liberation—, were at the same time infinitely far from grasping the latter's true interests. Over the next several decades the "discussion group intelligentsia" dwindled and in its place appeared the "kitchen intelligentsia," whose members once again discussed their own problems rather than questions of concern to the mass of the people.

But the interests of those gathering in kitchens instead of drawing-rooms were different. While at the beginning of the 1900s the intelligentsia was diverted by topics such as Marxism and the liberation of the people,[5] and pretended to despise erudition and their own mercantile interests, the later kitchen intelligentsia was preoccupied with different things altogether. By the time of *perestroika* the latter, together with the partocracy, had been very aptly renamed "new Russians.[6]"

But despite the many ways in which our present situation is distinct from that of the beginning of the twentieth century, it is still in certain respects reminiscent of that earlier one. This is borne out by the following quotations from a remarkable article of S. N. Bulgakov, which sound as if they had been written with today in mind, although the following words were actually written in 1906.

"Russia has survived the revolution.[7] This revolution has not yielded the anticipated results. It is generally held that any positive achievements of the liberalization movement remain at the very least problematical. Russian society, worn out by previous tensions and misfortunes, is in a state of numbness, apathy, spiritual confusion, and despondency. The Russian government shows no signs of seeking the renewal and consolidation so essential to it, as if our lethargic tsardom has been overcome by an irresistible somnolence. The Russian citizenry, demoralized by the extraordinary growth in the incidence of crime and the general coarsening of morals, has definitely regressed."

And a little further on: "There is indeed a reason for yielding to gloom and being dubious about Russia's long-term prospects."

Does it not seem to you, respected reader, as if these words had been written just today and addressed to all of us, citizens of a formerly great state?

* * *

I have several times before noted problems associated with the expression "Russian intelligentsia" as used in its widest sense. As I generally use the expression, it refers primarily to the stratum of our populace consisting of those people capable of looking beyond the framework of their own personal interests, reflecting on the fate of their country and the peculiariarities of its government, and looking for

[5] Since the serfs were liberated by Tsar Alexander II in 1861, this presumably refers to freedom from penury, perhaps, or from autocratic rule.

[6] That is, *nouveaux riches*.

[7] The 1905 revolution.

constructive—I emphasize the word—improvements of its structure and functioning. This definition embraces a much wider set of people than just those who do intellectual work. To take one example: throughout the war, my sergeant-major Eliseev, a simple collective farm driver from Ryzan′ Province, was always by my side. He functioned not just as my batman and driver, but also as papa and mama combined, being exactly twice my age. There have been very few people like him in my life, people with whom I could discuss problems plaguing me on equal terms and without fearing betrayal, and get intelligent answers.

Another example: In Skhodnya, where I lived in the 1920s, there was a stove-maker by the name of Ivan Mikhailovich Gryzlov,[8] who loved to visit us. He and my grandfather, a railway engineer, would talk over tea for hours about problems with the government. Once, just after he had left, my grandfather said: "A prime example of the intelligentsia: if only our commissars were like him!"

In this essay, however, I shall be using the word "intelligentsia" in a narrower sense, restricting it to the set of people who do intellectual kinds of work. It is important to note that I include in the intelligentsia in this narrower sense the upper layer of the working class, those working with forms of higher technology requiring special knowledge and intensive use of the mind.

The Structure and Fortunes of the Russian Intelligentsia

The present Russian intelligentsia is very inhomogeneous with respect to social class, core activity, aims, interests, and even national origins. It developed out of the serving nobility of the time of Tsar Peter I. A large part was played by German, or, more precisely, Swiss German, immigrants. They brought with them professionalism, and a culture of hard work. They Russified rapidly and in two or three generations had become indistinguishable from native Russians in their interests and way of thinking. They contributed to many worthy pages of Russian history. One should not forget also the contributions of Polish immigrants, especially in Siberia, whither they were exiled in the course of many failed attempts to free themselves from the oppressive Russian imperial yoke.[9]

Gradually the role of the nobility in the formative process of the intelligentsia was lessened. From the mid-nineteenth century, we find people from the merchant class, the priesthood, and even the peasantry entering the intelligentsia, giving rise to the "diversified" intelligentsia.

In my preface I talked initially of a relatively small section of the intelligentsia called by Berdyaev the "discussion group intelligentsia." It is precisely to these people that Russia owes the periods of instability and revolution in which the generations of the declining century[10] found themselves embroiled.

[8] See also Chap. 6.

[9] A large eastern swathe of Poland was governed by Russia from the late eighteenth century till 1918.

[10] The 20th.

S. N. Bulgakov makes the assertion, backed up with evidence, that the first Russian revolution[11] was "purely an achievement of the intelligentsia."[12] And of the role of the "discussion group intelligentsia" in the October revolution more than enough has been said! Furthermore, *perestroika*, begun by M. S. Gorbachov and his associates, was more than merely supported by a later version of the "discussion group intelligentsia." Members of the dissident movement of the "kitchen intelligentsia," the discussion group intelligentsia's heirs, whose discussions now took place in kitchens, became the "foremen of *perestroika*"! It's even possible that the basic ideology of *perestroika*, or at least some of its ideals, came from that movement. But what the dissident movement of the "kitchen intelligentsia" has led to we now know only too well from what we're going through at present: once more revolution and tragedy for the Russian people, whose fate was for the most part of very little concern to the dissident movement. The people were just as much merely experimental subjects for the kitchen intelligentsia as they were for the discussion group intelligentsia of Berdyaev's time. I shall return to this fact below. For the time being I confine myself to the remark that the activity of that section of the intelligentsia has always, independently of the government in power, had a nonconstructive or even destructive character: "Raze everything to the ground, and then . . .." But by "then" others of the intelligentsia were busy working since someone always has to do the work; in every historical period of our country, the discussion group intelligentsia has comprised but a tiny proportion of the overall intelligentsia. The main part of the intelligentsia worked, namely in the teaching and medical professions, in branches of culture such as literature and art, and a great many in engineering.

If below I write mainly of the scientific and technological intelligentsia, it's chiefly because that's the section of the intelligentsia I know best.

Thus, apart from the discussion group intelligentsia, to which the original anthology *Milestones* is devoted, there has always been the working intelligentsia, whose members taught or cured people, founded industries, or were simply conscientious in fulfilling their everyday duties, thereby contributing to the rapid development of our state and the welfare of the Russian peoples. Although they gave little thought to new-fangled economic or sociological theories, the developmental prospects and future of their country always figured in their field of interests. It is significant that the mass of the intelligentsia never constituted an opposition to the government, at least formally, although they had no qualms about being critical of their government's actions when appropriate.

Of course, each member of the intelligentsia had his or her own views and evaluations of the evolving situation. The famous aircraft designer Igor′ Ivanovich Sikorsky,[13] for instance, was a convinced monarchist, so his emigration following

[11] Of 1905.

[12] *Russian Thought. 1908, III.*

[13] Russian-American aviation designer of both helicopters and fixed-wing aircraft. Immigrated to the US in 1919, and founded the Sikorsky Aircraft Corporation in 1923. Designed and flew the first working American helicopter in 1939. Lived from 1889 to 1972.

the October revolution was inevitable. In much the same way, the inventor of modern television Vladimir Kuz′mich Zvorykin[14] was an active participant in the White movement to the extent of being an officer in the White Army, which circumstance determined his post-revolutionary emigration.

Over the last quarter of the nineteenth century, Russia underwent rapid development. Industry grew and urban centres ballooned; the population of Novonikolaevsk (now Novosibirsk), for example, even outran that of Chicago. And in parallel with this development the intelligentsia grew both in quantity and quality. In particular, in Russia there shortly appeared a great number of highly qualified engineers. Technical education in Russia was distinguished by its breadth, so that Russian engineers were able to master technological innovations relatively quickly. As we shall see below, this characteristic of Russian higher technical schools proved to be of great utility.

The revolution delivered a terrible blow to the intelligentsia. However, the fates of the intelligentsia overall as opposed to the tiny subset consisting of the discussion group intelligentsia were very different indeed.

The *émigré* contingent of the discussion group intelligentsia was relatively small. They weren't of much use to anyone in the West, since there they had trouble-makers enough of their own. So most of them opted to serve the Bolsheviks, who needed a large contingent of sufficiently literate personnel to organize the mass media and staff the bureaucracy—in particular the Cheka,[15] let's not forget.

I remember one of the earliest searches of our apartment, sometime around 1925, where the squad of Chekists doing the search was led by a former clerk in the railway, a bookkeeper or cashier. When they had departed, taking with them several finely bound books and a certain icon, a family relic which grandfather valued very highly, he said of the leader of the squad: "An appalling featherbrain. Just look at how such people get to be bosses! I remember when he worked for us how he was always organizing discussion groups. I had occasion to defend him to the police, explaining that there's no danger to be expected from such boneheads."

Thus in those early post-revolutionary times a significant portion of the "discussion group intelligentsia" had it quite good.

However, in the longer run the fate of the majority of representatives of that thin stratum of the wider intelligentsia tended to be tragic. Enthusiasm for leftist ideas and faithful service to the Bolsheviks were no guarantee of safety. Links to Trotskyites, social revolutionaries, or other leftist groups proved to be sufficient reason for the almost total annihilation of that set of people. But this was much later, towards the end of the 1920s and into the 1930s, when the consolidation of Stalin's dictatorship was under way.

---

[14] Russian-American inventor, engineer, and pioneer in the development of television technology using a design of P. T. Farnsworth. Immigrated to the US in 1918. Lived from 1888 to 1982.

[15] The All-Russian Extraordinary Commission (VChK or Cheka) was the first version of the Soviet secret police, established in December 1917 under Felix Dzerzhinsky, nicknamed "Iron Felix" for his ruthlessness.

The fate of the majority of the wider intelligentsia was altogether different. As early as the first post-October years a large portion of the scientific and technological intelligentsia emigrated. These were for the most part promising trainees or actively working specialists who saw no prospects for their professional line of work in post-revolutionary Russia. Ideological considerations also played a role, of course, or at least an unwillingness to accept the new order of things. This mass emigration constitutes a tremendous gift from Russia to the West. Above I mentioned Sikorsky and Zvorykin in this connection; let me expand on them a little further.

Igor′ Ivanovich Sikorsky built the first multi-engined plane, but perhaps the greatest achievement of this outstanding designer was the construction of the first helicopters in their modern form. (The Soviet Union would later purchase one of these for Khrushchov's use.) Vladimir Kuz′mich Zvorykin's contribution to American technology was no less significant: he gave television its modern technological shape, and founded the first television broadcasting company. And these were but two of the priceless gifts made by the "workers' and peasants' government" to its implacable foe Capitalism. How many others there were!

But the main migratory outflow of qualified personnel began later, in the second half of the 1920s, when the Soviet rulers began expelling members of the intelligentsia from the country, thereby perpetrating what amounts to treason at the highest level.

Towards the end of the 1950s, in the course of a visit to Paris, I had occasion to talk to Benoit.[16] This was a year or so before his death. We met at a reception held in the Château de Fontainebleau, where the curator of the museum was a certain Rozanov, a distant relative of the famous philosopher Vasily Rozanov.[17] We were discussing the contribution of Russian artists to French culture, when I said I didn't like Chagall's painting on the ceiling of the *Opéra*. Unexpectedly, Benoit, hitherto a stranger to me, replied that he also was not very taken with the painting. Emboldened, I dared put to the great *maître* the rather tactless question as to why he, curator of the Hermitage, had decided to emigrate. His answer was entirely unanticipated: "I am no emigrant. The Soviet government simply refused to give me a return visa." The fact is, that in those early years of Soviet rule anyone who left the country on official business had to obtain permission from the Soviet government to return to his Motherland! They simply didn't allow Benoit to go back home. He was thus excluded from his country.[18]

Something similar occurred in my family. One day my grandfather, Sergey Vasilievich Moiseev, received an invitation from the firm *Westinghouse* to take up the position of technical consultant or advisor to the firm.[19] A copy of the invitation

---

[16] See also Chap. 6.

[17] See Chap. 6.

[18] Many *émigré* Russians who *did* manage to return, soon regretted it. For instance, when the great poet Marina Tsvetaeva and her family returned in 1939, her husband was executed and she exiled to Elabuga, where she lacked means of support. She hanged herself on August 31, 1941, aged 48.

[19] Recounted also in Chap. 2.

was received by the People's Commissariat of Communication Channels (PCCC), where Grandfather held the rather prestigious post of chairman of the Finance and Accounting Commission. One evening—most likely in 1926 or 1927—Grandfather returned home from a meeting of the PCCC collegium in a terribly glum mood, and at the dinner table read out a document he had brought with him. It started as follows: "The collegium has considered the offer by the firm *Westinghouse*, and recommends that Moiseev, S. V., accept the offer and leave for America permanently with his whole family." The final few words were underlined.

But Grandfather rejected the offer. He said something that has since become my life-long motto: "The Bolsheviks have come and will go, but Russia will remain. To work!"

Those were the twilight years of the NEP, when Soviet Russia was gathering strength. It is perhaps not out of place to mention that in those days the Soviet Union was the only European economy to have returned to pre-World War I levels; villagers had never had it so good as in those late NEP years.

The GOELRO[20] plan inspired our technological intelligentsia, and they set to work. And how they worked! Sad to relate, however, the steam-roller of Stalinist repression rolled over this section of the intelligentsia also, regardless of all the work they had done for the benefit of the nation and the fact that they were by nature non-protesting in both thought and action. A portion of them were betrayed by their government and exiled from Russian in the late 1920s. I remember conversations at the dinner table: Such-and-such are leaving today, and tomorrow so-and-so, etc. And they left grief-stricken and in tears, but absolutely certain they would be able to return in a year or so! Who of them knew they were quitting Russia forever!?

Then all sorts of "trials" began: The Mines Affair,[21] the "Industrial Party Trial,[22]" and others. But the Russian intelligentsia somehow survived, and the relay baton was passed on.

---

[20] Acronym for State Plan for the Electrification of Russia. This was the first-ever Soviet plan for national economic recovery and development, implemented in 1920. It was the prototype of subsequent five-year plans.

[21] A public trial, held in Moscow in 1928, of 53 supervisors and specialists connected with mining operations in the Donbas. They were accused of wrecking and sabotage. Five of the accused were executed and others sentenced to ten years' imprisonment. Among those arrested were German specialists providing technical aid, and their arrest provoked a diplomatic crisis with Germany. The day following the trial Vladimir Mayakovsky's poem "Wrecker," echoing the verdict, was published in *Komsomolskaya Pravda*. In December 2000, the General Public Prosecutor of the Russian Federation officially rehabilitated the victims of the trial.

[22] "The Industrial Party Affair" is the name given to a 1930 show trial, at which it was announced that the ideological inspirer of a certain "Engineering Centre" was P. I. Pal′chinsky, president of the "Russian Technological Society," who had been shot in 1929 by special sentence of OGPU (that is, without trial). Pyotr Ioakimovich Pal′chinsky (1875–1929), outstanding Russian engineer and political activist, had never been shy, regardless of the government in power, of airing his point of view. Although he supported Soviet rule, he nevertheless criticized its projects if he thought them

The Relaying of the Baton

In this connection I shall discuss only the passing on of the traditions of the "working" intelligentsia, since the fate of the "discussion group intelligentsia," to which the original anthology *Milestones* is devoted, is clear enough. Its members filled vacancies in Cheka or the *nomenklatura*, and some of them ended up in the Gulag. But that was later. They neither preserved nor founded any traditions, and all trace of them had been erased by the early 1930s. They were simply forgotten.

But it's of interest that they, or rather a similarly thin slice of the intelligentsia, came to life during the Khrushchov thaw. I call this resurrected set the "kitchen intelligentsia" since, instead of meeting in semi-legal discussion groups, this new group now met fairly regularly in each other's apartments. It was from these meetings, which usually took place in kitchens,[23] that the famous dissident movement was born, which played a far from positive role in the fortunes of our country. At first glance it might appear that there was no "genetic" inheritance passed down to the "kitchen intelligentsia" from the "discussion group intelligentsia" described by Bulgakov. Members of the latter were interested in Marxism and various other leftist movements, whereas the former were mostly purveyors of very rightist ideas, seeing in capitalism a panacea for all troubles, especially their own. Even now I fail to understand why they called themselves democrats; theirs was a typical movement of the right.

But on closer examination, one finds several traits common to the "discussion group intelligentsia" which the anthology *Milestones* describes, and the dissident movement of the years 1960–1980. Most striking is the fact that they were both very distant from the mass of the people. They were unacquainted with, had no understanding of, and no wish to understand what that entity "the Russian people" might be. And they were both "destructivist" by nature. They were, as a rule, little concerned with the outcome of the Cold War or for that matter the fate of the Soviet Union. But they were *very* exercised by the matter of the freedom to prattle anywhere about anything. By this I don't at all mean to say that I'm against free

---

badly planned. In particular, he criticized three "shock" projects undertaken by the Soviet rulers: The Dnieper Hydroelectric Station (*Dneproges*), for its distance from consumers, the Magnitogorsk Steel Works, for the lack of nearby coal for smelting steel (coal had to be transported by train), and the White Sea-Baltic Canal, for the choice of the "eastern" rather than the "western" route, awarded a gold medal at the Paris World Exhibition of 1900. The "western" variant would have allowed passage of sea-going vessels along the canal, whereas the "eastern" variant ultimately built was too shallow. The point for Stalin was that the latter was 10% cheaper and didn't require mechanized labour, so could be constructed using labourers from the WhiteBaltCamp (*BelBaltLag*), [one of the camps of the Gulag]. Work on the White Sea Canal began in 1931. This project, much written about as a triumph of the Stalinist epoch of industrialization, proved unable to fulfill its strategic purpose, namely that of having a single area for combat duty and a single shipyard for the building and repair of vessels operating in both the White and Baltic Seas. At the present time this circumstance continues to act as a brake on projects to do with exploiting the Arctic. *Note by the editors of the Russian edition.*

[23] Perhaps to avoid being overheard on bugs planted in their apartments.

speech; it's just that the nation didn't and doesn't need that freedom limited to a certain group of people, as the situation is today, but *genuine* free speech.

I shall leave my discussion of the very significant role the dissident movement played in *perestroika* to the next section. I *do* wish to point out here, however, that there were several highly respected people participating in that movement— A. D. Sakharov, for instance.

In fact Sakharov's metamorphosis mystifies me. He and I had been acquainted since our student years. Although he was a little younger than I, we occasionally encountered each other at I. E. Tamm's theoretical physics lectures and in his seminar. Sometime in the mid-1950s, I happened to spend a few days in the town Arzamas-16,[24] where Sakharov, then at the peak of his career, was working. We met several times and discussed topics of mutual concern. I was struck by the strong similarity in our psychological make-ups. At the time we were both in the state of "worker euphoria" so widespread among the Russian intelligentsia postwar. He was working of problems to do with manufacturing the "filler," while my job was to get the "filler" to the place where it was needed. We agreed that the worst evil that could befall our country would be to lose the Cold War! But then we had no intention of losing it.

Many years later, at a guess about thirty, we once again met, this time in the office of Academician Belikov in the Presidium of the Academy of Sciences of the USSR. We discussed problems of "convergence," at that time in the forefront of the concerns of my colleagues, especially Academicians G. S. Pospelov and V. M. Glushkov, and myself, insofar as they might involve computer technology. I described several of our ideas, but Sakharov seemed unable to understand them, didn't even understand our terminology. As earlier, we wished to link the concept of "convergence" with a reorganization of the management of production that would allow our nation to insert itself in dignified fashion into the future world-wide system of division of labour, and also with improvements in our technology aimed at preventing defeat in the Cold War. But Sakharov understood the term completely differently. He was the same kind, intelligent, and upright person as before, but was concerned not so much for the nation as for the personal freedom of each of its individuals. I thought to myself sorrowfully: "Where is that Andrey Sakharov with whom in Arzamas I so agreeably discussed what needed to be done to neutralize the consequences of the Cold War?"

I shall return to the question of convergence in the next section; it is far too important to be dismissed in just a few sentences. For now, I return to the wider Russian intelligentsia.

One has to hand it to the Bolsheviks: despite the blows they meted out to the old Russian intelligentsia, they understood the latter's significance, in particular the importance of the Russian engineering tradition and of Russian higher education,

[24] Established in 1946 to accommodate Design Bureau 11. It was here that the first Soviet atomic and hydrogen bombs were designed. Renamed "Sarov," it is now the Russian Federal Nuclear Centre.

which they made sure was preserved. Only in Russia and perhaps Germany did there exist scientific schools as we understand the term.[25] These were groups of specialists centred around some idea or talented supervisor. A remarkable characteristic of such informal associations was the mutual responsibility accepted by each member for the others' fortunes. Our scientific seminars were very different from American ones. In the US a scientific speaker is most often advertising himself in the hope of securing a grant or a more prestigious university position, while with us the performer was primarily seeking advice, in the expectation that his auditors and colleagues wanted to help him, even if that advice should take the form of severe criticism.

The German system of scientific schools was destroyed under the fascists[26] and has not yet been restored. Our Soviet rulers had the wit to preserve our scientific schools even through the dreadful years of the Great Patriotic War,[27] and, furthermore, create a system of instruction of fresh cohorts of scientific and engineering intelligentsia, a system for passing on the baton of our scientific tradition and culture. The onus of that transmission lay squarely on my generation, one enduring the most extreme hardships, negotiating all the circles of hell, but managing to preserve not only itself but Russia as represented by it. My generation proved able not only to adhere to the traditions of Russian education but also to pass them on to following generations. It consisted, for the most part, of offspring of the old pre-revolutionary intelligentsia.

The reader will, I hope, indulge me if I now reproduce a few episodes from my own life, on the grounds that they were in many ways typical.[28]

I have already mentioned my grandfather Sergey Vasilievich Moiseev. My father, trained as an economist, worked in the People's Commissariat for Communication Channels, as did my grandfather. His job was that of rationalizing transportation along internal waterways. My mother was the adopted daughter of the well known railway agent Nikolai Karlovich von Meck, son of the famous Nadezhda Filaretovna, whose correspondence with P. I. Tchaikovsky became an epistolary classic. Nikolai Karlovich worked in the Supreme Council of the National Economy, on plans for the further development of railway transportation in the Soviet Union.

My father and grandfather, as well the engineers who used to visit us, were fired up by the GOELRO plan,[29] among other initiatives of the NEP period, and I recall the pride with which they talked of what they were involved in, namely the construction of a single state-wide energy system, something hitherto not even

[25] This may at first strike the reader as oddly neglectful of the great achievements of British science, not to mention the French, Italian and American varieties, in the nineteenth and twentieth centuries. But the author intends a narrow sense of "scientific school" as one where individuality is submerged in the commune.

[26] Nazis.

[27] World War II as fought in the Soviet arena from June 22, 1941; until then the Soviet Union was ostensibly in cahoots with Nazi Germany.

[28] Some of what follows is also related in Chaps. 2 and 3.

[29] The first Soviet state plan, this one for electrification of the Soviet Union; see above.

considered in the West. But if there was talk of "quirks" of one kind or another of the Party or government, Sergey Vasilievich would always respond with the same sentences: "The Bolsheviks have come and will go, but Russia will remain. My dear sirs, to work!"

Now, in the midst of the present difficult times, I often repeat what he used to say, slightly modified: "Your Gaidars[30] and your Chubaises[31] have come and will go, but Russia will remain. My dear sirs, to work!"—though our present situation is more complex and dangerous.

During the NEP period, my family enjoyed a relatively high standard of living and there was a healthy feeling that one was working for the general good, characteristic of most of the Russian intelligentsia of the time: Russia was on the rise and this gave people hope for the future, the repression of Stalin's dictatorship being as yet scarcely felt. As in the postwar years, people worked in good conscience rather than out of fear. But towards the end of the 1920s the atmosphere began to change, and my family was immediately affected.

In 1928 Nikolai Karlovich was arrested, declared a wrecker and shot. In 1930 my father was arrested in connection with the "Industrial Party Affair," and died in the Butyrka prison before his trial, from a heart attack, we were told. He was only forty-two.

Immediately following my father's death, Sergey Vasilievich retired, and, unable to bear the woe that had fallen on us, died—from the blow, as they used to say back then. Thus it was that, at the age of 14, I remained the only "man" of the formerly large family.

I pass over the deprivation and difficulties of the first half of the 1930s. My greatest ordeal occurred in 1935, when I decided to apply to enter the university. Despite the fact that my father had died before being tried, I was everywhere reckoned "the son of an enemy of the people," with all this implied.

Although every person has, of course, his or her different individual fortunes, the experiences of many representatives of my generation were nevertheless very similar. I don't know of a single one of my peers who, as youths and young men, had it more or less easy. On the contrary, we all negotiated the "circles of hell," and this required courage and faith—faith in our country, in Russia.

And we completed our mission: we managed to pass the baton on to the many millions of the new intelligentsia, thanks to which by the early 1960s our country had become the second greatest scientific and technological power in the world. We had succeeded where the German intelligentsia had failed. But that wasn't all: the Soviet Union had created a broad, several million-strong stratum of the populace

---

[30] For Gaidar, see Chap. 16.

[31] Anatoly Borisovich Chubais (born 1955), Russian politician and businessman of Jewish background, responsible, along with Gaidar and others of Boris Yel′tsin's administration, for the "shock therapy" of the 1990s, when the principles of a market economy were implemented in Russia.

capable of receiving the baton, demanding, in fact, to be provided with the means of absorbing the knowledge and traditions of the Russian intelligentsia.

Notwithstanding all our personal grief and strife, this fact reconciled us to the Bolsheviks, much as the GOELRO plan had reconciled the intelligentsia of the previous generation to them. For that reason there were almost no dissidents to be found amongst us, no "destructivists." But this in no way meant that we accepted everything that was happening unconditionally; it's just that most of us were "constructivists." All the same, although we understood a great deal, it's clear we didn't *do* enough! I shall discuss this in the following section.

So What Exactly Happened to the Soviet Union and What Part did the Intelligentsia Play?

There are many reasons given for the catastrophe that has befallen our country. I think one can say with authority—many take this view—that in the Soviet Union there arose a systemic crisis which the state was incapable of overcoming. So it perished! I believe this to be beyond argument. However, usually the phrase "systemic crisis" is not unpacked further, nor is an analysis undertaken as to how the crisis began and why it had such catastrophic consequences. I shall attempt here to convey my view of the matter.[32]

Every large system, be it a large company, corporation, or even a government, can function successfully if and only if at a minimum the following two conditions are satisfied.

First, it is essential that the system have clearly enunciated goals for further development and improved functioning. In other words, it must have a well-defined set of appropriate goals to realize which various projects are to be undertaken. One particular such goal is genetically crucial, so to speak, to every system, namely the preservation of the system's stability and integrity.

The second condition has to do with the system's management apparatus. This also is a kind of system, and one without which no complex system can function. This subsystem must be able to subordinate its own aims and interests—the ever-present personal aims of each worker and the independent aims of the entire subsystem—to those of the whole system.

In the 1950s and early 1960s the Soviet Union had a clearly defined goal: achieving parity with the USA in missile and nuclear armaments. This was our chief aim, dictated by the Party and the government, and considered an obligation on the whole nation. Not only were the people's personal interests sacrificed to this end, but also those of the whole of the ruling class. Beware the worker at any level who out of self-interest slackened in the fulfilment of his duties! The whole of the

[32] See also Chap. 17.

nomenklatura understood this in their bones and devoted themselves to solving the problems of realizing the overreaching goal.

One might have various attitudes to the system prevailing in the Soviet Union back then, in particular point out several grounds for criticizing what was going on, but it cannot be disputed that there was in the Soviet Union at that time no systemic crisis as I have defined it. The crisis began in the first years of the 1970s, when, at the end of the Khrushchov thaw, the desired parity was achieved. Our contemporary rulers proved unable to formulate new goals; their prattling about a future communist paradise could not possibly have provided a basis for any concrete action, and served only to irritate reflective people. Thenceforth the management apparatus busied itself looking after its own affairs.

When social scientists assert that the Soviet Union was indeed a socialist system, I think they're mistaken, to put it mildly. I would say rather that in our country in the postwar years a "single factory system" had been created and perfected, in which each branch of industry was like a component workshop with a monopoly on its output, if that term may be applied to such a system. There was a highly complex system of ownership, which would take me too far afield to describe here. But of course it should nevertheless be said that there were various features of socialism central to the system, namely in the areas of education, medicine, etc.

However one may evaluate that system, it has to be admitted that for some time it "worked." I like to compare the paths of economic development of the Soviet Union and Japan. In 1945 we and the Japanese both started from zero. Japan had the advantages of being beneficiaries of the Marshall Plan and enjoying favourable international relations, while for us it was the opposite, so to speak. Yet in 1952 our GDP *per capita* was between 10 and 12% ahead of Japan's, notwithstanding the fact that our system was in need of cardinal reconstruction if it was to avoid collapse. I emphasize here the word *reconstruction* as opposed to destruction. In what follows I shall try to support this point of view.

I registered the first signs of approaching trouble at the time of my first trips abroad in 1959 and 1961, allowing me to compare Western achievements in the use of computer technology with ours.

During the first of these two trips, I visited the computer centres of the firms *Siemens* and *Honeywell Bull*, and I have to say that the situation as I saw it then gave me grounds for a certain optimism. While on the one hand the technological equipment in those centres was approximately at the same level as that of the Computer Centre of the Academy of Sciences of the USSR, our ability to solve complex problems using computers was much greater, as was our software engineering.

But on my second trip, two years later, I saw that the situation had changed drastically, and our relative position had become cause for concern. It was the time of the changeover to transistors in computers, a huge advance in the technological revolution: unreliable vacuum tube computers, requiring highly qualified teams to operate them, were being replaced by machines using semiconductors, free from malfunction. As a result, computer technology took off and computers began to appear capable of being operated by anyone at all, no matter how unskilled. In the West computers were freed from their classified status, becoming familiar instruments for use no longer only within the military-industrial complex, but in the management of production, in business, and so on. The computerization era had begun, and we were totally unprepared for it.

On arriving home, I began agitating intensively for the need to radically alter the political situation as it impinged on the creation and most of all use of computer technology. It turned out that there were a great many like-minded people striving to change the *status quo*. First among these was Academician S. A. Lebedev, creator of our first electronic computer, the BESM-1. I told him of my impressions abroad and expressed the opinion that we needed to reduce production of specialized machines and move to building universal computers, with a view to using them in the management of production processes. My venerable interlocutor replied that he had been proposing this for some time to the ministry, but that the upshot of his lobbying might easily turn out to be the opposite of what he intended: the production of our famous BESM series of universal computers might shortly be shut down.

Alas, S. A. Lebedev's prediction was fated to come true.

But even before that I had begun to understand that our "single factory system" was overdue for radical reorganization, since it was simply unable to meet the challenge of the new scientific and technological revolution. A system of branch industries-cum-workshops is useful only when the overall industrial goal is rather basic, the variety of goods is limited, and the main concern is an ability to quickly change the sorts of goods produced. But we were then entering an age when the outcome of the Cold War would no longer be decided by parity of missile and nuclear technology, which in any case had been achieved, but by the general level of technological development of the state and the competitiveness of its industry. And it's precisely here that we came up against insuperable difficulties with our "single factory system." For some branch industries it had become objectively unprofitable to go over to producing complex universal computers, while for others incorporating computer technology into their round of activity would have required a tremendous amount of re-training, and the replacement of many workers by others!

But the most urgent indicator of coming stagnation was the failure of Kosygin's reforms,[33] all the more galling in that their promulgation had been sanctioned by the requisite number of Party decisions, so that their introduction had seemed certain. Their collapse was a sign that the *nomenklatura* was unable *in principle* to carry out reform. Their own special interests had taken precedence over the nation's best interests, and it was unlikely that our doddering elder heads of state understood the danger of what was happening. In fact, they were themselves not much interested in reform.

But the intelligentsia understood well enough the significance of what was taking place, and I would say that by the beginning of the 1970s they had taken up a detailed position on the matter, which, although generally adhered to, was nonetheless never made formally explicit. Here I have in mind not just my colleagues working on

[33] See Chap. 15.

problems of the use of computer technology for defence purposes, but the much larger stratum of working intelligentsia. It was at that time that I was elected to full membership of the LAAAS[34] and was collaborating with agricultural specialists in Stavropol′. And at the Computer Centre of the Academy of Sciences of the USSR, which I was in charge of, we were concurrently working on a computational model of the functioning of the biosphere, as a result of which I came into contact with many specialists from other scientific fields—not only engineers, but also biologists, agricultural scientists, etc. With all of these people I experienced a high level of mutual comprehension and found that they all had practically the same take on our political situation.

We saw two "chief dangers," which, given our political system as it was then and the worsening systemic crisis, many of us considered unavoidable.

The first of these was losing the Cold War, since this entailed our country's becoming an outsider as far as scientific and technological progress was concerned. It seemed to us, in other words, that the Soviet Union was drifting out towards the periphery of the developed world, and this was bound to have dire repercussions.

The second danger, a consequence of the first, was that of ineluctable internal changes of a revolutionary character. A nation which in a single century had endured several revolutions, a civil war, and two world wars, would surely be unable to survive yet another revolutionary cataclysm. It might very well be the death knell of our nation!

That is why all our efforts were directed towards finding a way to fix our political system that would allow us to circumvent the above-described dangers. Various approaches were aired. There was, for instance, that of G. S. Pospelov, who had created a programmable method for managing the national economy, based on fundamental ideas of contemporary computational planning and the creation of complex technical systems. To be frank, I thought his approach somewhat naive, but supported him as much as I could, reckoning that the implementation of a programmed method in government management practise would help not only to modify the system as it was and increase our ability to participate more efficiently in the world-wide process of scientific and technological development, but also, of special importance as it seemed to me, to limit the monopolies of our branch industries.

But the ideas closest to my own were those of V. M. Glushkov, then director of the Cybernetics Institute in Kiev. We often ran into one another at meetings of various committees of the Military-Industrial Complex. At that time the fate of the BESM series of computers was being decided, and Gluskov and I always found ourselves on the same side of the barricade. These meetings, which took place in the Kremlin, were frequently quite frictional, touching as they did on the very heart of the social structure of the nation, and we, the technologists, eventually lost the struggle to the government functionaries.

---

[34] The Lenin All-Union Academy of the Agricultural Sciences; see Chap. 11.

I was most of all concerned about the problem of "convergence," but not in the primitive form in which it was generally discussed back then. I had become convinced that both classical market capitalism and the "single factory system" of the Soviet Union were dead ends on the path of human development, so that we needed to find new principles of social development, *qualitatively* new ones! But the main thing was to avoid sudden revolutionary lurches.

On examining more and more closely the history of the NEP period, I began to feel, and in fact still feel, that a unique chance was missed back then of creating a really new kind of democratic society. It is no longer possible to repeat that experiment in view of the subsequent liquidation, in the 1930s, of the peasantry, the foundation of self-motivated labour. However, there is much else from the NEP era that might still be of use today.

Somewhere in late 1985 or early 1986, I was invited to a meeting chaired by M. S. Gorbachov, and during a break he approached me. A short but substantive conversation ensued. He asked: "Why have you stopped dropping in?" "It was one thing when you were Secretary for Agriculture, but quite another now you're General Secretary," I replied. He then asked what I had to say about what had been under discussion at the meeting, to which I responded verbatim as follows: "While you still have an emperor's power, you should liquidate the branch monopolies and attempt to revive something like the system of syndicates of the NEP era." Mikhail Sergeevich responded with: "Interesting. Write to me."

I wrote out my ideas in detail in a letter, which I handed in to his chancellery. It was not until two or three years later that I met Gorbachov again, at another meeting of some kind. I went up to him and asked about my letter, receiving the laconic reply: "I don't recall. Most likely I didn't receive it."

No comment, as they say. The government was not prepared to give us, the scientific and technological intelligentsia, a hearing.

The question of a gradual reformation of our "single factory system," and the consequent preservation of our system's social achievements in education, medicine, full employment, etc., allowing a citizen to feel secure about his or her future, finally became moot towards the end of the 1980s.

The Systemic Crisis Continues. Its Finale

The rulers of our country had their own consultants, and to these they *did* give a hearing. I mentioned above that, in addition to the "kitchen intelligentsia," there were in the dissident movement others much worthier of respect. Alas, none of them played first violin.

At the turn of the 1970s into the 1980s an acquaintance brought me a great bale of *samizdat*[35] literature. On perusing it, all illusions I might have had about the dissident movement vanished. I now knew that among the dissidents there were people with an intense hatred, not of the Bolsheviks, but of us, Russians, and of Russia as a whole.

I, together with those around me, were very cognizant of the failings of our system, as well as those of Russians generally, and found them acutely distressing and a source of great sorrow! We were always talking over our troubles and looking for a way out of the systemic crisis, whose nature we fully grasped. But at the same time we felt that our people were objectively worthy of the greatest respect, as well as empathy and concern.

Most of the region of the planet we live in, that vast northern stretch of Eurasia, is comfortless and harsh. Even the most northerly city of Canada is still only at the latitude of Kursk.[36] Our growing season is at least 100 days shorter than France's, for instance. Yet we have managed not only to survive in this rough clime, but to create a great culture and great science, and become a leading world power. And one shouldn't forget that we have always been surrounded on all sides by hostility: just read the Marquis de Custine[37] or listen to Brzezinski,[38] or recall the unending incursions from both east and west!

Yes, we're Europeans, but an alternative sort of Europeans from Western ones. It couldn't be otherwise! The protestant ethic and Western individualism could never have taken root with us. Under our prevailing climatic conditions they would have hampered our ability to survive. Even now we're struggling to survive! Collectivism, or communality, was essential to our survival, as was much else distinguishing Russians from West Europeans.

Perhaps I was just unlucky. In that great wad of *samizdat* pulp that someone passed on to me, I failed to find a single kind word about our people or our country, yet it had been written by some of my fellow citizens. It would be very hard to imagine any French citizen writing about his or her people or country the sort of negative things that I was reading about Russians and Russia.

[35] Portmanteau word for the underground self-publishing and distributing enterprise existing in the Soviet Union, by means of which illegal literary materials were secretly prepared and circulated among dissidents and those sympathetic to them.

[36] The author presumably had only Montreal, Toronto, Ottawa or Quebec in mind. Edmonton, capital of the province of Alberta, has almost the same latitude as Moscow.

[37] French aristocrat and writer best known for his travel writing, especially his account, entitled *La Russie en 1839*, of a visit to the Russian Empire during the reign of Nicholas I. Lived from 1790 to 1857.

[38] Perhaps this is Zbigniew Brzezinski (1928–2017), American diplomat, political scientist, and sometime US National Security Advisor.

It goes without saying that in using the word "pulp" above I don't have in mind the writings of people such as A. I. Solzhenitsyn or V. Maksimov[39] or other writers of that stature forced out of their country.

As it happened, it was precisely the "kitchen intelligentsia" who became the ideologists of *perestroika* and especially post-*perestroika*, and not the working intelligentsia, to whom our state was beholden for its technological might and who had adopted a generally constructive outlook. How did this come about? How was it that an ideology cooked up in dissidents' kitchens had filtered up to the level of government? Here I can only hazard a guess.

I had met M. S. Gorbachov much earlier, when he was working in Stavropol′, where, in the 1970s, the Computer Centre of the Academy of Sciences of the USSR was helping local specialists set up a computer system.[40] I had by that time met many secretaries of Party regional committees, and Gorbachov stood out from the rest; he was more democratic and, what was most important, knew how to listen to others and attend to their reasoning. I was very glad when he was made General Secretary, and had great hopes for the future under his leadership.

But then I observed, with disappointment, that, rather than begin formulating a large-scale strategic plan, he became engaged in haphazard actions such as the anti-alcohol campaign. But what I found most disquieting was his enthusiasm for "openness." Openness[41] is, of course, essential, however not as a root source of reform, but rather as an emergent consequence of the enfolding system's main features. As I had written to Gorbachov in 1986, what was really needed was a root-and-branch restructuring of our "single factory system." It was on the wave of this "openness" that the dissident intelligentsia rose to the top, but not under the leadership of its best representatives. A rabble of "foremen of *perestroika*" came to the fore. This hypertrophied openness became a source of national discord and disunity.

However, it must be admitted that the wave of restructuring captivated a much wider section of the intelligentsia. There was the phenomenon of "Gorbachov euphoria." Even I was taken in to the point of writing an article titled "The Restructuring Type," to which I now plead guilty! And am ashamed of.

But this euphoria for restructuring was short-lived amongst the intelligentsia; we very soon understood that *perestroika* was not about to halt the systemic crisis. As earlier, no clear goals for the future development of the nation were being adumbrated. No one knew whither we were going nor why. The *nomenklatura* continued running affairs, but chiefly in their own interests, quietly gathering to themselves in one way or another the people's national wealth. Our country

[39] Vladimir Emelyanovich Maksimov (1930–1995), Russian writer, poet and publicist. After some years in exile and in the Gulag, returned to Moscow in 1956. Forced to emigrate in 1974 for his anti-communist writings.

[40] This period is described in detail in Chap. 11.

[41] In Russian *glasnost′*, meaning openness to the public.

was becoming more and more impoverished and famished, and scientific and technological progress was at a standstill.

One could see the nation faltering and giving way. The government had clearly lost control. The situation was getting to that stage the anticipation of which had most worried us over the previous two decades, and of which I have written above: we had already lost the Cold War, and would surely soon be subject to unendurable revolutionary shocks. The system was self-destructing—not undergoing restructuring, but imploding: the catastrophic collapse of our country was already inevitable. But we of the engineering intelligentsia nonetheless kept on talking about what might be done to deflect the approaching disaster.

By early 1991, we had come to a clear appreciation of a possible catastrophe. But, to be frank, we didn't think the situation would degenerate so rapidly or the government turn out to be so helpless. I had hoped that we had another two or three years in reserve, and, God willing, that, given the robust souls I saw around me in society and the encouraging result of the referendum showing the people to be against the disintegration of the USSR, we'd be able to endure the approaching period of agony.

But in one thing at least we had been absolutely correct: the collapse was the fault of the *nomenklatura*, those wielding power in the Party, who had taken advantage of our enfeebled state to engage in their own private chicanery, while the "kitchen intelligentsia" assumed responsibility, expressed in pretty words, for the various kinds of abomination perpetrated on our people and nation.

The systemic crisis reached its logical conclusion: the system collapsed. A new system rose out of its ashes, one with its own problems.

The Intelligentsia and the Post-*perestroika* Period

The new system, arising on a truncated region of our former Great State and beginning to be called Russia, to my mind inappropriately since around 25 million Russians, enough for quite a large country, remained outside its borders,[42] immediately descended into its own systemic crisis. But before explaining what I mean by this, I wish to bring to mind certain facts.

The dissident movement came to a safe end with the launching of Gorbachov's *perestroika*. The openness the dissidents were so insistent on was granted. Some of the "kitchen intelligentsia" left the country and began penning libellous squibs about the Russian nation and its people, a significant number of others obtained positions in the government, and a sizeable segment of them, together with a portion of the

[42] It might be objected here that the existence of so many ethnic Russians outside Russia proper was the result of the perhaps deliberate Soviet, and even pre-Soviet, (neo)-imperialist policy of colonization of formerly independent countries or ones desiring their independence, such as the Baltic countries and Ukraine.

nomenklatura, became "new Russians," having acquired, by means unclear to me, significant capital. The smallish set of decent people among the dissidents was now absorbed into the rest of the intelligentsia, sharing their dismal fate.

Thus a certain subset of the intelligentsia ended up in the halls of power. I sometimes call them "Gaidaromorphs." The coming to power of such people was a terrible misfortune for our country. As a rule they are not very highly educated, but enormously ambitious and with even bigger appetites. But the most frightening of their traits is an overweening self-sufficiency, revealing of the defects of their intellect and upbringing.

Alas, the wider intelligentsia proved unable to oppose to these any effective action: they were crushed, demoralized, and also somewhat out of touch with the essence of what was going on. And they didn't have access to the mass media. Nevertheless, some of them tried to do something. I shall now relate one such attempt.[43]

Towards the close of 1991, the final General Meeting of the Academy of Sciences of the USSR was held, presided over by its last president G. I. Marchuk.[44] While we were debating the future of the Academy, primarily the question as to whom it should belong, I stepped up with a little speech which went approximately as follows: "The question of belonging is moot: The Academy of Sciences of the USSR is heir to the former Russian Academy of Sciences, so now it can revert to its former name. But there is a much more important question: how can we utilize the intellectual potential of the Academy in our nation's present difficult, even critical, times? During the Great Patriotic War we had the program 'Science to the Frontlines.' Don't we need something similar now?"

Marchuk commented on my speech as follows: "Initiative always comes with a penalty. You, Nikita Nikolaevich, should write a letter to Yel′tsin."

I wrote an appropriate letter next day, and placed it on the desk of the secretary to the President of the Academy. Over the following week it was signed by around twenty academicians, and I then took it to Old Square, where at that time the president of the Russian Federation had his chancellery. A few months later we were all surprised when Yel′tsin ordered that a Council for the Analysis of Crises be struck. It was to include twelve members of the Academy, and I was appointed Chair. But an overseer of the council was also appointed, in the person of G. E. Burbulis, then the second most powerful person in the government.

A few days later I had a remarkable conversation with our overseer. I told him that although the work of our council would be carried out on a voluntary basis, in order for it to function effectively it should have, in the first place, suitable premises, and, secondly, a scientific secretary at a decent salary, a few technical assistants, and office equipment, including a computer and a direct link to the government.

To this Gennady Eduardovich responded, practically word for word, as follows: "Now, Nikita Nikolaevich, your council contains twelve wise men from the

[43] Related also in Chap. 14.

[44] Related also, in somewhat less detail, in Chap. 14.

Academy. Such matters are therefore best referred to the Academy, and all the more so in that much of the Academy's office space has been freed up." So I went along to Yu. S. Osipov, who had by then succeeded Marchuk as President of the Academy, and repeated my request to him. He answered, again almost word for word, as follows: "Now, Nikita Nikolaevich, we are, after all, not specialists in such matters." As if in the whole wide world there are specialists in the transition from socialism to capitalism! So he also gave us nothing, and for our meetings and associated work we mostly used the office of Academician V. S. Stepin, Director of the Institute of Philosophy. I note by the way that, although Yu. S. Osipov was also a member of our council, he attended not a single meeting.

In 1992 our council was granted two hearings before the Russian Supreme Council, and even published two reports. But then we began gradually bringing our work to a close, and not only because it was practically impossible to continue without technical support. The fact was that the two hearings before the Supreme Council had left me and many of my colleagues with a sense of the futility of our enterprise. The hearings were quite well attended, but not out of any interest in what we had to say; the members of the Supreme Council regarded us as cranks, people so eccentric as to busy themselves not with politics, or making money, or lobbying, but with some sort of higher matters. Our auditors were immersed in their politics, or, more precisely, in their politicking. They put no questions to us, and there was no discussion. They listened and then left!

The attempt of our council to influence the train of events was not the only one to be made by the scientific and engineering intelligentsia. But at the government level no one really listened to us; we couldn't get through to anyone in power! And meanwhile the Gaidaromorphs continued with their anti-people activity.

Everyone belonging to the scientific and engineering intelligentsia was talking about the same thing, albeit expressed variously. What exactly was it then, that we, the "non-professionals," as the President of the Academy Yu. S. Osipov called us, were discussing? I shall now reveal our subject matter, but couched in the words of an outsider.

Almost seven years had elapsed when, on December 31, 1998, the *Independent Newspaper* published an assessment by Jeffrey Sachs,[45] Gaidar's chief advisor. Here is his appraisal as reported by that newspaper: "What mainly undercut us was the colossal gap between the reformers' rhetoric and their actions. ... And, as it seems to me, in its actions the Russian leadership exceeded the most fantastic imaginings of the Marxists about capitalism: they considered it their duty to be of service to a small coterie of capitalists, stuffing their pockets with as much money as possible as quickly as possible This is not so much "shock therapy" as an intentionally evil,

[45] Jeffrey David Sachs (born 1954), American economist, public policy analyst, and former director of the Earth Institute at Columbia University. Known as a leading expert on economic development and the fight against poverty. Advisor to several governments (Bolivia, Poland, Russia) on the transition from Marxism-Leninism and other economies to market-based ones.

well-thought-out action, having as its goal a massive redistribution of wealth in the interests of a tiny set of people."

That's what we were talking about, only less acerbically and more politely, since we were still hoping that those calling themselves democrats would turn out to be decent people and that the fortunes of their people were, as with us, of foremost concern!

* * *

At the beginning of this section I mentioned that the systemic crisis did not come to an end with the emergence of the Russian Federation as an independent state. It would have been more accurate to say that the people and the nation continued to live through a time of degradation and disintegration. There is, of course, a continuing crisis, though not a *systemic* one in the precise sense I earlier gave to the phrase. But it's a crisis nonetheless, since, as Sachs described it, we are witnessing the willful destruction of the nation.

And the terrible thing is that the "reformers" may achieve their goal. They control the mass media, are propped up by a corrupt bureaucracy, and feel secure in the support of the West—just read the interview with Brzezinski published in that same *Independent Newspaper*! At the beginning of the post-*perestroika* period I tried to take the following tack: "People like Gaidar and Chubais come and go but Russia will remain; meanwhile one must work." I'm afraid that this line of reasoning is no longer valid. The "reformers" are much more dangerous than the Bolsheviks: although they will indeed be gone at some stage, by then Russia will no longer exist. She will disappear along with them, or perhaps even before them.

That is why the intelligentsia must today speak out and show the way forward to the preservation of Russia, its people, and its culture and traditions. This will be the subject of the final section of my essay.

The Intelligentsia as the Backbone of the Electorate

I urge you, highly respected reader, to read the book *From Moscow to St. Petersburg* by Valery Pisigin.[46] At some time in the post-*perestroika* period, this author followed the path trod long ago by Radishchev,[47] only in the opposite direction. He stayed *en route* at small towns and settlements, conversed with doctors and teachers, and attended their modest celebrations. When I read this small book, I was overcome by a feeling for which I am very grateful to that author: I felt that Russia was still

[46] Valery Fridrikhovich Pisigin (born 1957), Russian writer and historian, and researcher into British and American folk music. He actively supported *perestroika* and, in the 1990s, democracy in Russia.

[47] Alexander Nikolaevich Radishchev (1749–1802), Russian writer and social critic. His depiction of socio-economic conditions in Russia in his book *Journey from St. Petersburg to Moscow*, published in 1790, led to his arrest and exile by Catherine the Great.

viable, and that its intelligentsia was still there, the true Russian intelligentsia, not that offshoot that had come to power with the second-level *nomenklatura*, but the genuine article, living amongst the people, sharing their thoughts, feelings, and love of country, and doing its eternally necessary work for the sake of Russia.

From impressions gathered from my seminars for teachers and from talking to people from the remote countryside over the past two or three years, I have assembled a picture of the hopes and expectations of the Russian intelligentsia, usually unexpressed, or, if given expression, then very laconically. Perhaps what these people are waiting for most of all is a few words of gratitude and respect. But they are also in dire need of real work and a return of a sense of self-worth. Without a revival of their feelings of self-respect it will be very difficult to achieve that surge of energy so necessary to the nation, and the intelligentsia most of all, in order for our country to return to a state of peaceful development based on real labour.

But it's surprising how, although beset by worry about surviving on a pitiful salary, moreover paid irregularly, our Russian intelligentsia is still capable not just of thinking of the future but of living it! This is a characteristically Russian phenomenon and one to be reckoned with. The present relative calm prevailing is due not only to the inherent long-suffering natures of our people, but also to genuine love of the Motherland—although, on the other hand, the potential for hatred is building and may one day blaze up to such an extent that it's impossible to quench. This also should be kept in mind.

Once there came to my home a certain "provincial representative of the intelligentsia" from Kostroma Province. He wanted me to autograph the copy of a book of mine he had bought somewhere in Moscow. We talked long and interestingly. I asked him what party he favoured, and he replied: "None of them. Those party types are uninterested in us. They have their own agendas; let them sort themselves out! We'll somehow get by without them." Respected reader, think over these words: perhaps they hide history's essence, that of an historical epoch awaiting our people!

Today the intelligentsia, in the sense I have been using the word, constitutes an enormous multitude of people, between ten and twelve million, perhaps. They have a common fate, and well they know it. One senses already a feeling of mutual understanding and community. They really do know that there is no future for them under present conditions and that none of the existing political parties is about to concern itself with them.

The communists apply certain standards to their thinking, and these they will never abandon, but those who are now calling themselves democrats are totally indifferent to the concerns of the Russian intelligentsia. My guest from Kostroma was right on the mark!

But one thing everyone understands: there's no returning to the past!

Although it's hardly worthwhile discussing here the possible variants, scenarios and programs of future development, I shall proffer a possible sequence of milestones.

The first thing I would like to offer for the reader's consideration is the following fact, already become banal: the planet is in the process of becoming a single system, both economic and social, outside which no state can survive. The main question, therefore, to be answered not by any single party but by the whole of our country, is what we should do to ensure we occupy a worthy place in that complex, variegated, yet unified world.

A single economic organism is taking shape. A process of self-organization is under way, unregulated and ungovernable. I call this process the formation of a World of Transnational Corporations, or WTNC.[48]

Thus for us to ensure we occupy a strategic position in the WTNC—keeping in mind that there will be no existing outside it—, we must have our own powerful corporations. When I reflect on the possible forms such organizations should have, images of the the syndicates or combines formed in the NEP era involuntarily come to mind, those combines thanks to which our economy was resuscitated. Of course, there is no question of recreating the combines of the 1920s in their original form, but we should nonetheless seriously consider implementing certain of their special features, primarily their links to the government. Although the combines' capital was state-owned, they functioned independently of the State, while at the same time working closely with the Supreme Council of the National Economy and Gosplan[49] on a contractual basis. Basic Government revenue was then determined by the efficiency of these combines: the repletion of the national budget depended on their productivity rather than individuals' income tax. I feel that at the present time the State should once again have a controlling interest in our leading corporations, with their profits accounting for the lion's share of national income.

My second idea has to do with agriculture. In view of our harsh natural environment, we cannot expect our agricultural produce to be competitive, at least on the world market. But a nation, especially one as large as Russia, can survive in the WTNC only if it is at least able to feed itself. We must, therefore, have a special production program. It should be subsidized by the State and special measures taken so that our agriculture is competitive internally, among other things.

But I would like to point out that ideas of a pure market (and the associated kind of thinking) are by themselves inadequate for solving complex agricultural problems, since these depend at base on the mutual relations of Nature and Society, so require for their solution a deep understanding of their non-market essence. I give one example of many, this one concerning the extent of mechanization.

In the USA the productivity of a single agricultural worker is around a hundred times greater than that of a Chinese peasant. But your American farmer could scarcely imagine the harvest a Chinese peasant reaps from a single hectare. And one must not forget, furthermore, that before all else agriculture is fundamental to our civilization. The invention of agriculture heralded the appearance of a completely

[48] It's true that this name is not completely accurate since it does not capture an important feature of the concept. The leading role in this formative process is perhaps played more by financial speculation than by production. I shall keep to the name chosen, however, since it is suitably short. *Author's note.*

[49] The overarching Soviet economic planning organ.

new sort of human relations with Nature, based on the creation of an artificial recycling of matter.

Finally: Only that nation has a chance of a decent future that is capable of establishing systems of higher education and morality. This thesis is especially actual in Russia's case since in our harsh land only by means of "higher technology" can we hope to secure a sufficiently high level of well-being among our people.

That is why there must lie at the basis of our polity, whatever form it may take and under whatever conditions, the ancient peasant principle of "the preservation of the stock of seeds": no matter how hungry one is in Winter, the stock of seeds must not be touched since it is our hedge against the future. That is why I reckon the greatest crime of the powerful of this world before their peoples to be the drop in the level of education and the catastrophic reduction in scientific potential.

* * *

Today our nation is overcome by apathy and a feeling of hopelessness. And even the "backbone of the electorate"—the intelligentsia—so far barely reacts to the party leaders' spiel. This is likely to continue till such time as a party or movement is formed able to offer a word of thanks to our people for having in the NEP era resurrected the nation, for being able, even following the Stalinist genocide, to emerge victorious in the Great Patriotic War and then restore our Great State. That movement or party should announce to the nation that all is not lost, that it has ideas for opening perspectives of revival, and show them the way.

I think it essential to establish a social council, to be completely independent of the government and political parties, with the mandate of exploring prospects for national development.[50] Its actions should be made public and open to debate. Such an institution would enable the "backbone of the electorate," to whom this essay is dedicated, to decisively influence the fortunes of our country for the better.

1999

[50] The social council of which N. N. Moiseev is speaking here should not be confused with the Social Chamber of the Russian Federation, established in 2005, as is clear from the avowed aims of the latter. *Note by the editors of the Russian edition.*

Index of Names

N. N. Moiseev, *How Far It Is to Tomorrow*. . . ,
https://doi.org/10.1007/978-3-030-96651-5